D1154690

"Michael Harris writes with all-absorbing exuberance and intensity about what it feels like from the inside to do mathematics, and he succeeds, for the uninitiated like myself, in conveying the motives and the pleasure that have impelled him and his precursors and peers to seek to understand. But *Mathematics without Apologies* is many things besides: it combines thoughtful personal memoir, sharp social chronicle, entertaining literary analysis, and jeux d'esprit reflecting on formulae for love or on the hidden structures in the fiction of Thomas Pynchon. Most importantly, however, Harris issues a clarion call for the autonomy of research in our time. He defends—fiercely and lucidly—the pursuit of understanding without recourse to commercial interests or other principles of utility. This is an original and passionate book; Michael Harris has fashioned much-needed luminous arguments for the cause of intellectual independence."

    —Marina Warner, professor of English and creative writing, Birkbeck, University of London, and author of *Stranger Magic: Charmed States and the Arabian Nights*

"Michael Harris opens the doors and gently guides you into a magic world. Once inside, you can't help but feel mesmerized, eager to see how deep the rabbit hole goes. And no wonder: a major thinker of our time is talking to you about math and so much more, like you've never heard before."

    —Edward Frenkel, author of *Love and Math: The Heart of Hidden Reality*

"Harris offers a unique and passionate view of the life and work of contemporary mathematicians. Rich in detail and marvelously broad in scope, *Mathematics without Apologies* gives an unforgettable account of the frustrations, elations, and sheer wonder of doing mathematics."

    —Barry Mazur, Harvard University

"Becoming a mathematician is like becoming a musician. The apprentice must master the technique, but along the way he also has to develop an aesthetic sense. Only then can he become a master in his own right. In this lively book, Harris examines the mathematician's craft from every angle, from the elementary and familiar to the sophisticated and exotic, and questions the ethics of using mathematics in finance. *Mathematics without Apologies* is a very personal book dealing with timeless questions."
　　—Ivar Ekeland, mathematician and economist, author of *The Best of All Possible Worlds* and *The Cat in Numberland*

"*Mathematics without Apologies* is a work of relentless intelligence that depicts Harris's experience of mathematics, but it is not at all a mathematician's autobiography. It is a madly erudite and creative reflection on the mathematical life."
　　—Colin McLarty, author of *Elementary Categories, Elementary Toposes*

"Harris vividly conveys what it is to work as a researcher in pure mathematics today. Through a series of novel and unexpected perspectives, he transforms readers' preconceptions of this activity. What we encounter here are the reflections of an erudite mathematician, uncommonly well read outside his field, on the nature and purpose of his subject."
　　—David Corfield, author of *Towards a Philosophy of Real Mathematics*

"Mathematical high culture collides with pop culture and all hell breaks loose! Harris takes us on a wild ride—never a dull moment!"
　　—Gregory Chaitin, author of *Proving Darwin: Making Biology Mathematical*

mathematics
without
apologies

mathematics
without
apologies

portrait
of a
problematic
vocation

michael harris

princeton university press

princeton and oxford

Copyright © 2015 by Princeton University Press
Published by Princeton University Press, 41 William Street, Princeton, New Jersey 08540
In the United Kingdom: Princeton University Press, 6 Oxford Street, Woodstock, Oxfordshire
OX20 1TW

press.princeton.edu

Cover illustration by Dimitri Karetnikov

ISBN 978-0-691-15423-7

Library of Congress Control Number: 2014953422

British Library Cataloging-in-Publication Data is available

This book has been composed in Times New Roman and Archer display

Printed on acid-free paper ∞

Printed in the United States of America

10 9 8 7 6 5 4 3 2 1

To Béatrice,
who didn't want to be thanked

contents

Wer das Buch schriebe, hätte die Vorrede Schritt für Schritt zurückzunehmen, aber sie ist das Beste daran, das Einzige was wir können, wir Modernen...

Ich will aus solchen Vorworten zu ungeschriebenen Büchern ein Buch machen, ein modernes Buch. Und ich schrieb eben—das Vorwort dazu.

—Paul Mongré, *Sant' Ilario**

When this book was nearly done and my colleagues started asking me what it is about, I found it simplest to answer that it's about how hard it is to write a book about mathematics. That's the short answer; the unabridged version involves a few pages of explanation. Here are those pages.

Of course people are writing books about mathematics all the time—and not only for expert audiences. The most effective of these books strip away the technical jargon to convey the magical sense that pure thought can conjure a second life, a virtual world of shapes and numbers and order and rules where not only do we know that everything is as it should be but we are also satisfied that we know *why*. Knowing why is the specialty of mathematical reasoning, but the virtual world of *pure* mathematics, not designed for any practical application, is remote from our first and authentic life; those of us dedicated to that world feel (or are made to feel) obliged to justify our indulging in an activity that is charming and engrossing but that appears to bring no benefit beyond the pleasure of knowing why.

These attempts at justification are the "apologies" of the title. They usually take one of three forms. Pure research in mathematics as in

* "Whoever would write the book would have to undo its preface step by step, but it's the best, the only thing we can do, we Moderns. . . . I want to make a book of such prefaces to unwritten books, a modern book. And thus I wrote—the preface to this book" (Hausdorff 2004).

other fields is *good* because it often leads to useful practical consequences (Steven Shapin calls this the *Golden Goose* argument); it is *true* because it offers a privileged access to certain truths; it is *beautiful*, an art form. To claim that these virtues are present in mathematics is not wrong, but it sheds little light on what is distinctively *mathematical* and even less about pure mathematicians' *intentions*. Intentions lie at the core of this book. I want to give the reader a sense of the mathematical life—what it feels like to be a mathematician in a society of mathematicians where first and second lives overlap. But during this guided tour of what I want to call the *pathos* of mathematics, we will repeatedly see our intentions misrepresented, and we will be reminded how hard it is to explain what impels us along this peculiar path.

Rather than rely on apologies, this book pieces together fragments found in libraries, in the arts, in popular culture, and in the media, to create a composite portrait of the mathematical vocation. The sequence of chapters very roughly follows the trajectory from the vocation's awakening, through struggles with various kinds of temptation, to its consolidation, followed by a conclusion consisting of inconclusive reflections on what we know when we "know why" and what it all means. Although I have consulted actual transcripts or recordings of mathematicians talking, my sources consist mainly of writings about mathematics, especially by participants—so the portrait is largely a self-portrait, though not of the author, of course!—but also by (usually, but not always, sympathetic) observers. Preconceptions and misrepresentations are fair game but are usually identified as such. I have paid special attention to writing or speaking by mathematicians whose manifest content may concern truth or utility or beauty, but which exhibit an aiming at something else, the values and emotional investment—the pathos, in other words—involved in pursuing the mathematical life.

Writing the book was a process of assembling and organizing this material in connection with selected themes and unifying perspectives. The process of assembling suggested virtues rather different from those usually invoked. Alternatives I explore in this book include the sense of contributing to a coherent and meaningful *tradition*, which entails both an attention to past achievements and an orientation to the future that is particularly pronounced in the areas of number theory to which my work is devoted; the participation in what has been described, in other settings, as a *relaxed field*, not subject to the pressures of material gain and productivity; and the pursuit of *pleasure* of an elusive, but nevertheless specific, kind.

The alert reader will have noted, correctly, that these alternatives are no more distinctively mathematical than *good, true,* and *beautiful.* I certainly don't think they offer definitive solutions to the riddles of mathematical pathos; but they did make it possible for me to hint at a vision of the mathematical good life that I find more reliable than the standard account. Another author, presented with the same material, would assemble it in a different way and would likely reveal a different set of habits, virtues, and goals. This is only natural; I try to make the diversity of the community of pure mathematicians visible by recording their distinctive opinions, and there's no reason to assume they come to the field sharing identical motivations. At most, I hope that the reader will see coherence in my personal assemblage.

The "problematic" subtitle alludes to the problems that define the intellectual landscape where the mathematical life makes its home. It's conventional to classify mathematicians as "problem solvers" or "theory builders," depending on temperament. My experience and the sources I consulted in writing this book convince me that curiosity about problems guides the growth of theories, rather than the other way around. Alexander Grothendieck and Robert Langlands, two elusive costars of the present book's narrative, count among the most ambitious of all builders of mathematical theories, but everything they built was addressed to specific problems with ancient roots. Entering the mathematical life is largely a matter of seeking an orientation among such outstanding problems. In this way, as in every other way, the mathematical life is a running dialogue with human history.

The mathematical life is problematic in other ways. Trade books about mathematics typically follow a quest narrative. They share with the currently dominant model of science writing an attachment to a simple moral economy in which the forces of light and darkness are clearly delineated. The quest may be embodied in the protagonist's need to overcome external obstacles or to meet an intellectual challenge; its happy ending takes the form of a triumph over a hostile or unpromising environment or the rewarding of the protagonist's unique talents—or both at once. The reality is not so simple. The most interesting obstacles are less straightforward and more problematic. These can never be overcome: they are inseparable from the practice of mathematics itself. There is the need to guarantee a constant supply of the material underpinnings of our practice, what the moral philosopher Alasdair MacIntyre calls "external goods." We might feel deeply ambivalent about what we provide in exchange. We set aside our ethical compass and contract Faustian bargains (Faust is a recurring figure in this book). We promise Golden Geese, immutable truths, ineffable beauty. We collude in the misrepresentation of our values and our intentions, in "the alienation from oneself that is experienced by those who are forced to describe their activities in misleading terms."*

But any burdens left on our conscience when we contract bargains, Faustian or otherwise, can be separated from mathematics, at least conceptually. It's not the least of the paradoxes explored in this book that the pathos of mathematics grows darkest and most problematic at its moments of greatest success. Satisfaction in solving a problem can be intense, but it is short-lived; our pathos is driven by what we have not yet understood. André Weil, one of the twentieth century's dominant mathematicians, described this as "achiev[ing] knowledge and indifference at the same time." We never understand more than a finite amount of the limitlessness of what mathematics potentially offers to the understanding. If anything, the situation is even more frustrating: the more we learn, the more we realize how much more we have yet to understand. This is also a kind of Faustian bargain—Goethe's Faust got to keep his soul until he reported to the Devil that he was satisfied with what he had seen. The mathematical soul, embodied in a historical tradition oriented to a limitless future, can rest secure in the knowledge that its dissatisfaction is guaranteed.

And yet there are enough of us who find the attractions of the mathematical vocation irresistible to fill any number of books this size. Is this because the vocation is so problematic, or in spite of that? In fact, this book narrates a quest after all, an abstract quest, to explain how being a pure mathematician is possible and, no less importantly, to explain why the reader should wonder why an explanation is needed—all

---

* The quotation is from (Collini 2013), about the subordination of higher education to market priorities—thus primarily concerned with the "good"; but it applies equally well to the two other ways of describing mathematics to which my title refers.

of which is just a roundabout way of saying that the book is about how hard it is to write a book about mathematics.

APOLOGIES NONETHELESS

The title is a transparent allusion to G. H. Hardy's *A Mathematician's Apology*, published in 1940, the closest thing pure mathematicians have to a moral manifesto. As such it tends to be read as a timeless declaration of principles, especially by outsiders writing about the ethos of mathematics. In chapter 10, I restore it to its historical setting and examine how the ethical obligations to which its author subscribes measure up against contemporary circumstances. "Apology" for Hardy meant an attempt to justify the choices he made in the course of his life according to the ethical standards of antiquity, as updated by G. E. Moore's *Principia Ethica*. Hardy's essay spurns useful Golden Geese and instead formulates the classic justification of mathematics on the grounds that it is *beautiful*.

This book does not offer an alternative apology (hence the title), nor does it claim that society owes pure mathematicians a comfortable (though hardly extravagant) living in our relaxed field. The point is rather that apologizing for the practice of pure mathematics is a way of postponing coming to terms with what its practitioners are really up to—what society gets in exchange for allowing pure mathematicians to thrive in its midst. But apologies are due, nonetheless, to my peers in other disciplines, as well as to the reader. In writing this book, I have drawn extensively, if not systematically, on the work of certain historians, philosophers, and sociologists, for whom mathematics is a matter of professional concern, and the reader may suspect me of making unjustified claims of disciplinary expertise. Separate apologies, or explanations, are therefore due to each of these professions.

To historians, I apologize if I have not been sufficiently diligent in insisting that what counts as mathematics—and what it means to be a mathematician—are in constant flux.* When the text alludes to mathe-

---

* The professional experience of mathematicians who teach at universities, which is where most research in pure mathematics takes place, has evolved considerably during my lifetime. Extended postdoctoral periods are now the norm in much of the world, for example, and attendance

matics of the past, including the distant past, I am not suggesting that I know what it shares, if anything, with the mathematics of my contemporaries. I am, rather, implicitly acknowledging the very different contention that a relation to the mathematics of the past is integral to the experience of being a contemporary mathematician. A professional historian can provide evidence that what a practicing mathematician believes about the mathematics of the past often involves significant distortions. Because my purpose is to give a snapshot of the mathematics I know, distorted ideas are no less valid than accurate ideas, and for the most part I try not to distinguish between them; they play an equivalent role in shaping the professional understanding of my peers. This does not mean I am necessarily accurate in presenting what my colleagues believe (accurately or not) about the history of mathematics, but I do try to be; on the other hand, I do not pretend to be professionally equipped to provide historical accuracy, much less to take sides in the extremely lively, sometimes bitter, controversies that persist in history of mathematics. Several chapters in this book, especially chapters 6, 8, and 10, contain material that may be relevant to questions that historians of mathematics seem not to have addressed; but this material was not gathered systematically, and I make no claims as to its significance.

By taking issue with the premise that mathematics is of interest to philosophy primarily by virtue of its relation to the *true*, I necessarily take sides in a debate that divides philosophers of mathematics. In the text I distinguish what I call philosophy of Mathematics from philosophy of mathematics. The former, practiced by the majority of the (relatively small) community of specialists, at least in English-speaking countries, concerns itself with traditional questions of truth, proof, and ontology and typically pays little attention to what mathematicians actually do, unless they happen to be logicians. Philosophers of (small-m) mathematics, in contrast, can be very attentive to the practice of mathematicians; indeed, there is a considerable overlap between my philosophy of (small-m) mathematics and the branch of philosophy called *philosophy of mathematical practice*. It's not surprising that when mathematicians do interact with philosophers, it's mainly with those of

at research and training conferences is much more common than it used to be, so that today a typical graduate student in a major research department is likely to have traveled to Paris, Berkeley, Princeton, Bonn, and some of the numerous centers in China.

the second tendency. Philosophical questions recur throughout the book but are especially prominent in chapters 3 and 7. If there were anything as structured as an "ism" in my approach to philosophy, it would probably be *naturalism*,* which as far as I can tell means the attempt on the part of philosophers to account for mathematics as mathematicians see it. But that's too obvious to mention; in any case, I have not been professionally trained to defend a philosophy of any kind; and I apologize to both branches of the philosophy of M/mathematics if at any point I appear to be overstepping my disciplinary prerogatives.

Readers will recognize that the framing of chapter 2 as an essay on the sociology of mathematics is mainly a conceit that allows me to introduce the main themes of the book—and most of the main *characters* as well. I nevertheless owe an apology to professional sociologists of mathematics (an even smaller community than the corresponding group of philosophers), inasmuch as my themes—how one acquires the values consistent with a certain appreciation (accurate or not) of the history of mathematics, for example, or the disposition to use words such as existence or proof or to say things such as "Mathematics is an Art"—are not the ones they generally prefer to address. Chapter 2 is more than mildly critical of sociologists of what I call the "genealogical" tendency for their focus on the Mathematics of the philosophers rather than the mathematics of mathematicians; but upon reflection, I realize I owe them a specific apology. It's thanks to the Sociology of Scientific Knowledge (SSK) and similar movements that I came to believe that the philosophical norms to which mathematicians (often implicitly) adhere arise from mathematical practice and not the other way around. This belief informs the entire book, and it would be perverse on my part not to acknowledge my debt to the genealogists.

Each of the three chapters in part 2 explores mathematical culture from a somewhat unconventional angle, and each one includes lengthy detours through a philosophical tradition—Russian Orthodox, Indian, and (medieval) Islamic—not usually cited in books about mathematics. This was by no means part of my plan when I started writing these chapters, but it turned out to fit my purpose. One of the most exciting

---

* It could just as well be called *Spinozism*, in reference to the opening of the *Political Treatise*, where Spinoza ridicules philosophers who "conceive of men, not as they are, but as they themselves would like them to be."

trends in history of mathematics is the comparative study across cultures, especially between European (and Near Eastern) mathematics and the mathematics of East Asia. These studies, which are occasionally (too rarely) accompanied by no less exciting comparative philosophy, is necessarily cautious and painstaking, because its authors are trying to establish a reliable basis for future comparisons. I have no such obligation because I am not trying to establish anything; my knowledge of the relevant literature is secondhand and extremely limited in any event; and the allusions in the middle chapters do nothing more than provide an excuse to suggest that an exclusive reliance on the western metaphysical tradition (including its antimetaphysical versions) invariably leads to a stunted account of what mathematics is about. In the same way, the occasional references to sociology of religion or to religious texts are NOT to be taken as symptomatic of a belief that mathematics is a form of religion, even metaphorically. It rather expresses my sense that the way we talk about value in mathematics borrows heavily from the discourses associated with religion.

And to all my colleagues in other disciplines, I want to apologize again for the introduction of idiosyncratic terminology (like mathematics versus Mathematics) or my use out of context of terms (such as *charisma* or *relaxed field*) with well-defined meanings. This is not intended as a sign of disrespect. I chose these words to give names to conceptual gaps I find particularly problematic, and for this very reason I would be disappointed if anyone else used them; that would be a failure to recognize that the gaps deserve rigorous analysis.

The remaining apologies are addressed to readers. First and foremost, I need to explain my own first-person appearances in the book. My mathematical life, while by no means exemplary, does offer some illustrative examples; but the main function of the "I" who occasionally surfaces to interrupt a general or abstract discussion is to provide narrative unity. I drew on my personal experience when it served to illustrate a larger point (when it was *ideal-typical* in the Weberian sense I try to mobilize in chapter 2). This is especially true in chapter 9: the story is as literally accurate as my memory allowed but the reader should recognize that in dramatizing the creative process the chapter makes use of familiar narrative conventions, including the quest narrative. For such conventions a protagonist is indispensable. In chapter 9 and elsewhere I occasionally fill that role, but other protagonists make brief appearances. They are taken from the rosters of prominent mathematicians past and present, real or fictional—or both at once, as is notoriously the case of Evariste Galois, prototype of romantic idealism in mathematics as well as forerunner of many of the themes that dominate contemporary number theory.

Although they were inspired by a (very brief) true incident, the series of chapters entitled "How to Explain Number Theory at a Dinner Party" are the least autobiographical as well as the most technical in the book. Each one, labeled by a Greek letter, is divided into two parts, the first an exposition of a basic notion in number theory, presented in approximate order of complexity, and the second a fictional dialogue designed to address questions that might have occurred to the reader while watching the mathematical material flow heedlessly to its conclusion. The thought was that there ought to be some number theory in a book about mathematics written by a number theorist, but since number theory has largely to do with equations, which lay readers reportedly find repellent, the equations and the surrounding number theory have been cordoned off in an alternative route, a separate tunnel from one end of the book to the other that can easily be visited but no less easily skipped.

One of the author's alter egos in the dialogues repeatedly invites the reader to think of a number neither as a thing nor as a platonic idea but rather as *an answer to a question*. Outside the fictional dialogues, I wouldn't put a lot of energy into defending this way of thinking—not because it's wrong but because the kind of number theory with which the book and its author are concerned cares less about the numbers themselves than about the kinds of questions that can be asked about them. The habit of questioning questions, rather than answering them, pervades contemporary pure mathematics. In this book, and especially in chapter 3, the habit spills over to question some of the questions that come to us from outside, the ones that seem to compel us to answer along the lines of "good," "true," or "beautiful." Chapter 3 is written in a more academic style than most of the others and is easily the most superficially "serious" part of the book. Readers who want to know what mathematics *is* and who don't need to be convinced that the subject is not exhausted by the good-true-beautiful trinity can safely skim or even skip chapter 3.

The expositions of mathematical material in the "Dinner Party" sequence are as scholarly as anything in the book, but it seemed to me that to maintain a serious tone in the dialogues that follow would make them impossibly stilted. Why this is the case inevitably raises the question of just how seriously mathematics is to be taken. My confusion on this point is such that I sincerely can't tell whether or not the question itself is a serious one. The position of this book is that what we understand as pure mathematics is a necessarily human activity and, as such, is mixed up with all the other activities, serious or otherwise, that we usually associate with human beings. The three chapters in part 2 try to make this apparent by depicting mathematicians in unfamiliar but recognizably human roles: the lover (or theorist of love), the visionary, and the trickster. Modes of mathematical seriousness come in for extensive examination throughout the book, especially in the final chapter of part 2. I can assure the reader who finds this unconvincing that my intentions, at least, are serious.

The contrast between seriousness and frivolity overlaps thematically with the opposition of high culture to popular culture. The presumption that mathematics is a high cultural form is one of those unexamined preconceptions that can stand to be examined; much of chapter 8 is devoted to precisely this question, but it comes up in other connections as well. I draw on popular culture because I find that it serves as a reliable mirror of the cultural status of high culture, and the representations of mathematics in popular culture have as much to do with sustaining the field's high cultural pretensions as do its (often-clichéd) representations by professionals.

Finally, although I have made an effort to list the many sources I consulted during the writing of this book and, although I have double-checked translations (sometimes with the help of native speakers) when they looked like cultural projections, the reader should not mistake this book for a work of scholarship. I tried to make sense of the material I did find, reconciling it as far as possible with mathematicians' accounts of their experience and with my own experience, but another author could well have consulted other sources and other experiences and come to very different, but no less valid, conclusions.

acknowledgments

The idea of writing a book of this kind came to me gradually, during the period between the Delphi meeting on *Mathematics and Narrative*, organized in 2007 by Apostolos Doxiadis and Barry Mazur, and the 2009 conference in Hatfield, England, on *Two Streams in the Philosophy of Mathematics*, organized by David Corfield and Brendan Larvor. All participants in both conferences encouraged me, directly or indirectly, consciously or not, to write this book, but without the consistent support in the early stages of the project of the four organizers of the two conferences and of Peter Galison, who was also at Delphi, this book would surely never have existed.

Several chapters are expanded versions of texts that have been published or presented elsewhere. An earlier and shorter version of chapter 5 was scheduled for presentation at the 2009 meeting of the Society for Literature, Science, and the Arts in Atlanta; I thank the organizers for accepting my proposal and Arkady Plotnitsky for presenting the paper in my place when a schedule conflict forced me to cancel. Chapter 7 is the third incarnation, much expanded, of a paper that was first presented at the 2009 Hatfield conference and again as a twenty-minute "After Hours Conversation" at the Institute for Advanced Study in 2011. I am grateful to Helmut Hofer and Piet Hut for making the 2011 version possible and to all the participants in both meetings for their generous attention and their challenging questions. The Hardy thread in chapter 10 originally appeared, in streamlined form, as a stand-alone essay in Issue 50 of *Tin House*, under the title *Unexpected, Economical, Inevitable*. Working with *Tin House* editor Cheston Knapp to get the article into shape, overcoming surprisingly deep differences in habits and expectations, was an immensely enjoyable and instructive experience, and it taught me lessons that I have tried to put to good use while writing this book.

The book began to take its present form began in the spring of 2011, while I was a Member of the Institute for Advanced Study in Princeton. The IAS has been one of the world's capitals of mathematical research since its founding in the 1930s, and it should inevitably figure prominently in any account of the culture of contemporary mathematics. If the IAS gets more than its strictly mathematical share of attention in this book, it's because I took advantage of (some of) the time I spent there to talk to scholars outside the School of Mathematics. I am grateful to Richard Taylor for inviting me to spend a term at the IAS and to then-director Peter Goddard for maintaining an environment conducive to this sort of interdisciplinary interaction.

I was fortunate to make use of the libraries at Princeton University and the IAS at the time and, later, the rich but scattered holdings in Paris in conjunction with preliminary searches on the Internet. Readers with access to the NSA's PRISM data center can probably reconstruct my Google searches, but even with the proper security clearance, current technology won't allow you to reconstruct the deep satisfaction I felt when, after finding a reference to an interesting item that had not yet been published or that was hidden behind a paywall, I was able to obtain the text in question from the author. There were many such authors, and practically all of them quickly responded to my requests. Invariably, the initial contact was continued with extended exchanges. Some authors were kind enough to look over the relevant chapters and make suggestions. Their names are listed below, alongside those of old friends and a great number of colleagues. Any mistakes that remain, of course, are my responsibility alone.

I thank Don Blasius for providing photographs of the array of mathematicians on the walls of the UCLA mathematics department, one of which was used in figure 2.1. Catherine Goldstein was kind enough to allow me to reproduce her father's drawing of the *Formule de l'amour prodigieux*, which appears here as figure 6.4. Alberto Arabia gave generously of his time to help me contact the officials of the Museo Nacional de Bellas Artes in Buenos Aires responsible for authorizing the reproduction of Luca Giordano's *Un matemático* (figure 8.2). I also thank Monica Petracci for permitting the inclusion of stills from *Riducimi in forma canonica* (figure 6.6) and William Stein for creating the Sato-Tate graph (figure δ.3).

Because the choice of wording or of individual words has been crucial in orienting the perception of mathematics by outsiders as well as practictioners, I owe special thanks to Alexey Nikolayevich Parshin, Marwan Rashed, Yuri Tschinkel, and Yang Xiao, who helped me decide among alternative translations of ancient and modern texts and who answered numerous other linguistic questions. In a handful of places I chose to ignore their advice and to include imperfect or tendentious translations when they were better suited to the narrative, but I have kept this sort of thing to a minimum. I also thank Bas Edixhoven, Kai-Wen Lan, and Teruyoshi Yoshida for telling me what mathematicians call "tricks" in their respective languages and Jean-Benoît Bost for helping me to understand how the word *astuce* came to fill that role in French.

For their thoughtful reading of early drafts of several chapters, for their challenging questions and suggestions, and for their constant encouragement, I am deeply grateful to Don Blasius, Harold Bursztajn, Emmanuel Farjoun, Edward Frenkel, Catherine Goldstein, Jeremy Gray, Graeme Segal, Yuri Tschinkel, Marina Warner, and Yang Xiao. My son Emiliano deserves thanks for many reasons, not least for the healthy skepticism he brought (and continues to bring) to the project and for finding several highly cogent quotations. I am especially indebted to David Corfield, Ian Hacking, Brendan Larvor, and Colin McLarty for their careful and critical reading of several versions of the philosophical sections and for the many references they provided.

In addition to the people just mentioned, a great many colleagues, friends, and virtual acquaintances supplied references, anecdotes, quotations, crucial background material, and kind words at difficult moments; asked questions; and read individual chapters, either making suggestions to improve clarity or coherence or simply reassuring me that the project was legitimate. Each deserves individual thanks, and I hope they will forgive me for instead simply listing them in an alphabetical roster: Amir Alexander, Stephon Alexander, Judy Arnow, Derek Bermel, Jean-Pierre Boudine, Martha Bragin, Olivia Chandeigne-Chevallier, David Daim, Chandler Davis, Sonia Dayan, François de Blois, Pierre Deligne, Joseph Dreier, Laurence Dreyfus, Reidar Due, Ivar Ekeland, Matt Emerton, Nina Engelhardt, Toby Gee, Neve Gordon, Christian Greiffenhagen, Mathieu Guillien, Rob Hamm, Peter Hogness, Juliette Kennedy, Irving Lavin, Pierre Lochak, Yuri Ivanovich Manin, Loïc Merel, Kumar Murty, Gaël Octavia, Frans Oort, John Pittman, Marwan Rashed, Catherine Rottenberg, Norbert Schappacher, Winfried Scharlau, Leila Schneps, Joan Scott, Rohini Somanathan, Glenn Stevens, Boban Velicković, T. N. Venkataramana, and the hosts and all the guests at that dinner in New York in 2008.

Even before I knew I was writing the book, Jean-Michel Kantor, Barry Mazur, Nathalie Sinclair, and Vladimir Tasić were already reading it. Their questions and suggestions have changed the book materially, always for the better. For this and for their friendship I am deeply grateful to them.

PART I

chapter 1

Introduction: The Veil

Who of us would not be glad to lift the veil behind which the future lies hidden; to cast a glance at the next advances of our science and at the secrets of its development during future centuries? What particular goals will there be toward which the leading mathematical spirits of coming generations will strive? What new methods and new facts in the wide and rich field of mathematical thought will the new centuries disclose?

—David Hilbert, Paris 1900

The next sentence of Hilbert's famous lecture at the Paris International Congress of Mathematicians (ICM), in which he proposed twenty-three problems to guide research in the dawning century, claims that "History teaches the continuity of the development of science."[1] We would still be glad to lift the veil, but we no longer believe in continuity. And we may no longer be sure that it's enough to lift a veil to make our goals clear to ourselves, much less to outsiders.

The standard wisdom is now that sciences undergo periodic ruptures so thorough that the generations of scientists on either side of the break express themselves in mutually incomprehensible languages. In the most familiar version of this thesis, outlined in T. S. Kuhn's *Structure of Scientific Revolutions*, the languages are called *paradigms*. Historians of science have puzzled over the relevance of Kuhn's framework to mathematics.[2] It's not as though mathematicians were unfamiliar with change. Kuhn had already pointed out that "Even in the mathematical sciences there are also theoretical problems of paradigm articulation."[3] Writing in 1891, shortly before the paradoxes in Cantor's set theory provoked a *Foundations Crisis* that took several decades to sort out, Leopold Kronecker insisted that "with the richer development of a science the need arises to alter its underlying concepts and principles. In this respect mathematics is no different from the natural sciences: new phenomena [*neue Erscheinungen*] overturn the old hypotheses and put others in their place."[4] And the new concepts often meet with resistance: the great Carl Ludwig Siegel thought he saw "a *pig* broken into a beautiful *garden* and rooting up all flowers and trees"[5] when a subject he had done so much to create in the 1920s was reworked in the 1960s.

Nevertheless, one might suppose pure mathematics to be relatively immune to revolutionary paradigm shift because, unlike the natural sciences, mathematics is not *about* anything and, therefore, does not really have to adjust to accommodate new discoveries. Kronecker's *neue Erscheinungen* are the unforeseen implications of our hypotheses, and if we don't like them, we are free to alter either our hypotheses or our sense of the acceptable. This is one way to understand Cantor's famous dictum that "the essence of mathematics lies in its freedom."

It's a matter of personal philosophy whether one sees the result of this freedom as evolution or revolution. For historian Jeremy Gray, it's part of the *professional autonomy* that characterizes what he calls *modernism* in mathematics; the imaginations of premodern mathematicians were constrained by preconceptions about the relations between mathematics and philosophy or the physical sciences:

> Without . . . professional autonomy the modernist shift could not have taken place. Modernism in mathematics is the appropriate ideology, the appropriate rationalization or overview of the enterprise. . . . it became the mainstream view because it articulated very well the new situation that mathematicians found themselves in.[6]

This "new situation" involved both the incorporation of mathematics within the structure of the modern research university—the creation of an international community of professional mathematicians—and new attitudes to the subject matter and objectives of mathematics. The new form and the new content appeared at roughly the same time and have persisted with little change, in spite of the dramatic expansion of mathematics and of universities in general in the second half of the twentieth century.

Insofar as the present book is about anything, it is about how it feels to live a mathematician's double life: one life within this framework of professional autonomy, answerable only to our colleagues, and the other life in the world at large. It's so hard to explain *what* we do—as David Mumford, one of my former teachers, put it, "I am accustomed, as a professional mathematician, to living in a sort of vacuum, surrounded by people who declare with an odd sort of pride that they are mathematically illiterate"[7]—that when, on rare occasions, we make the attempt, we wind up so frustrated at having left our interlocutor unconvinced, or at the gross misrepresentations to which we have resorted, or usually both at once, that we leave the next questions unasked: What *are* our goals? *Why* do we do it?

But sometimes we do get to the "why" question, and the reasons we usually advance are of three sorts. Two of them are obviously wrong. Mathematics is routinely justified either because of its fruitfulness for practical applications or because of its unique capacity to demonstrate truths not subject to doubt, *apodictically* certain (to revive a word Kant borrowed from Aristotle). Whatever the merits of these arguments, they are not credible as motivations for what's called *pure mathematics*—mathematics, that is, not designed to solve a specific range of practical problems—since the motivations come from outside mathematics and the justifications proposed imply that (pure) mathematicians are either failed engineers or failed philosophers. Instead, the motivation usually acknowledged is *aesthetic*, that mathematicians are seekers of beauty, that mathematics is in fact art as much as science, or that it is even more art than science. The classic statement of this motivation, due to G. H. Hardy, will be reviewed in the final chapter. Mathematics defended in this way is obviously open to the charge of sterility and self-indulgence, tolerated only because of those practical applications (such as radar, electronic computing, cryptography for e-commerce, and image compression, not to mention control of guided missiles, data mining, or options pricing) and because, for the time being at least, universities still need mathematicians to train authentically useful citizens.

There are new strains on this situation of tolerance. The economic crisis that began in 2008 arrived against the background of a global trend of importing methods of corporate governance into university administration and of attempting to foster an "entreprenurial mindset" among researchers in all potentially useful academic fields. The markets for apodictically certain truths or for inputs to the so-called knowledge economy may some day be saturated by products of inexpensive mechanical surrogate mathematicians; the entrepreneurial mindset may find mathematics a less secure investment than the more traditional arts. All this leaves a big question mark over the future of mathematics as a human activity. My original aim in writing this book was to suggest new and more plausible answers to the "why" question; but since it's pointless to say why one does something without saying what that something is, much of the book is devoted to the "what" question. Since the book is written for readers without specialized training, this means it is primarily an account of mathematics as a way of life. Technical material is introduced only when it serves to illustrate a point and, as far as possible, only at the level of dinner-party conversation. But the "why" will never be far off, nor will reminders of the pressures on professional autonomy that make justification of our way of life, as we understand it, increasingly urgent.

The reader is warned at the outset that my objective in this book is not to arrive at definitive conclusions but rather to elaborate on what Herbert Mehrtens calls "the usual answer to the question of what mathematics is," namely, by pointing: "This is how one does mathematics."* And before I return to the "why" question, I had better start pointing.

* *Ich gebe damit auch die übliche Antwort auf die Frage, was Mathematik sei: So macht man Mathematik* (Mehrtens 1990, p. 18).

chapter 2

How I Acquired Charisma

*J'ai glissé dans cette moitié du monde. pour laquelle l'autre
n'est qu'un décor.*

—Annie Ernaux

My mathematical socialization began during the prodigious summer of
1968. While my future colleagues chanted in the Paris streets by day and
ran the printing presses by night, helping to prepare the transition from
structuralism to poststructuralism; while headlines screamed of upheav-
als—The Tet offensive! The Prague spring! Student demonstrations in
Mexico City!—too varied and too numerous for my teenage imagination
to put into any meaningful order; while cities across America burst into
flames in reaction to the assassination of Martin Luther King and continued to smolder, I was enrolled in the Temple University summer program in mathematics for high school students at the suggestion of Mr.
Nicholas Grant, who had just guided my class through a two-year experimental course in vector geometry. It was the summer between tenth
and eleventh grades and between the presidential primaries and the unforgettable Democratic National Convention in Chicago. Like many of
my classmates, I was already a veteran of partisan politics. Over the
course of the summer, the certainty of the religious, patriotic, and famil-
ial narratives that had accompanied the first fourteen years of my own
life were shaken, in some cases to the breaking point. How convenient,
then, that a new and timeless certainty was ready and waiting to take
their place.[1]

At Temple that summer, I discovered the Men of Modern Mathematics poster that I subsequently rediscovered in nearly every mathematics
department I visited around the world: at Swarthmore, where the marvelous Mr. Grant drove me during my senior year to hear a lecture by
Philadelphia native L. J. Mordell, an alumnus of my high school and

G. H. Hardy's successor at Cambridge to the Sadleirian Chair of Pure Mathematics; near the University of Pennsylvania mathematics library, where I did research for a high school project; and through all the steps of my undergraduate and graduate education. The poster was ubiquitous and certainly seemed timeless to my adolescent mind but had, in fact, been created only two years earlier by IBM. Its title alludes to Eric Temple Bell's *Men of Mathematics*, the lively but unreliable collection of biographies that served as motivational reading that summer at Temple. You will have noticed at least one problem with the title, and it's not only that one of the "men" in Bell's book and (a different) one on the IBM poster are, in fact, women. Whole books can and should be devoted to this problem, but for now let's just be grateful that something (though hardly enough) has been learned since 1968 and move on to the topic of this chapter: the contours and the hierarchical structure of what I did not yet know would be my chosen profession when I first saw that poster.

It was designed, according to Wikipedia, by the "famous California designer team of Charles Eames and his wife Ray Eames," with the "mathematical items" prepared by UCLA Professor Raymond Redheffer."[2] Each "man" is framed by a rectangle, with a portrait occupying the left-hand side, a black band running along the top with name and dates and places of scientific activity, and Redheffer's capsule scientific biography filling the rest of the space, stretched to the length of "his" lifespan. As my education progressed, I began to understand the biographies, but at the time most of the names were unfamiliar to me. With E. T. Bell's help, we learned some of the more entertaining or pathetic stories attached to these names. That's when I first heard not only about the work of Nils Henrik Abel and Evariste Galois (see figure 2.1) in connection with the impossibility of trisecting the angle and with the problem of solving polynomial equations of degree 5—they both showed there is no *formula* for the roots—but also how they both died at tragically young ages, ostensibly[3] through the neglect of Augustin-Louis Cauchy, acting as referee for the French *Académie des Sciences*. What surprised me was that Abel and Galois both had portraits and biographies of standard size, while Cauchy, of whose work I knew nothing at the time, belonged to the very select company of nine Men of Mathematics entitled to supersized entries. The other eight were (I recite from memory)

Figure 2.1. Portraits of eminent mathematicians on the wall of the UCLA mathematics department, with Galois in front. (Photo Don Blasius)

Pierre de Fermat, Sir Isaac Newton, Leonhard Euler, Joseph-Louis Lagrange, Carl Friedrich Gauss (the "Prince of Mathematicians"), Bernhard Riemann, Henri Poincaré, and David Hilbert.

Maybe Bell's book and the IBM poster should have been entitled *Giants of Mathematics*,[4] with a special category of *Supergiants*, including the nine just mentioned plus Archimedes and a few others from antiquity (the poster's timeline starts in AD 1000). The hierarchy admitted additional refinements, the Temple professors told us. It was generally agreed—the judgment goes back at least to Felix Klein, if not to Gauss himself—that Archimedes, Newton, and Gauss were the three greatest mathematicians of all time. And who among those three was the very greatest, we asked? One of our professors voted for Newton; the others invited us to make up our own minds.

> The field of mathematics has a natural hierarchy. Mathematicians generally work on research problems. There are problems and then there are hard problems. Mathematicians look to publish their work in journals. There are good journals and there are great journals. Mathematicians look to get academic jobs. There are good jobs and great jobs. . . . It is hard to do mathematics and not care about what your standing is.
>
> In Wall Street every year bonus numbers come out, promotions are made and people are reminded of where they stand. In mathematics, it is no different. . . . Even in graduate school, I found that everyone was trying to see where they stood.[5]

That's hedge-fund manager Neil Chriss, explaining why he quit mathematics for Wall Street. But his analogy between finance and mathematics doesn't quite hold up. For mathematicians, the fundamental comparisons are with those pictures on the wall. "To enter into a practice," according to moral philosopher Alasdair MacIntyre, "is to enter into a relationship not only with its contemporary practitioners, but also with those who have preceded us. . . ."[6] Adam Smith, writing in the eighteenth century, found these relations harmonious:

> Mathematicians and natural philosophers . . . live in good harmony with one another, are the friends of one another's reputation, enter into no intrigue in order to secure the public applause, but are pleased when their works are approved of, without being either much vexed or very angry when they are neglected.[7]

Two centuries later, one meets a more varied range of personality types:

> In the 1950's there was a math department Christmas party at the University of Chicago. Many distinguished mathematicians were present, including André Weil. . . . For amusement, the gathered company endeavored to . . . list . . . the ten greatest living mathematicians, present company excluded. Weil, however, insisted on being included in the consideration.
>
> The company then turned to the . . . list of the ten greatest mathematicians of all time. Weil again insisted on being included.

Weil soon moved to the Institute for Advanced Study (IAS) in Princeton, and when, in the mid-1970s, a Princeton University graduate student asked him to name the greatest twentieth-century mathematician, "the answer (without hesitation) was 'Carl Ludwig Siegel.' " Asked next to name the century's second-greatest mathematician, he "just smiled and proceeded to polish his fingernails on his lapel."[8] Fifteen years later my colleagues in Moscow proposed a different ranking: A. N. Kolmogorov was by consensus the greatest mathematician of the twentieth century, with a plurality supporting Alexander Grothendieck for the second spot.

Hierarchy and snobbery are, naturally, not specific to mathematics. "Democracy should be used only where it is in place," wrote Max Weber in the 1920s. "Scientific training . . . is the affair of an intellectual aristocracy, and we should not hide this from ourselves."[9] In the nineteenth century, Harvard professor Benjamin Peirce, perhaps the first American

Figure 2.2. "Word Cloud" of twentieth-century mathematicians. Generated by José Figueroa-O'Farrill for the StackExchange Web site MathOverflow, based on frequency of references in (Gowers et al. 2008). See http://mathoverflow.net/questions/10103/great-mathematicians-born-1850-1920-et-bells-book-x-fields-medalists/10105#10105. Weil makes a respectable showing on the left, but Kolmogorov is barely visible on the right between Minkowski and Banach. Note the prominence of Einstein, who was not a mathematician.

mathematician to enjoy an international reputation, could "cast himself . . . as the enemy of sentimental egalitarianism . . . a pure meritocrat with no democracy about him."[10] Nowadays, of course, mathematicians are no less committed to democracy than the rest of our university colleagues. But we do seem peculiarly obsessed with ordered lists.[11] A lively exchange in 2009 on the collective blog MathOverflow aimed at filling in the gap between the last Giants of Bell's book and the winners in 1950 of the first postwar *Fields Medals*, awarded every four years to distinguished mathematicians under 40 and still the most prestigious of mathematical honors. The discussion generated several overlapping lists of "great mathematicians born 1850–1920"[12] and at least one novel graphic representation, as shown in figure 2.2.

The MathOverflowers—who mainly treat the blog as a forum for exchanging technical questions and answers—favored the word *romanticizable* to qualify candidates for inclusion in the list. You read that right: mathematicians are not only individually fit subjects for romantic idealization—we'll see a lot of that in chapter 6—but romanticizable *collectively*, as befits the theme of this chapter. This may seem odd if you haven't read Bell's book and even odder if you have run through the list of Fields medalists—or winners of prizes created more recently, such as the *Wolf Prize* (since 1978), the *Abel Prize* (administered since 2003 by the Norwegian Academy of Science and Letters as an explicit substitute for the missing Nobel Prize in mathematics), or the *Shaw Prize* (the "Nobel of the East," awarded in Hong Kong since 2004).[13] Few prizewinners are obvious candidates for biographical treatment in Bell's romantic vein, but the same could have been said of the majority of Bell's subjects before he got hold of them. Leaving aside the tragic cases of Galois and Abel (after whom the prize is named), perhaps the most authentically romantic figure in Bell's book was Sofia Kovalevskaia— "scientist, writer, revolutionary" is the subtitle of one of her biographies—the first woman to receive a PhD in mathematics; she traveled to Paris to witness the Commune of 1871 and devoted herself as energetically to prose and political activism as to her work on differential equations. Among prizewinners, Fields medalist Laurent Schwartz (1950) surely qualifies as romantic for his long and courageous commitment to human rights; so do Fields medalists Grothendieck (1966) and Grigori Perelman (2006) and (Wolf Prize winner) Paul Erdős, for reasons we will explore at length.[14] And Cédric Villani, with his "romantic" dress code, has been a fixture of Paris talk shows since he won the Fields Medal in 2010 (he's on the radio again as I write this, talking about ideal gases and how most efficiently to transport croissants to cafés). As for the others, their native romanticism has found no public outlet. This is not an accident.

Weber, writing a few years before Bell, had a romantic vision of science: "Without this strange intoxication, ridiculed by every outsider; without this passion . . . you have no calling for science and you should do something else." Did this change after the war? Not for impressionable teenagers, in any case. The romance of a vocation and the mystique of hierarchy intertwined when the undergraduate adviser received me, a first-year student still under the spell of a recent encounter with Cauchy's residue formula, in one of the alcoves of the Princeton mathematics department common room. His deep voice and distant demeanor made me think of fate incarnate.[15] We talked about course requirements and such, but my only distinct memory of the conversation is that at one point he said, "You want to be the world's greatest mathematician"; my

Figure 2.3. (a) A "romantic" mathematician of the nineteenth century: Sofya (Sonia) Kovalevskaya on a 1951 Soviet postage stamp. (The text reads: S. V. Kovalevskaya, outstanding Russian mathematician.) (b) A "romantic" mathematician of the twenty-first century: Cédric Villani in one of his trademark *lavallière* ties.
(© Hervé Thouroude)

immediate reaction was to wonder, How can he tell? Even at age 16, though, I had enough sense to recognize this as a sentence he had used before, and this meant there would be competition. My mother used to get birthday cards addressed to the World's Greatest Grandma, and while there is no more a World's Greatest Mathematician than there is an international ordered ranking of Grandmas, competition within the acknowledged hierarchy of mathematicians is fundamental to the elaboration of professional self-image, as Neil Chriss understood, in a way it's not for Grandmas. The aspiration to be something like the World's Greatest Mathematician is thus one of the concrete forms taken by the initiation into what I am going to call *charisma*.

Shortly after that common-room meeting I was *converted* to number theory—there's no other way to describe the experience—when a professor explained Weil's idea to count solutions to polynomial equations using *topology*, the geometric theory of shape. The equations of interest to Weil are like the equation in figure 2.4 (see p. 20) that was the object of Galois' attention, but they involve lots of variables, not only the single variable x, and thus describe geometric objects.[16] I had already learned what it meant to solve polynomial equations like figure 2.4 in my high school algebra class. I even knew what it meant to solve equations in more than one variable—that was L. J. Mordell's specialty and the subject of his talk at Swarthmore. The reader, however, may not know or may have forgotten what it means, in which case it is advisable to flip ahead and look at the first few pages of chapter β, which includes a sampling of polynomial equations in two variables, along with pictures (graphs) of their solutions. The use of geometry to understand solutions to polynomial equations and the use of algebra to solve problems in geometry comprise the field of *algebraic geometry*, whose history can be traced back through Descartes to medieval Islamic algebraists—and specifically to the Persian Omar al-Khayyām, the very first Giant on the IBM chart.[17] Weil's idea was that solutions to certain kinds of equations could be *counted*[18] by interpreting them as *fixed points*, the points that stay put under a motion of a sort of space; these fixed points could then be counted using a *fixed point theorem*. I had seen some fixed point theorems in one of the books Mr. Grant had advised me to read as a high school junior,[19] but those had to do with spaces that looked like balls or tori that had nothing to do with Weil's equations, and it was the very incongruity of the visual analogy that I found irresistible.

Be assured that this is not a series of clippings from my autobiography. "When the studies of a philosopher, and especially of a mathematician, have been described, his discoveries recorded, and his writings considered, his history has been written. There is little else to say of such a man: his private life is generally uninteresting and unvaried."[20] Too true! I can't even begin to imagine what might make for an interesting private life. The "I" of this chapter's title is not the hateful "I" of Blaise Pascal's *Pensées* but rather the hypothetical "I" of a Weberian *ideal type*. "Type of what?" Maybe we'll know by the end of this book. In this chapter it's the type associated with the possession of a specific degree of *routinized charisma*, with one curious feature: its subject crossed the threshhold relatively late, granting him an unusual and, he likes to think, unbiased perspective on life on both sides of the charisma divide.

The word *charisma* colloquially means a kind of personal magnetism, often mixed with glamour, but Weber chose the word to designate the quality endowing its bearer with *authority* (*Herrschaft*, also translated *domination*) that is neither *traditional* nor *rational* (legally prescribed). Charisma is (in the first place) "a certain quality of an individual personality, by virtue of which he is set apart from ordinary men and treated as endowed with supernatural, superhuman, or at least specifically exceptional powers or qualities," whose legitimacy is based on "the conception that it is the duty of those subject to charismatic authority to recognize its genuineness and to act accordingly."[21]

For mathematicians "acting accordingly" consists in participation in a *research program*. For example, the *Langlands program*, established by Robert P. Langlands—one of the great research programs of our time, very much in the way this term was understood by Imre Lakatos[22]—benefited from its founder's meticulous elaboration of the program's ultimate goals, too distant to serve as more than motivation, as well as a remarkably precise vision of accessible intermediate goals and the steps needed to attain them. In both these respects, the Langlands program, developed in several stages during the 1970s and thus part of my generation's collective memory, resembled the program promoted by Alexander Grothendieck in the 1960s, whose ultimate goal was to realize the implications of Weil's ideas about fixed point theorems.

It's for its subjective, romantic, and not altogether rational associations that I prefer charisma to words like *prestige*, *status*, *standing*, or *visibility* currently in use in the sociology of science.[23] Weber's primary target was religious leaders—Jesus, Mohammed, or the Buddha were extreme cases—but even mathematical charisma does not derive from "objective" external considerations. After Weil made his "topological" insight work for algebraic curves—a ten-year undertaking that required a complete rethinking of what he called the *Foundations of Algebraic Geometry*, the title of the most elaborate of the three books he wrote for the project—he formulated in three conjectures the outlines of a new geometry that would place topology at the center of number theory. For the next twenty-five years, the Weil conjectures served as a focus of charisma, what I would like to call a *guiding problem* in number theory and algebraic geometry, a challenge to specialists and a test of the rightness of their perspective. Grothendieck, who called them his "principal source of inspiration"[24] during his most active period of research, was the first to find a geometry that met Weil's specifications, proving all but one of Weil's conjectures and reorienting much of mathematics along the way, number theory in particular, before withdrawing from active mathematical research.

> Charisma . . . is imputed to persons, actions, roles, institutions, symbols, and material objects because of their presumed connection with "ultimate," "fundamental," "vital," order-determining powers. This presumed connection with the ultimately "serious" elements in the universe and in human life is seen as a quality or a state of being, manifested in the bearing or demeanor and in the actions of individual persons; it is also seen as inhering in certain roles and collectivities.
>
> What is alone important is how the individual is actually regarded by those subject to charismatic authority, by his "followers" or "disciples."[25]

Of course some mathematicians possess charisma in the colloquial sense as well as charismatic authority within the field.[26] Grothendieck was by all accounts such an individual. His close colleague, the French mathematician Jean-Pierre Serre, who received the Fields Medal in 1954, has for more than sixty years been one of the world's most influential mathematicians. In addition to his original research that literally reshaped at least four central branches of mathematics, he is an exceptionally gifted lecturer and an incomparable clarifier, whose books have been required reading since before I was a graduate student. The *Matthew Effect*, sociologist Robert K. Merton's name for the familiar tendency of prizes and honors to accrue to those who have already been honored,[27] works both ways: by awarding him the first Abel Prize in 2003, the Norwegian Academy borrowed Serre's charisma to secure their own legitimacy and to confirm the new prize's compatibility with community norms.

Serre is known for his effortless personal charm,[28] and in France, where politicians routinely point to the many French Fields Medalists as a mark of national glory, he could easily have become a media favorite. But he has limited his field of action to mathematics, apparently agreeing with Weber that the scientist who "steps upon the stage and seeks to legitimate himself through 'experience,' asking: How can I prove that I am something other than a mere 'specialist' ... always makes a petty impression." It "debases the one who is thus concerned. Instead of this, an inner devotion to the task, and that alone, should lift the scientist to the height and dignity of the subject he pretends to serve."[29]

Cambridge professor Sir Timothy Gowers (Fields Medal 1998, knighted in 2012) is not the sort to "step upon the stage," but unlike Serre, he has chosen to devote time and energy to exploring mathematical communication in a variety of forms. For professional mathematicians and the mathematically informed, he edited the encyclopedic *Princeton Companion to Mathematics* and runs a popular blog that covers topics of broad interest; for specialists he has pioneered "massively collaborative mathematics" online; and for the general public he has written a *Very Short Introduction to Mathematics*. He's very good at communication—he was chosen to give one of the talks in Paris at the meeting organized by the Clay Mathematics Institute for the 100th anniversary of Hilbert's 1900 Paris lecture—and he has also shown considerable courage, since it's commonly felt, at least in English-speaking countries, that a mathematician who willingly spends time on anything other than research must be short of ideas.[30]

"In every age," according to Northrop Frye, "the ruling social or intellectual class tends to project its ideals in some form of Romance." Class struggle is a poor guide to the history of mathematics;[31] it's more accurate to say that the romantic ideal is projected onto charismatic figures and any "ruling" is as likely as not to be posthumous. What Weber called the *routinization of charisma* helps explain why mathematical romanticism is mainly to be sought in a legendary (nineteenth-century) past. A professional mathematician will regularly enjoy the privilege of consorting, consulting, lunching, and even partying with Giants of Mathematics. This is consistent with the dynamics of the routinization of charisma, which does not necessarily divide mathematics socially into segregated Weberian "status groups" (*Stände*): "The very effort of a charismatic elite to stabilize its position and to impose a charismatic order on the society or institution it controls entails ... spreading the particular charismatic sensitivity to persons who did not share it previously. This means a considerable extension of the circle of charisma."[32] If we ignore charged verbs such as impose and controls that are misleading in the mathematical context, this sentence helps to understand how charisma can propagate from the "specifically exceptional" individuals to those (like the author of these lines) who occupy "certain roles" or participate in "collectivities."

"The charismatic leader," writes Pierre Bourdieu, "manages to be for the group what he is for himself, instead of being for himself, like those dominated in the symbolic struggle, what he is for others. He 'makes' the opinion which makes him; he constitutes himself as an absolute by a manipulation of symbolic power which is constitutive of his power since it enables him to produce and impose his own objectification." It may sound as if Bourdieu is saying that the charismatic leader, and the charismatic academic, in particular, exercises political power over the group. Whether or not this applies to mathematicians, it's not what I have in mind.[33] The bearer of mathematical charisma, consistently with Bourdieu's model, contributes to producing the objectification—the reality—of the discipline, in the process producing or imposing the objectification of his or her own position within the discipline. The reader can judge whether or not this is compatible with democracy—more material will be provided as the chapter progresses—but I want to stress that there is nothing deliberately misleading or sinister or even mysterious about this process, which is manifested in practice as well as in the perception of the field as a whole:

> In mathematics, many details of a proof are omitted because they are considered obvious. But what is "obvious" in a given subject evolves through time. It is the result of an implicit agreement between the reseachers based on their knowledge and experience. A mathematical theory is a social construction.[34]

The symbolic infrastructure of mathematical charisma is, likewise, a social construction, the result of an implicit agreement. But it is also the "objectification" of mathematics: the common object to which researchers refer, which in turn drives the evolution of aspects of mathematical discourse like the "obviousness" just cited. Does this mean that mathematicians can share only an understanding of mathematical theories and discoveries associated with charismatic individuals? No, but insofar as a mathematical theory is a social construction, one measures the importance of one's own contributions in terms of an accepted scale of values, which in turn is how charisma is qualified.

In other words, it's not just a theory's contents that are defined by a social understanding: so are the value judgments that organize these contents. Caroline Ehrhardt captures this process well in her account of the construction of *Galois theory*, decades after Galois' tragic and romantic death (my emphasis added to highlight effects of charisma):

> [W]riters of university textbooks not only played the passive role of collectors of research ideas . . . they also **created** mathematical knowledge in that when they introduced students to Galois theory, they offered **an organisation and a hierarchy of its constitutive elements** which were anything but established within the initial, fragmented landscape of local memories. In this way, these authors **structured the mathematical field**; they **redistributed symbolic capital between the authors**, they **defined which objects are legitimate**, which orientations took precedence, and they enabled the constitution of a community of specialists who had received the same kind of training. Mathematical content and practice thus **defined the social space** corresponding to Galois theory at the end of the 19th century within the mathematical field.[35]

Galois, you'll recall, showed that there is no formula for finding the root of a polynomial equation of degree 5 or more, like the one shown in figure 2.4. So did Abel. But Galois, before he died in a duel at the age of 20, did much more for equations: he invented a *method* for understanding their roots, even in the absence of a formula—the *Galois group*, which governs what, for want of a more precise term, one would now call the *structure* of the roots of an equation. Together with his successors who "defined the social space" of Galois theory, he also created a new *point of view*: that what's interesting is no longer the centuries-old goal of finding a root of the equation, but rather to understand the structure of all the roots—the Galois group. This is a stage in an ongoing process of abstraction; it is also a change in perspective. Today's mathematicians, especially number theorists, have taken this process one step further: rather than focus on the Galois group of a single equation, number theorists look at all the Galois groups of all equations simultaneously as a single structure. Is this a paradigm shift or a priority shift?

$$x^5 + 40x - 3 = 0$$

Figure 2.4. There is no formula for the solutions x of this polynomial equation. But they have a *structure* determined by the Galois group.

Once it is routinized, charisma can be quantified. These days a popular name for quantifiable academic charisma is *excellence*, the recent pursuit of which has spawned an immense secondary literature and bureaucratic infrastructure, notably in Europe. The exact day I and all my colleagues at the Université Paris-Diderot acquired *excellence* is recorded in the French *Journal Officiel*. The news arrived in my inbox on February 3, 2012, that my university, together with its seven partners, had been chosen as one of eight *Initiatives d'excellence*. These are the French contribution to an effort to improve the position of European universities in international rankings:

> A metric that purports to show the stock and presumed rate of creation of new intellectual capital in those institutions, universities, whose role it is [to] conserve and create it, becomes the equivalent of a stock market quotation, and also an index of national intellectual prestige.[36]

The *League of European Research Universities* (LERU) cautioned against attaching too much importance to these metrics, writing that they "promote [. . .] the idea of the university as a supermarket selling modular products that happen currently to be in vogue" and that "It's about measuring a brand, like Gucci or Chanel. It almost paints universities as a fashion accessory."[37] But excellence is the professed goal, and a system of material rewards and penalties is the means chosen to encourage us to strive for excellence.

The rules of bureaucratic rationalization therefore require quantitative metrics. For scientists the typical metrics count citations, weighted for "impact." The market in impact factors, jealously guarded by Thomson Reuters and similar commercial services, is rooted in early attempts by Mertonian sociologists to quantify scientific "visibility." The brothers Jonathan and Stephen Cole, both students of Merton, demonstrated in their statistical analysis of the reward system in physics that *honors* correlate more highly with *quality* than with *quantity* of research; but to reach this conclusion they needed an objective measure of quality, and citation indices were their metrics of choice.[38] Mathematicians, whose papers remain relevant for decades, tend to dismiss citation metrics adapted to the experimental sciences, such as the h-index,[39] as poor indicators of mathematical influence. Andrew Wiles, for example, proved Fermat's Last Theorem (see chapter δ) as well as a long list of theorems of the highest importance in number theory (see note 52 for the first of the list) and is undoubtedly the most famous of all living number theorists; but his h-index of 13 would just barely qualify him for tenure if he were a physicist. On the other hand, I doubt my colleagues would have much sympathy for the view, promoted in the 1960s by sociologist Warren Hagstrom and still occasionally cited, that mathematics is Durkheimian rather than Weberian, characterized by *anomy* [*sic*] rather than charisma—that mathematicians are so narrowly specialized that we don't read one another's papers and couldn't understand them if we did.[40]

Chapter 10 shows how "excellence"—it is no coincidence that "excellence" is a translation of the Greek word *aristos*, as in Weber's "intellectual aristocracy"—colludes with market forces to shape European research budgets. Innovations with immediate practical applications, such as radar, image processing, or cryptography, are obviously subject to external judgments of excellence or profitability or whatever norms are favored by those in a position to impose them. What's specific to *pure* mathematics is that it is obliged to create its own value system—and that it has the means to do so. "The essence of mathematics," Cantor might have written, "lies in its *socially constructed* freedom." And because this freedom is inseparable from value judgments, its starting point is the charismatic hierarchy it tends (with revisions) to reproduce.

Of course charisma in mathematics, as in other academic disciplines, brings power in the conventional sense: power to organize one's time, power to set the research agenda, power to attract talented students and to place them in prominent positions, as well as material perks, including the generous salary that (in the United States, at least) helps distinguish a "great job" from a "good job." Departmental rankings are broadly charisma based, so that a professor at one of the top U.S. mathematics departments will be perceived as charismatic, and a professorship like Weil's at the IAS, the pinnacle of pure mathematics research practically since its creation, is presumptive evidence of Gianthood.[41] The Cole brothers argued that "there is substantial overlap in the groups having power and those having prestige" (charisma in my sense). The two overlapping groups are the "scientists who have earned recognition for their outstanding contributions to knowledge and those who hold key administrative positions." The first group "extends the circle of charisma" by distributing what Bourdieu calls "symbolic capital": Chriss's "great jobs," publication in "great journals". . . ; the second group distributes material resources ("[t]hey can determine what specific research areas are to receive priority, and what individuals are to receive support for their research programs").[42]

Let's leave the second group to professional sociologists and focus on the first group. A mathematician derives *authority* from being an *author*—the two words have the same root—but if one asks, with Michel Foucault, "What is an author?" one begins to get a sense of how peculiar the constellation of power around a charismatic mathematician looks, compared to the authorial aura of a charismatic scientist. Unlike articles in particle physics or biomedicine that can be signed by a thousand authors or contributors, it's unusual for articles in pure mathematics to have more than two or three authors, and single-author articles are common. Mathematics research, as opposed to mathematics publishing, is, nevertheless, intensely collaborative, and breakthroughs are always prepared by years of preliminary work, whose authors are usually given full credit by specialists and are quickly forgotten by nonspecialists. This *communalism*, in Merton's sense, is especially relevant in highly visible research programs like those of Langlands and Grothendieck. Langlands himself has observed that

> when the theorem in which the solution is formulated is a result of cumulative efforts by several mathematicians over decades, even over centuries . . . and when there may have been considerable effort—the more famous the problem the more intense—in the last stages, it is not easy, even for those with considerable understanding of the topic, to determine whose imagination and whose mathematical power were critical.

Langlands contrasts those responsible for "the novelty and insight of the solution" to mathematicians "whose contributions were presented with more aplomb and at a more auspicious moment."[43] So the question of how credit—and, therefore, authority—are apportioned has real conse-

quences, and it is a shame that its sociological and philosophical underpinnings are so poorly understood. Foucault has left a hint. Alongside a
mathematical treatise's historically determinate author and the "I" who
serves as the subject of the proofs, with whom the reader identifies by
accepting the rules in force, Foucault alludes to a "third self, one that
speaks to tell the work's meaning, the obstacles encountered, the results
obtained, and the remaining problems . . . situated in the field of already
existing or yet-to-appear mathematical discourses."[44] This authorial self
places each new work in one or more of the discourses instituted by
those figures peering down from the (metaphorical) wall. Highly charismatic mathematical authors have a more fundamental responsibility:
not only are they in dialogue with the portraits on the wall, whether or
not they have a scholarly interest in the details of history; they—and not
"nature" —mediate this dialogue for the rest of us, including those who
contribute to their research programs.

• •

The way a graduate student, barely past 20, chooses a thesis adviser, and,
in so doing, nearly always determines a permanent career orientation,
has always fascinated me. Is the choice based on an ineffable preexisting
harmony or is it true, as Pascal thought, that "chance decides"? Did I ask
Barry Mazur to supervise my Harvard PhD because I admired his research style that unites methods and insights in novel and unexpected
combinations or because he was (and still is) one of the few people I've
met who can without hesitation speak engagingly and cogently on practically any topic—because of his personal charisma in the familiar
sense? Does it even make sense to separate his personal style from his
research style? However I made my choice, I soon found myself caught
up in the thrill of the first encounter between two research programs,
each of a scope and precision that would have been inconceivable to
previous generations, each based on radically new heuristics, each experienced by my teachers' generation as a paradigm shift. Hilbert, whose
quotation opens the previous chapter, pioneered the conception of mathematics as a practice oriented to the future, whose meaning is defined
less by what we can prove than by what we expect to be able to prove.[45]
But nothing in Hilbert's list of twenty-three problems could compare
with the detailed predictions of Robert Langlands's program for number

theory in the light of *automorphic forms* or of the conjectures deriving from Alexander Grothendieck's hypothetical theory of *motives*, both on the brink of spectacular expansion when I entered the field, both too complicated to do more than shadow my narrative.

The IHÉS (Institut des Hautes-Études Scientifiques) was created outside Paris in conscious imitation, on a smaller scale, of Princeton's IAS; Grothendieck was one of the first permanent professors. His IHES seminars in the 1960s, collected in a series of volumes known as SGA— *Séminaire de géométrie algébrique*, or Seminar of Algebraic Geometry[46]—were both symbolically and practically the founding acts of a school of experts, mainly Grothendieck's own students, who wrote most of the text and continued the research program after the leader's hermetic withdrawal from active research in 1972. Langlands, who joined the IAS as Hermann Weyl Professor that same year, promoted his unifying vision for automorphic forms and number theory for two decades, serving as the focus of research for widening circles of specialists before taking a break to think about physics; over the past fifteen years, he has returned to his guiding role with a "reckless" new approach to his conjectures.

Steven Shapin points out in *Scientific Life* that the word *charisma* is "the word used by both participants and commentators" to describe the "personally embodied leadership" exercised by scientific entrepreneurs. Quoting the economist Richard Langlois, Shapin argues that "charismatic authority 'solves a coordination problem'... in circumstances of radical normative uncertainty" and that the use of the word *charisma* in this way "is a consequential, reality-making usage."[47] Leaders of mathematical research programs, such as Grothendieck and Langlands, exercise their charismatic authority in a very similar way, even in the absence of entrepreneurial incentives.

Langlands and Grothendieck are both (at least) Giants by any measure, and both were consciously successors of Galois. In his attempt to prove the last of Weil's conjectures—solved instead by his most brilliant student, 1978 Fields medalist and IAS Professor Pierre Deligne—Grothendieck imagined a theory of elementary particles of algebraic geometry, called *motives*. The theory, still far from complete, represents one vision of a Galois theory of algebraic geometry, the geometry of equations. The Langlands program is another vision of Galois theory: his

Figure 2.5. Grothendieck lecturing at the Séminaire de Géométrie Algébrique (SGA), IHES, 1962–1964.

unexpected insight was that the structure of the theory of *automorphic forms*, rooted in a different geometry, the geometry of mechanics, is in large part determined by Galois' theory of polynomial equations. In a famous article entitled *Ein Märchen*—a fairy tale—Langlands speculates about a reformulation of number theory, also still incomplete, in terms of elementary particles (*automorphic representations*) that mirror Grothendieck's (hypothetical) motives but include additional mysterious particles that—unlike motives—the Galois group is not equipped to detect.[48]

Grothendieck, whose relations to the material world (and to his colleagues) are far from straightforward (see chapter 7 and his underground memoirs, *Récoltes et Sémailles*), may well qualify as the last century's most romantic mathematician; his life story begs for fictional treatment. Langlands' life has been by no means as extravagant as Grothendieck's, but his romanticism is evident to anyone who reads his prose; the audacity of his program, one of the most elaborate syntheses of conjectures and theorems ever undertaken, has few equivalents in any field of scholarship. While neither of them has ever "ruled" in Frye's sense, their research programs, like Hilbert's problems, typify charisma by focusing attention and establishing standards of value for specialists. They also provide a way for colleagues in other branches of mathematics to judge the importance of someone's research: proving a central conjecture of a named research program often suffices to establish one's charisma in the eyes of such colleagues, who may only dimly apprehend the program as a whole. Three of the seven *Clay Millennium Problems*, announced at the Clay meeting in Paris in 2000 by an all-star roster of speakers, including Gowers as well as John Tate (Abel Prize 2010) and Sir Michael Atiyah (Fields Medal 1966, Abel Prize 2004), in a conscious echo of Hilbert's problems—but each offering a million-dollar reward for its solution—predate but are at least loosely connected with one or both of these research programs.

Number theorists began building bridges back from the Langlands program to motives in connection with problems like the *Conjecture of Birch and Swinnerton-Dyer* (BSD), one of the three Clay problems related to number theory,[49] which served as the guiding problem for my own entry into research. Just as Weil's conjectures were about counting solutions to equations in a situation where the number of solutions is known to be finite, the BSD conjecture concerns the simplest class of polynomial equations—*elliptic curves*—for which there is no simple way to decide whether the number of solutions is finite or infinite. Here are the equations of two elliptic curves:

$$y^2 = x^3 - x, \tag{E1}$$

$$y^2 = x^3 - 25x. \tag{E5}$$

We will see these curves again in chapters γ and δ, along with a whole family labeled EN (one for every number N). It can be shown (but not in a simple way) that equation (E1) has only finitely many solutions, while (E5) has infinitely many solutions.

Each of these equations belongs to Grothendieck's research program—it can be seen as a motive—and to Langlands' program—it can be seen as an automorphic form.[50] The BSD conjecture states roughly that the number of solutions can be determined by properties of this automorphic form. That is not how Birch and Swinnerton-Dyer initially formulated their conjecture, but a few years later, just before I arrived at Harvard, the Langlands perspective was gaining ground. John Tate and Barry Mazur, Harvard's two number theorists at that time, had both been working on the BSD conjecture; they were also both actively studying Langlands' work, and their PhD students were encouraged to work on problems connected to Langlands' program (see chapter 9).

Between 1976 and 1993, there were successive striking partial solutions to the BSD and closely related problems by six groups of authors. Although the full BSD conjecture remains unsolved (and its solution is still worth $1,000,000), each of these authors was immediately and dramatically rewarded—not materially, though material rewards eventually came to some of them—but with an appropriate degree of *charismatic authority* within the community, in accordance with the general principles of continuity of Kuhn's "normal" science:

> Because the criteria of scientific adequacy and value within any research area are so specialized and subtle, the task of deciding which contributions are acceptable and which are important tends to be given to those who are already recognized as having contributed significantly to the field . . . scientists who wish to advance their careers and/or to produce acceptable contributions to knowledge must comply with the cognitive standards set and exemplified by these leaders.[51]

It goes without saying that their work was published in "great journals," and (as far as national circumstances allowed) they were offered "great jobs" at institutions where they could hope to advise "great students."[52] The gates of authority opened to welcome them to conference committees, journal editorial boards, and prize panels, alongside or replacing their former mentors in accordance with the informal rules maintaining the organizational balance of the discipline's distinct branches. Hiring and promotion committees solicited their advice—and Deans wanted additional information—in a process that offers valuable insight for a future sociology concerned with the reproduction of the mathematical elite, since the authors in question had without exception spent some of or all their careers studying or teaching at the leading poles of number theory in their respective countries: Oxford, Cambridge, Harvard, Princeton, Bonn, the École Normale Supérieure, Moscow State University, and the University of Tokyo. They spoke at the most important international seminars, where their work was examined and its lessons were absorbed by the most qualified specialists, and at colloquia at peripheral as well as central mathematics institutes, where students and colleagues from all branches of mathematics were invited to share the dreams and aspirations of number theory for the space of an hour. A few of the authors even had the supreme honor of addressing a plenary session of the International Congress of Mathematicians (ICM)—as Hilbert had done in Paris in 1900, when he announced his list of problems. Their names, hyphenated in the case of joint work, became shorthand for breakthroughs and served as titles for intensive study groups around the world, where a new generation of students, by seeking to eliminate unnecessary hypotheses and to streamline arguments, transformed each landmark article into the founding document of a new research program, branches of which persist today as semiautonomous subcultures within the broad scope of the Grothendieck or Langlands program or both at once.

The Langlands program remained in the background in all the work I've just mentioned, since less elaborate methods sufficed for the authors' purposes. But they would not do for the sequel, which is why I gradually moved away from a direct involvement with BSD and began thinking about—thinking *by means of*—the Langlands program. This lateral movement across subdisciplinary boundaries would have been conceptually impossible without the help of my colleagues—I've always been better at learning from conversations than from reading research articles, which is probably not uncommon in the profession—but it would have been materially impossible had I not been encouraged and tutored by a series of professors who occupied leading positions within the respective subdisciplines and were thus in an authoritative position to represent my work as significant from their charismatically enhanced standpoints.[53] Concretely, my mentors and validators wrote the letters that convinced the hiring and promotion committees—and ultimately the Deans—that I was on the way to acquiring the charisma I would need to function effectively within the routine at the level of the hierarchy to which I was soon successfully appointed.

• •

The Cole brothers' use of citation indices to quantify *visibility* and *influence* is characteristic of the Mertonian school that dominates what, taking a hint from philosopher David Corfield, one might call the *encyclopedic* tendency in sociology of science. Corfield borrowed the term to adapt Alasdair MacIntyre's tripartite division of "moral inquiry" to the philosophy of mathematics. Alongside the encyclopedic approach, associated with various strains of positivism, MacIntyre places the *genealogical* (inspired by Nietzche or Foucault but often conflated with historical relativism) and the *tradition-based*, which he and Corfield favor in their respective fields. Sociology of science can be similarly divided. Encyclopedic sociologists of science (to oversimplify) follow their philosophical counterparts in leaving unexamined the assumptions of practitioners—or the assumptions philosophers attribute to the latter—and concentrate on factors they are prepared to consider objective, which in practice means quantifiable. For the genealogists, on the other hand, specifically those associated with the *Sociology of Scientific Knowledge* (SSK) program, everything practitioners say is suspect.[54] David Bloor, for example, has written that "[t]he unique, compelling character of mathematics is part of the phenomenology of that subject. An account of the nature of mathematics is not duty-bound to affirm these appearances as truths, but it is bound to explain them as appearances." Bloor continues, "It is a notable characteristic of some philosophies of mathematics"—he has the encyclopedic approach in mind—"that they uncritically take over the phenomenological data and turn them into a metaphysics." This implies that "there can be no sociology of mathematics in the sense of the strong programme. What is required is a more critical and naturalistic approach."[55]

Appearances notwithstanding, my intention in this chapter is not to indulge in amateur sociologizing; but it's true that the questions I've raised are among the topics I'd like to see covered in a sociology of mathematics—a sociology of *meaning* as opposed to encyclopedists' exclusive emphasis on the implementation of preexisting consensual norms or the genealogists' focus on effects of power. Questions of meaning fit best with the tradition-based perspective—or so a participant in the tradition might be expected to say—but the sociology (as opposed to philosophy) of mathematics doesn't seem to have a tradition-based wing. I find it striking that among the philosophers and the (rare) sociologists who write about the subject, neither encyclopedists nor genealogists actually have much to say about mathematics. Instead, mathematics tends to serve as an excuse either to promote—or, as in the previous Bloor quotation, to strike a blow against—a questionable ideology, promoted by certain philosophers, most long dead.

(Controversies among philosophers, especially encyclopedist philosophers, surrounding the word *meaning*, which has just made its third appearance in this chapter and which will reappear repeatedly throughout the book, are as lively as they get. My intention in introducing the word is not to enter into these controversies, but rather to give the reader a sense of how it is used by pure mathematicians, even if the reader happens to be a philosopher. One theme of chapter 7 is that this use is fraught with paradox—which doesn't make it less meaningful.)

The sociological texts I've seen specifically devoted to mathematics mainly belong to the genealogical tendency,[56] but with few exceptions the encyclopedists and genealogists are united by their grounding in what I will be calling philosophy of Mathematics.[57] Capital-*M* Mathematics is a purely hypothetical subject invented by philosophers to address (for example) problems of truth and reference, which these days presupposes an outsized concern for logical formalization, with very little attention paid to what matters to mathematicians. The encyclopedic and genealogical tendencies are for the most part variants of the philosophy of Mathematics, the main difference being their differing degree of credulity versus skepticism regarding truth claims.[58] Philosophy of (small-*m*) mathematics, on the other hand, takes as its subject of inquiry the activity of mathematicians and other mathematically motivated humans—the basis, one would have thought, of what Bloor calls the "phenomenological data."[59]

Genealogists are especially concerned with the achievement of *epistemic consensus*, how mathematicians come to agree on the truth despite the paradoxes and antinomies that plague reason, but ignore *charismatic consensus*, whereas the encyclopedists treat the latter as a functional question, without looking into content. Christian Greiffenhagen, a sociologist trained in the SSK perspective, observes in a literature review that, thirty years after Bloor wrote the words quoted earlier, "There is almost a complete absence of anthropological or sociological studies of professional mathematics." By this he means there have been few "laboratory studies" of "science in the making" of the sort pioneered by SSK, and he proposes explanations as to why this might be the case. His own "video ethnography" of "situations in which mathematical competence is accountably visible," concerned as it is with mathematical communication, necessarily addresses questions of meaning.[60]

As for the third tendency, when Corfield and MacIntyre use the word *tradition* they mean it to be read descriptively; MacIntyre makes a point of divesting it of its associations with (traditional) conservatism. Within the tradition-based perspective, Corfield emphasizes the meaning attached to *value* rather than *truth*, to ethics rather than epistemology: "[I]f we wish to treat the vital decisions of mathematicians as to how to direct their own and others' research as more than mere preferences," he writes—in other words, if we want to treat these decisions as following from "rational considerations"—then value has to be admitted as a matter for rational analysis. "[I]t is the notion of progress toward a *telos*"—a goal—"that distinguishes genealogy and tradition."[61]

An example may help illustrate how a sociology of meaning can illuminate what Corfield means by *telos*. If Langlands was alone with a few of his friends and associates during his program's Act 1, extending, refining, and revising the outlines of his conjectures through calculations and a series of letters, Act 2 mobilized an international cast to verify these conjectures in a restricted situation called *endoscopy*, where Langlands thought his preferred technique would suffice if elaborated with sufficient diligence and attention.[62] Twenty years of intense work completed most of the steps required to bring Act 2 to the conclusion Langlands had marked out as an intermediate *telos*. But one vexing bottleneck remained, a series of calculations of the form

$$\text{quantity } G = \text{quantity } H \qquad \text{(FL1)}$$

conjectured by Langlands and Diana Shelstad and called the *fundamental lemma*. The earliest examples of (FL1) could be checked by fairly simple calculations of both sides of the equation, but the calculations quickly became tedious and then intractable, and by the year 2000 experts were convinced that the only sensible way to prove (FL1) was by deriving it from an identification of the form

$$\text{object } \mathscr{G} = \text{object } \mathscr{H} \qquad \text{(FL2)}$$

where $\mathscr{G}$ and $\mathscr{H}$ are objects in Grothendieck's geometry. Since Grothendieck's and Langlands' programs both emanate from Galois theory, it's not surprising that they overlap. Pierre Deligne, the Russian Vladimir Drinfel'd, and Langlands himself had been applying Grothendieck's methods to the questions raised in Act 1 since the 1970s, and two Fields Medals had been awarded for such work (to Drinfel'd in 1990 and to Laurent Lafforgue in 2002) by the time the Vietnamese mathematician Ngô Bảo Châu realized that the geometric objects needed for (FL2) could be built out of the repertoire of mathematical physics.[63] Here, by the way, Weil's and Grothendieck's original priorities are reversed: instead of using the results of counting to say something about the geometric objects—to pin them down as *motives*, for example—Grothendieck's methods provide the quantities of interest from the objects, and the identification (FL2) is used to avoid calculation of the complicated quantities in (FL1)—which brings Act 2 to a close, since in the applications all one needs to know is the equality in (FL1) and not the quantities themselves.

To explain why it was at this point generally agreed that Ngô's work was worthy of the Fields Medal—which he received in 2010—requires a vocabulary of value, of meaning, of *telos*. Had he devised the objects $\mathscr{G}$ and $\mathscr{H}$ and proved (FL2) without reference to the *telos* of the Langlands program—without the goal of deducing (FL1)—it would have been deep and difficult but unlikely to have attracted much notice; the Grothendieck tradition provides the means for proving any number of identifications like (FL2) but would not have recognized this one as furthering its specific goals. On the other hand, if Ngô had simply proved the experts wrong by displaying enough stamina to calculate the two sides of (FL1) directly—and before you set out to flaunt your own stamina I should warn you that (FL1) is shorthand for infinitely many separate calculations that would somehow have to be carried out simultaneously—he would have completed the *telos* of Act 2. Rapturous applause and offers of "great jobs" would have greeted the news. But Ngô would probably not have won the Fields Medal. This is because such a calculation has no *meaning* outside Langlands' endoscopy project, whereas (FL2)—precisely when it is seen against the background of the Langlands program—is the starting point for a new and open-ended research program (overlapping with Act 3 of the Langlands program, the "reckless" new initiative mentioned previously).

Ngô's prize-winning achievement was to create a *synoptic* proof, a proof that made the result "obvious" to those trained to see it, that served a specific *telos* within the overlapping traditions of the Grothendieck and Langlands research programs, and these programs link him both historically and conceptually to that poster on the wall and specifically to the figure of Galois. The danger of a tradition-based ethic is that it limits both understanding and participation to what David Pimm and Nathalie Sinclair call an "oligarchy" (Weber's "aristocracy"), whose rule, exercised "through the explicit notion of 'taste,'" excludes anyone outside the tradition. Asking "[I]n what sense . . . can mathematics be considered a democratic regime . . . " open to all, Pimm and Sinclair quote (Supergiant) Henri Poincaré to the effect that "only mathematicians are privy to the aesthetic sensibilities that enable" the decision of "what is worth studying."[64] The article, published in a journal for educators, is motivated by the "view that mathematics can do something for me in a humanistic sense that repays the careful attention and deep engagement it may require; one that may expose students to a fundamental sense and experience of equality . . . and provide them with another sense of human commonality."

The concern that mathematics is structured as a gated community to which only professional mathematicians are admitted calls to mind William Stanley Jevons's advocacy of "the better class of dance music, old English melodies, popular classical songs" as a vehicle for the moral improvement of nineteenth-century Britain's untutored working classes, who lacked the "long musical training" needed to appreciate "great musical structures," or the more recent defenders of elite standards against the encroachment of mass culture.[65] This is a debate mathematicians cannot afford to ignore, but this chapter is concerned only with a more specialized question: whether the hierarchy of values is compatible with democracy within the profession.

Rational authority in mathematics is vested not only in national (and in Europe supranational) administrative bodies responsible for research funding and orientation, but especially in national professional organizations. Officers of these organizations are elected by the membership; they, in turn, elect the leadership of the International Mathematical Union (IMU), whose functions include the organization every four years of an International Congress and defending the perceived interests of mathematics. The IMU and certain national professional and governmental bodies, including the AMS, SIAM, and NSF[66] in the United States, devote substantial resources to improving access for mathematicians from developing countries and for women and (at least in the United States and Britain) minorities in the metropolitan centers—promoting inclusion in mathematics as presently constituted (the term in the United States is *diversity*)—and fostering the growing awareness of the historical importance of long-standing mathematical traditions outside the European cultural sphere, notably in China, India, Japan, and the Islamic world. The main thrust of this work has been to insist on the underlying unity of mathematics, even in traditions that could not influence one another, rather than to suggest (as sociologists of the genealogical tendency might) that traditional Chinese and Greek mathematics, say, are based on norms that are fundamentally different and, presumably, equally valid. The success of the inclusion of mathematicians from all over the world is ascribed to the field's universality rather than the victory of the European worldview.

Specialized mathematics institutes and conference centers[67] display their rational authority by organizing the day-to-day and year-to-year activity of mathematical research, setting priorities by devoting time and resources to areas and problems identified as important by their national or international scientific advisory boards. I started writing this book during a three-month membership at the IAS in Princeton, as a participant in a year-long program on *Galois representations and automorphic forms*. Unofficial bodies can also exercise a kind of semirational authority. The most celebrated example is the ostensibly secret Bourbaki association, founded in Paris in the 1930s by André Weil and a few of his contemporaries to reform mathematics according to a clear conceptual "architecture." The Bourbakistes, including Serre, Tate, Grothendieck, and Deligne, just to list those mentioned in this chapter, promoted their program through the collective authorship and publication over sixty years of a series of books entitled *Eléments de mathématique*—the singular *mathématique* itself an expression of their vision of the field's unity—and through the choice of topics at their three-times-yearly (now four-times-yearly) seminar in Paris. Seminars identified with prominent mathematicians in Russia used to fulfill a similar function.

Charismatic and rational authority overlap, but they are not identical. Both came to public attention in the spring of 2012 when Timothy Gowers announced on his blog his decision to boycott mathematical journals published by Elsevier. Commercial firms, relative latecomers to scientific publishing, are driven by economic goals increasingly incompatible with the budgets of the libraries on which the researchers depend in order to produce the articles that publishers sell back to them at a profit. Gowers listed a few of Elsevier's commercial practices he found inexcusable, and within days a Web site had collected thousands of pledges by mathematicians and scholars in other fields to refuse to publish in or otherwise cooperate with Elsevier's journals. A statement of purpose (SoP) signed by thirty-four international specialists from all branches of mathematics—including Gowers and fellow Fields medalists Terence Tao and Wendelin Werner, several laureats of other prestigious prizes, and (in their private capacity) past and present top officers of the IMU and the (U.S.-based) SIAM—was posted on the internet and published in several languages in the official organs of national mathematics associations.[68]

Coming as it did on the heels of the prodigious events of 2011, the boycott inspired talk of an *academic spring* aimed at democratizating the access to knowledge, just as the *Arab spring* a year earlier was a revolt against the undemocratic regimes of the Middle East. But while Arab demonstrators stopped far short of calling for the "withering away of the state," the Elsevier boycott left mathematicians wondering whether the Internet had not made traditional scientific journals obsolete. Two years before Gowers launched the Elsevier boycott, an article entitled "Why hasn't scientific publishing been disrupted already?" had already pointed out that "While the journal was a brilliant solution to the dissemination problems of the 17th century . . . dissemination [i.e., making the results of scientific research widely available] is no longer a problem that requires journals." Mathematicians have for years been posting advance versions of their articles on their home pages or on open Web sites, especially the prepublication server at arxiv.org. Quoting the "disruption" article in one of the discussions spawned by the boycott, Terence Tao—like Gowers, Tao hosts a wide-ranging and popular blog—suggested that the journal form remained necessary only for "validation (certifying correctness and significance of a paper) and designation (providing evidence of research achievement for the purposes of career advancement)." With regard to attempts to incorporate these functions into what some have called a "Math 2.0" framework, Tao wrote that "so far I don't see how to scale these efforts so that a typical maths research paper gets vetted at a comparable level to what a typical maths journal currently provides, without basically having the functional equivalent of a journal."[69]

The still unresolved case of Elsevier raises fascinating questions about the possibility of reconciling the goals of science with the material organization of society. A mathematical tradition unified around charismatic norms maintained by overlapping research programs hardly fits comfortably in the intellectual property regime monitored by the supervisors of globalization. What I find intriguing is the apparent consensus that publishing retains one indispensable function that can't be automated—the maintenance of the charismatic structure of the field. Given the preponderant role of informal communication even in the 1970s, long before the Internet made delivery of preprints instantaneous, the sociologist Bernard Gustin is convincing when he calls scientific publication a form of "Traditional Ritual."[70] "Validation"—the process of peer review—has two aspects in mathematics, as indicated before: "certifying correctness," the painstaking reading of an article, line by line if possible, to make sure all the arguments are valid; and certifying "significance," irrelevant to the "Mathematics" of logical empiricist philosophy, but the very life of the "mathematics" of mathematicians. The "good journals" of Chriss' comment differ from "great journals" by the relative "significance" of what they choose to publish; "designation" in turn relies on the "significance," as validated by more or less great journals, to assign candidates to the "good" or "great" (or lousy) jobs they are deemed to merit.

The commercial publishing issue points up the arbitrariness or absurdity of inherited social forms: there is no reason to assume traditional forms (like the research journal) are rational. I find it striking that most mathematicians take the logic of the current system for granted. The SoP acknowledges that junior colleagues place themselves at risk when boycotting expensive commercial publishers that hold a monopoly on the crucial fountains—"great journals"[71]—that dispense the charisma hiring committees (and those omnipotent Deans) need to evaluate before agreeing to award tenure; the SoP encourages senior mathematicians supporting the Elsevier boycott to "do their best to help minimize any negative career consequences."

"In his later writings," writes Robert Bellah, "Durkheim identified 'society' not with its existing reality but with the ideals that gave it coherence and purpose."[72] Some colleagues judge that participation in such networked collective activities as blog discussions and MathOverflow represents a Sociability 2.0, close to mathematical ideals. MathOverflow's constantly updated register lists the most active contributors, and those who take the time to answer questions are rewarded by a system of "badges" and reputation points, as on amazon.com or Facebook, but much more elaborate (that's the 2.0 part); so are those who manage to find the right ways to frame questions that are or should be on everyone's minds. Modesty regarding one's own standing while displaying one's expertise is an unwritten rule of good manners on this and similar blogs (there are also written rules of conduct). Gowers leads the gold-badge competition (he has nine) and clocks in with a very impressive 16,608 reputation points (as of November 3, 2012), which still places him well behind logician Joel David Hamkins of CUNY, the current reputation champion (64,313 points!).

"Reputation," explains the MathOverflow FAQ page, "is earned by convincing other users that you know what you're talking about." A few participants in the Math 2.0 blog, launched soon after Gowers announced his Elsevier boycott, wonder why publishing can't be reorganized along similar lines, eliminating the profit motive and replacing the less democratic features of the charismatic hierarchy by a permanent plebiscite.[73] One blogger wrote "some people like 'elite communities,' [others] prefer more democratic communities," suggesting that the latter are in the majority. In contrast, SoP signatories were for the most part explicitly committed to maintaining the "élitist" functions of journals—validation and designation—in what they hoped would be a post-Elsevier era of mathematical publication; some were actively working to create new vehicles for this very purpose. If the content of mathematics is bound up, as I've argued, with a hierarchical charismatic structure, something of this sort is inevitable; if journals relinquish these functions, other institutions will take them up.

· ·

By granting me tenure at the age of twenty-seven, Brandeis University ratified my permanent admission to the community of mathematicians. Thus I was endowed with the routinized charisma symbolized by my institutional position and fulfilled the (rational-bureaucratic) obligations[74] incumbent upon one enjoying the privileges befitting my charismatic status. These privileges included and still include regular invitations to research centers like the IAS (where I began writing this book in 2011), the IHES, or the Tata Institute of Fundamental Research (TIFR) in Mumbai (Bombay); to specialized mathematics institutes like the Fields Institute in Toronto, the Mathematical Sciences Research Institute (MSRI) in Berkeley, or the Institut Henri Poincaré (IHP) in Paris, where Cédric Villani is now director; or to speak about my work at picturesque locations like the *Mathematisches Forschungsinstitut Oberwolfach*, a conference center nestled high in the hills of the Black Forest in Germany, that for nearly seven decades has hosted weekly mathematics meetings.[75]

I could, therefore, end this chapter right now and consider that I have respected the bargain announced in its title. Eleven years after obtaining tenure, however, an unexpected event—a mystic vision, no less—started me on the path to being bumped up a few rungs on the charisma ladder. The incident, recounted in chapter 9, resulted directly in item 34 on my publication list, a premonitory sign of my impending symbolic promotion. The process continued with item 37 and was clinched with the much more substantial item 43, a collaboration with Richard Taylor in the form of a 276-page book published in the highly visible *Annals of Mathematics Studies* series edited by the mathematics departments of Princeton University and the IAS (where Taylor is now a professor).

My book with Taylor concluded with the proof of a conjecture in a named research program—the *local Langlands conjecture*, a step in Langlands' program. Like every mathematical breakthrough, this one was prepared by the extensive work of numerous colleagues, especially my Paris colleague Guy Henniart, who turned a qualitative prediction by Langlands into a precise and optimal quantitative conjecture. Henniart, whose office was across the hall from mine when I began thinking about the question in 1992, was also instrumental in making sense of my first results in the field. So it's only fitting that his name should be attached to the solution of the problem, and so it is, but—because the rules of charisma almost always severely discount work that stops short of the perceived *telos*—it's there on the grounds of the separate proof he found shortly after my announcement with Taylor.[76]

Ideal-typical honors—the details are unimportant, and anyway some of them are listed on the book jacket—flowed from this and subsequent collaborations with Taylor. Consistently with the Matthew Effect, the prize the French *Académie des Sciences* awarded my research group, including a substantial research grant, is routinely but inaccurately described as a prize for me alone. I'm sometimes invited to put an ordinary face on the Langlands program's charisma for the sake of mathematicians in other fields—Langlands can't be everywhere, and while none of his surrogates can really represent his perspective, he has been a good sport about it. Students with only the vaguest idea of my motivations think they want to work with me; indefatigable Deans solicit my opinions before proceeding with hiring decisions; colleagues I visit often hasten to meet even my tentative requests, as if the Matthew Effect had outfitted me with a built-in megaphone.

And, to cap it off, this book. A friend whose bluntness I cherish told me, "Of course if you're able to publish this book, it's because of the kind of mathematician you are." This is misleading. We'll see throughout the book quotations by Giants and Supergiants in which they conflate their own private opinions and feelings with the norms and values of mathematical research, seemingly unaware that the latter might benefit from more systematic examination. One of the premises of this chapter is that the generous license granted hieratic figures is of epistemological as well as ethical import.

My own experiments with the expression of what appear to be my private opinions resemble this model only superficially and only because they conform to the prevailing model for writing about mathematics. My friend's point was that even my modest level of charisma entitles me not only to say in public whatever nonsense comes into my head—at a philosophy meeting, for example, like the one for which I wrote the first version of chapter 7—but even to get it published. Or, to quote Pierre Bourdieu and Jean-Claude Passeron, "There is nothing upon which [the charismatic professor] cannot hold forth . . . because his situation, his person, and his rank 'neutralize' his remarks."[77] This book's ideal (-typical) reader, on the other hand, will know how to neutralize this kind of charisma, will have already recognized its ostensible narrator as little more than a convenient focal point around which to organize the text, and will be attentive to what is being organized.

chapter α

How to Explain Number Theory at a Dinner Party

Researcher: So your image of a mathematician—describe a mathematician to me apart from the fact that they're a maths academic. What are they like physically? What are they, I mean, are they . . . ?
Babs: An old bloke with a big grey beard.[1]

INTRODUCTION

During the spring of 2008 I was invited by the Columbia University mathematics department to deliver the Samuel Eilenberg lectures—a perfect illustration of the Matthew Effect described in the previous chapter. The appointment involved living away from my family for several months. Working late in the department one Friday evening, I must have looked even more forlorn than usual, because a colleague passing my open door decided on the spot to invite me home to dinner. Several other mathematicians had been invited, along with a neighbor from another department and the neighbor's visiting friend, a young British woman who turned out to be a performer between jobs—a real, professional actress, with an agent and a long string of film and TV credits as well as a steady and successful career on the stage. She talked about the trials of being an actress, hinting that not all her peers suffered quite so much as she did. The younger men and women among the mathematicians alluded to their own career anxieties, while their tenured colleagues offered reassuring but noncommital replies. The actress glowed enigmatically during this part of the conversation, but when it came time to serve dessert, she turned to me without warning and asked, "What is it you do in number theory, anyway?"

The other mathematicians looked at me in unison, holding their collective breath. I had stumbled into the awkward moment every mathematician dreads, my predicament highlighted by the questioner's quiet radiance. "If you have ever found yourself next to a mathematician at a dinner party," Tim Gowers once wrote, "you have probably, out of politeness, or perhaps desperation, asked what he or she works on." Variants of this scenario take place in cocktail parties, on long-distance flights, more rarely at singles bars. "If you do not have a mathematics degree," Gowers continued, "you will almost certainly have received a disappointing answer such as, 'I work in Iwasawa theory, but it would take too long to explain to you what that is.' "[2]

All of us in the room were in our own ways performing artists, of course, but the mathematicians were trapped in the invariable role of emissaries from a distant and reputedly inhospitable planet. In 1952 Helmut Hasse lamented that he often heard ("especially from ladies") remarks like these: "You mathematicians are all cold, sober rationalists. How can you find satisfaction in an activity that offers nothing for the heart, nothing for the soul?"[3] My answer to the actress and the conversation that ensued lasted at most ten minutes. The consensus in the department over the following days was that I didn't do so badly, given the hopelessness of my situation. But one always feels one could have done better. This chapter, which has been broken up into short sessions in order to impede the accumulation of formulas—each of which, according to publishing wisdom, is enough to scare off 10,000 potential readers—contains what should have been my answer. Those intrepid readers who continue to the end will see that I work my way up to the *guiding problem* of the Birch-Swinnerton-Dyer conjecture, which concerns equations of degree 3 in two variables, through problems of increasing complexity: equations of degree 2 in one variable in the first session, of degree 3 (or 4) in one variable or degree 2 in two variables in the second session, with a detour through congruences in the third session in order to hint at the substance of the guiding problem in the last session. Parts of what I actually said and of the actress's reactions have been incorporated into the first of the (otherwise entirely imaginary) dialogues that conclude each session.

FIRST SESSION: PRIMES

One theorem found in nearly every popular book on mathematics, whether or not it is really relevant to the topic at hand, is the irrationality of the square root of 2. This can be stated as a theorem:

Theorem: There is no fraction p/q, where p and q are whole numbers, with the property that

$$\left(\frac{p}{q}\right)^2 = 2. \tag{Q}$$

In other words, equation (Q) in the unknown quantities p and q is impossible to solve if p and q are required to be whole numbers. The theorem is easy to state, its proof fits in a few lines, and it's accompanied by a story about the crisis the discovery of such numbers provoked among the ancient Greeks who concerned themselves with such questions.

There is also a picture: figure α.1.

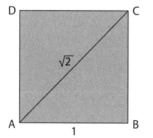

Figure α.1.

If $ABCD$ is a square with sides of length 1, then the Pythagorean theorem tells us that

$$AC^2 = AB^2 + BC^2 = 1+1 = 2.$$

Thus the diagonal AC has length $\sqrt{2}$, which the theorem tells us is an irrational number, that is, a number that is not a fraction. Here, therefore. is a quantity that one can perceive, or believe one perceives, but that cannot be apprehended rationally, or so the argument goes. The idea is that the rational number p/q can be understood in terms of acts that can be completed in finite time—cutting a unit length into q equal parts and then stringing p of these parts together; maybe this requires cutting several (or several billion) unit lengths in the first step. But $\sqrt{2}$ cannot be understood in such terms. It can be understood in other terms, say as the diagonal AC of the unit square in figure α.1, or as the number x that satisfies an equation of degree 2 in one variable:

$$x^2 = 2. \tag{I}$$

Comparing the ways one can understand this number to the way that the theorem tells us is off limits helps to attenuate the mystery that may be attached to $\sqrt{2}$. Figure α.1 suggests that $\sqrt{2}$ can be embedded in what Ludwig Wittgenstein called a "language game" that can be used to talk about geometry, whereas equation (I) tells us that $\sqrt{2}$ can be embedded in a language game that can be used to talk about algebraic equations. The theorem tells us that the language game of rational fractions has to be expanded in order to accommodate $\sqrt{2}$. The square root of 2, and the many numbers like it, thus serve to establish the relations among various language games that arise from different aspects of our experience with mathematics—measuring, counting, dividing up space into geometric figures, and comparing numbers with one another. The square root of 2 is old news, literally ancient history, but it remains true that mathematicians and those drawn to mathematics pay special attention to notions that can serve to link distinct language games—the more distinct, the better.

I borrow Wittgenstein's metaphor "language games" not because I believe experience is neatly divided up into such games (Wittgenstein certainly believed no such thing) but because it draws attention to the habit mathematicians have of using the word *exist* in regard to the objects we study when they become the subjects of the stories we tell about them. I have already begun to tell three stories about the square root of 2. The natural continuation of the first of the three stories is the proof of the theorem. Here's how it goes. We can write the fraction p/q in lowest terms. That means that we can assume that any number that divides p doesn't also divide q; otherwise, we can remove this factor from both numerator and denominator and simplify the fraction. "Simplifying" fractions is one of the skills we learn as part of learning to use fractions. There is an implicit value judgment that 2/3 is simpler than 10/15 even though they are two ways of writing the same number. Learning that a given number can be written in (infinitely) many ways is a very important step in learning to use numbers at all and contributes to forging their meaning.

In our case, all we really care about is that p and q can't both be even. What would be the point in writing $\sqrt{2}$ as the quotient of two even numbers? We can divide both numerator and denominator by 2 as many times as it takes, simplifying over and over again until one or the other is no longer even.

Now we rewrite equation (Q):

$$p^2 = 2q^2.$$

All we have done is to convert the equality of fractions into an equality of whole numbers. But now the right-hand side is $2q^2$, which is even. Since this is an equality, p^2 must also be even.

Now comes the *crucial step*. If p were an odd number, then p^2 would also be an odd number. We know that because when you multiply two odd numbers, the result is odd. How do we know that? Undoubtedly, a teacher told us this, we confirmed it by experience if we were curious, and if we were very curious. we eventually cobbled together something resembling a mathematical proof of the fact.

Let's just accept that we know this fact and resume the arguments. If p were an odd number, then p^2 would also be odd, but we know this is not the case; hence p must be even. Note the strategic use of that word *hence*! This is argument by contradiction, a form of argument we all use but that leads to somewhat problematic, or at least controversial, conclusions if applied incautiously.

Anyway, p is even, so it can be divided by 2; say, $p = 2n$. Now rewrite equation (Q) again:

$$(2n)^2 = 2q^2;$$

or again:

$$4n^2 = 2q^2;$$

or, finally, dividing both sides by 2:

$$2n^2 = q^2.$$

Now we are nearly done. The right-hand side is now even, so q^2 is even; just as before, this must imply that q is even.

We started out by arranging that p and q would not both be even, but here we have shown that p and q must necessarily both be even. This is a contradiction, which means we must have made a false step somewhere along the way. But our reasoning has been flawless. The false step must therefore have come at the very beginning, when we assumed equation (Q). This shows that equation (Q) has no solution, and that is what the theorem asserts.

I want to return to what I called the *crucial step*. There were between five and ten steps, depending on how you count. What makes that particular step more crucial than any of the others? You might say it's less obvious than the others, except that by the time we get to the stage of reading this proof, it's hard to perceive this difference in obviousness. The crucial step is crucial to what you might call the *deep structure* of the proof: *deep* relative to proofs of other theorems that bear some resemblance to the original theorem. One can generate other potential theorems by tinkering with the statement. For example:

Theorem: There is no fraction p/q, where p and q are whole numbers, with the property that

$$\left(\frac{p}{q}\right)^2 = 3. \tag{Q3}$$

That one is certainly true (and is attributed to the ancient Greek mathematician Thaeatetus, who gave his name to one of Plato's dialogues), whereas

"Theorem": There is no fraction p/q, where p and q are whole numbers, with the property that

$$\left(\frac{p}{q}\right)^2 = 4 \tag{Q4}$$

is certainly false, because $4 = 2^2$: take $p = 2$ and $q = 1$ (of course, the fraction $2/1$ is also the whole number 2). The step I identified as crucial is the one that can be imitated to give a proof that (Q3) has no solutions but cannot be twisted to give a proof of the false theorem that (Q4) has no solutions. One needs to replace our notion of "odd" by one adapted to 3. Just as odd is a synonym for not a multiple of 2, the appropriate notion to show that $\sqrt{3}$ is irrational is not a multiple of 3.

The crucial fact, then, is that if neither of two numbers is a multiple of 3, then their product is again not a multiple of 3. This fact is less familiar than the corresponding fact for odd numbers, but it is true nonetheless. In contrast, neither 2 nor 6 is a multiple of 4, but $2 \times 6 = 12$ is also 4×3.

Armed with this crucial fact for 3, the proof proceeds as before: a solution to (Q3) would yield

$$p^2 = 3q^2,$$

forcing us to conclude that p is a multiple of 3—this is what fails if 3 is replaced by 4—and then a few steps later that q is also a multiple of 3, which contradicts the hypothesis that p and q have no common factors.

We can, therefore, divide up the whole numbers (positive integers) into those, like 2 and 3, for which the crucial fact is valid and those, like 4, for which it is not. Say that a positive integer p is *prime* if, whenever a and b are two numbers that are not divisible by p, the product ab is again not divisible by p. This is not the familiar definition of a prime as a (positive) integer that has no divisors other than 1 and itself, but Euclid already knew that the two definitions are equivalent, and it is the former to which our proof has drawn attention. The first few primes are 2, 3, 5, 7, 11,. . . .

For any prime p, our proof shows that $\sqrt{p}$ is not a rational number. That is as far as our proof can stretch without modification. For example, $6 = 2 \times 3$ is not a prime, but $\sqrt{6}$ is still irrational. The proof requires additional properties of primes that need not be mentioned here, but they are not particularly novel; they were also known at the time of Euclid.

Number theory gets started as a theory of prime numbers, and prime numbers are on the front lines in analyzing any number-theoretic question. One has known at least since Euclid that there are infinitely many prime numbers (another theorem found in nearly every popular book on mathematics).[4] Another basic fact proved by Euclid is that every number can be factored as a product of prime numbers—for example, $238 = 2 \times 7 \times 17$—and that the factorization is unique.

Most numbers are not primes—no even number other than 2 is prime, for instance—but experiment shows that they are rather evenly distributed among numbers in general. One of the major mathematical problems of the nineteenth century was to prove that the number of primes less than a given number N was roughly $N/\log(N)$, which is, even more roughly, N divided by three times the number of digits in N. This claim, the *prime number theorem*, was proved by two French mathematicians at the turn of the twentieth century and has been re-proved in countless ways since then, but it's not the last word. One can ask how well the quantity $N/\log(N)$ approximates the actual number of primes less than N. The claim that it is statistically an optimal approximation is the import of the *Riemann hypothesis*, the subject of at least four books for the nonspecialist public,[5] considered by most mathematicians to be the outstanding unsolved problem in number theory and indeed in any branch of mathematics; it is one of the seven *Clay Millennium Problems*, for the solution of which the Clay Mathematical Foundation has offered a $1,000,000 prize.

At least two other well-known problems, simpler to state and therefore more popular with amateurs, concern prime numbers: the *twin primes conjecture* to the effect that there are infinitely many primes p such that $p + 2$ is also prime (examples of twin prime pairs: 3, 5; 11, 13; 29, 31; $65,516,468,355 \cdot 2^{333333} \pm 1$; and so on); and *Goldbach's conjecture* that every even number can be written as the sum of two primes (for example $8 = 3 + 5$, $28 = 11 + 17$; $100 = 97 + 3$; . . .), the subject of a popular novel by Apostolos Doxiadis.[6]

> PERFORMING ARTIST: I've always wondered why the square root of 2 deserved to be called a number. When you count 1, 2, 3, . . . , you don't want to stop for the square root of 2, otherwise you would have to let all sorts of other square roots in and you'd never get to 10.
>
> NUMBER THEORIST: You could make the same objection to fractions. Before you get from one fraction to the next you have to pass the one halfway between.
>
> P. A.: Yes, Tom Stoppard made this point in *Jumpers*:
>
> > *It was precisely this notion of infinite series which in the sixth century BC led the Greek philosopher Zeno to conclude that since an arrow shot towards a target first had to cover half the distance, and then half*

the remainder, and then half the remainder after that, and so on ad infi-
nitum, the result was, as I will now demonstrate, that though an arrow
is always approaching its target, it never quite gets there, and Saint Se-
bastian died of fright.

But we're used to using fractions as numbers. The fraction of this cake
that's being divided—not very evenly, by the way—among nine people,
for example. What's the use of the square root of 2?

N. T.: Well, there's figure $\alpha.1$, for a start.

P. A.: But that's a *length*. Why does every length deserve to be called a
number?

N. T.: You sound suspiciously like a logician. When logicians started worrying about using infinite numbers to count things, they convinced
themselves they already didn't know how to explain what they were
doing when they counted finite collections of objects—even one object,
for that matter. But then they realized that just because they didn't
know how to count one object, they did, in fact, know how to count
zero objects, because that's how many objects you have counted before
you have figured out how to start counting. On that insight, starting
with zero the logicians built a whole theory of counting[7] that includes a
place for the square root of 2 as well as any other number you can think
of. Some people still claim this is the only reliable way to talk about
numbers. But I find it most helpful to think of a number as an answer to
a question.

P. A.: I don't find that at all helpful. Most questions can't be answered by
numbers.

N. T.: Of course not. A number is an answer to a question about numbers.

P. A.: You could just as well explain trains by saying they are answers to
questions about trains.

N. T.: I grant that it's a circular explanation, but trust me, you won't get a
better one, at least not at a dinner party.

P. A.: Remind me when I might have asked a question about numbers.

N. T.: Well, you may wonder how many people attended your last performance. That appears to be a question about people, but it is so only incidentally, since you are considering them only quantitatively and dis-
regarding their individual characteristics. That is a *counting* question
and is in the spirit of the logicians' approach. But we can also ask ques-

tions about *measurement*—how much space your audience occupied in the theater. Because we can do arithmetic with the answers—because we can add lengths, or multiply them to get areas, for example, and because the arithmetic of measurements follows the same rules as the arithmetic of counting numbers—we decide that the answer to a question about measurement is also a number. If you were to ask the length of the diagonal in figure α.1, the answer would be $\sqrt{2}$, which thereby qualifies as a number.

P. A.: My impression is that one hardly needs an advanced degree to answer that sort of question. For that matter, each of your examples looks to me like a question about something else that you, in your self-serving way, have chosen to interpret perversely as a question about numbers.

N. T.: Very much to the point. We could also ask

Question 1: What number, multiplied by itself, yields 2?
This is a question purely about numbers and its answer is, again, the square root of 2. As we have already observed, that number is also the answer to a question about length. But now suppose we asked

Question 2: What number, multiplied by itself and added to 1, yields 0?
Or, at the risk of losing another 10,000 readers, we can ask for the solution to the equation $x^2 + 1 = 0$. The answer, which is usually called i, or $\sqrt{-1}$, can't be used to count anything, and it's certainly not the length of anything; all it does is answer the question.

P. A.: But what makes it a number??

N. T.: It's just as I said, i is the answer to a question about numbers, namely, question 2. May I ask whether you are a devotee of Leopold Kronecker, who once claimed that *Die ganzen Zahlen hat der liebe Gott gemacht, alles andere ist Menschenwerk?**

P. A.: May I ask whether you are trying to distract me from the circularity of your reasoning by bringing up irrelevant theological considerations in German?[8]

N. T.: Not so irrelevant, unfortunately. But you might find my reasoning a little less circular if I mentioned that, just as with lengths, we can do

* "The Good Lord made the whole numbers, all the rest is the work of man."

arithmetic with numbers like i. Strictly speaking it's a new kind of arithmetic: when we want to multiply i by 3 and add, say, $\sqrt{2}$, the answer is just $\sqrt{2} + 3i$. But when we allow ourselves to do that, we find we can answer questions such as

Question 3: What number solves the equation $x^4 + 14x^2 + 121 = 0$?
When you get home you can check that $\sqrt{2} + 3i$ is an answer to question 3.

And, while we're on the subject of circles, your homework assignment for our next session is to understand the relation between numbers in the language game of *arithmetic*—for example, whether 1 can be written as the sum of squares of two rational numbers or whether the square root of 2 is rational; numbers in the language game of *equations*—for example, $x^2 + y^2 = 1$ (or $x^2 = 2$); and numbers in the language game of *pictures*—for example a circle (or whatever picture you can find to go along with $x^2 = 2$).

P. A.: So to summarize what you've been saying, a number theorist sits at a desk and answers questions about numbers all day, just like the person at the railway station information desk answers questions about trains.

N. T.: Actually, number theorists are not especially interested in answering questions about numbers. We really get excited when we notice that answers seem to be coming out a certain way, and then we try to explain why that is. For example, the equation $x^4 + 14x^2 + 121 = 0$—the question, what number solves that equation?—has not one but four answers: $\sqrt{2} + 3i$, $\sqrt{2} - 3i$, $-\sqrt{2} + 3i$, $-\sqrt{2} - 3i$. There's a pattern: you can permute $\sqrt{2}$ with $-\sqrt{2}$ and $3i$ with $-3i$. What does it mean? What does it tell us about solutions to other equations?

When our ideas about possible explanations are sufficiently clear, we set ourselves the goal of finding the correct explanation and then justifying it.[9] In other words, our goal as number theorists is to solve *problems*, such as the Riemann hypothesis, but in general much easier.

P. A.: I'm not convinced railway staff are especially interested in answering questions about trains either. But I think I understand what you mean by *problems*. Most of my work as a performer involves solving problems.

N. T.: What kind of problems?

P. A.: About characters, mostly. It's my job to create the character as a credible presence for the duration of the performance. You may think the playwright has created the characters, and all we have to do is deliver the lines in the right order. If that were all it involved, someone could program a computer to take my place. Someone like you, for instance. As Marilyn Monroe said, "I'm not a Model T . . . An actor is not a machine, no matter how much they want to say you are."

N. T.: A familiar dilemma! If you glance at some of the earlier chapters, you might agree that we have quite a bit in common, after all. As it happens, I can't program a computer to do anything, and in any case I doubt any computer could reproduce your stage presence, even for the duration of this brief sketch. But now it's your turn to provide an example of what kind of problem you have to solve.

P. A.: For example, when I played Nora in *A Doll's House*, there's a problem that has as much to do with the audience as with the character. Everyone comes to the theater knowing that Nora is trapped by the hypocrisy of the conventions of marriage, and when she slams the door at the end, she is signaling her escape from all that. But the audience doesn't follow her out the door. One of my main problems is to use the character to create an image of what Nora expects to find on the other side of the door and to make the image real for the audience.

N. T.: That reminds me of a problem I have writing this book.

P. A.: A number theory problem?

N. T.: More like a writing problem.

P. A.: Are you a writer? I thought you were a number theorist.

N. T.: Let's say it's the kind of problem I would have if I were a writer.

P. A.: Maybe I can help.

N. T.: I'm afraid I'll have to solve that one by myself. It's easy to argue that both the mechanizers and promoters of the "knowledge economy"[10] are missing what is most precious about mathematics: namely, that it is a *human* activity and one of the few remaining human activities not driven by commercial considerations. But to complain that these images of mathematics are stifling and confining, like Nora's marriage, is merely negative. There's not much sympathy to be gained in saying what mathematics is *not*.

On the other hand, maybe you can help me with a smaller problem I need to solve right now, which is to explain complex numbers—expressions like $\sqrt{2} + 3i$—without turning my readers to stone. Barry Mazur, my thesis adviser, published a 288-page book about a single complex number, the square root of -15, and some people still found it too difficult to grasp.

P. A.: I suppose you want me to ask what question a complex number is designed to answer . . .

N. T.: More questions than you can possibly imagine!

P. A.: We'd better stick with the questions I can imagine, so I suggest you find a way to connect your complex numbers with literature.

N. T.: Mazur talked about poetry, but he made a convincing case that there are real similarities between what we do when we imagine the square root of -15 and the imagination inspired by reading poetry. I have no such case to make.

P. A.: Even if your literary references are artificial, your readers will put down the book feeling they haven't been wasting their time, because they'll at least have seen something that interests them.

chapter 3

Not Merely Good, True, and Beautiful

> *[I]f this is a "We-system," why isn't it at least thoughtful*
> *enough to interlock in a reasonable way, like They-systems*
> *do?*
>
> —Thomas Pynchon, *Gravity's Rainbow*

If it has become urgent for mathematicians to take up the "why" question, to which we are now ready to turn, it is because the professional autonomy to which we have grown attached is challenged by at least two proposals for reconfiguration. The first proposal is neither new nor specific to mathematics. The economic crisis that began in 2008 placed universities under enormous stress. Cutbacks have been especially severe in humanities departments, where the elimination of entire programs has become routine, but the crisis has also intensified existing pressure to subordinate scientific research in Europe and North America to commercial applications. It is telling that universities in Britain are now the responsibility of the Department of Business, Innovation, and Skills.[1] Here is how Gary Goodyear, Canada's Minister of Science and Technology, put it in a recent television interview:

> [W]e are also asking our scientists to appreciate the business side of science ... as the Prime Minister has said many times, science powers commerce. ... We have to move forward on applied science, the commercialization end ... we are the first to say that there is far too much knowledge left in the laboratory, and *knowledge that is not taken off the shelf and put into our factories is actually of no value*" [my emphasis].[2]

Goodyear is unusually blunt, but he is hardly the first to talk this way. Already in 1907, Clarence Birdseye had written that "our colleges have become a part of the business and commercial machinery of our country, and must therefore be measured by somewhat the same standards."[3]

Quoting Birdseye and other voices of what we might call "market in-strumentalism," Joan Wallach Scott pointed out in 2008 that "Business-men and politicians, then as now, have had little patience with the ideal of learning for its own sake. . . . Today the sums may be larger and their impact on university research operations greater, but the pressure to bring universities in line with corporate styles of accounting and man-agement persists."[4]

There is now a massive literature on the pressures facing university laboratories.[5] These books mostly ignore mathematics, where stakes are not so high and opportunities for commercial applications are scarce, especially in the pure mathematics that is the subject of the present book. Nevertheless, even pure mathematicians, when presenting the case for research support to the "Powerful Beings"* who often go by the name of "decision-makers," find it necessary to point to potential commercial or industrial spinoffs or at least to resort to what Steven Shapin calls the Golden Goose argument, to the effect that BECAUSE the potential ben-efits of scientific research are so often unpredictable, THEREFORE, Powerful Beings have every interest in funding research motivated by curiosity alone.[6] Weimar Germany's Kaiser-Wilhelm Institutes, sup-ported by public and private funds, put the Golden Goose argument to practical effect by allowing scientists to "focus on basic research in re-sponse to problems posed by industry without concern for immediately applicable results."[7] Abraham Flexner, founder of Princeton's Institute for Advanced Study (IAS), devoted a 1939 *Harper's* article, entitled "The Usefulness of Useless Knowledge," to the Golden Goose argu-ment. After listing some of the unexpected historical benefits of purely theoretical research in the sciences, including pure mathematics—and announcing plans for the construction of Fuld Hall, the main building at the IAS—Flexner wrote that "we cherish the hope that the unobstructed pursuit of useless knowledge will prove to have consequences in the future as in the past."[8]

Mathematical societies today see promotion of the Golden Goose argument as one of their primary responsibilities. Chapter 10 reviews some recent examples in France, Germany, Britain, and the United States. There is obviously no reason not to hope that the ways of thinking

---

* Explanation to follow.

that arise in pure as well as applied mathematics might contribute to solving a few of the problems facing humanity.[9] Leibniz's thoughts as mathematician and philosopher tended to be abstract, but his motto was *Theoria cum Praxi*—"theory with practice." That's a very different ethical stance from equating our worth with our potential as Golden Geese, especially when the Powerful Beings to whom we promise our gold are primarily concerned with short-term economic goals to which we may not subscribe. Flexner promoted the IAS not as a Golden Goose refuge but "as a paradise for scholars who, like poets and musicians, have won the right to do as they please and who accomplish most when enabled to do so." The physicist Peter Goddard, director of the IAS in 2011, when I was beginning this book, vigorously defended Flexner's vision against unspecified but persistent challenges. Goddard argued that, with universities facing the "introduction of business principles and methods"—"systems of standardization, accountancy, and piecework," he wrote, quoting Thorstein Veblen—the reasons for creating institutes like the IAS, of which there are now hundreds, "have greater force than they did a century ago."[10] It is not only dishonest but also self-defeating to pretend that research in pure mathematics is motivated by potential applications. Pure mathematics is a living community, and we'll see that the motivations of pure mathematicians are rich and varied, but there is no evidence that market instrumentalism ranks prominently (or anywhere at all) among them.

The second proposed paradigm shift poses less of an immediate material challenge but is more interesting in that it arises from within the profession. The proposal stems from an observation with which no one would disagree. Proofs in some areas of mathematics are growing increasingly elaborate, involving in some cases hundreds of pages of calculations or expertise in too many areas to be fully understood by any one mathematician. Traditional methods exhaust the physical limits of human beings as presently configured and for that reason are deemed inadequate to confer legitimacy on some of the most remarkable theorems of the past few decades. The most notorious examples are the 1976 computer-assisted proof of the *four color theorem* by Kenneth Appel and Wolfgang Haken, which launched the ongoing debate on the epistemic status of computer-assisted proofs;[11] the *classification of finite simple groups*, whose proof, completed in the 1980s but still undergoing adjustment, is estimated to fill five thousand pages; and the proof of the *Kepler conjecture*, completed in 1998 by Thomas Hales and Samuel Ferguson with the help of around 3 gigabytes of computer code, which no committee of human referees could validate in its entirety.[12]

Some mathematicians, disturbed by the uncertainty surrounding these proofs, envision a future in which the human theorem prover is yoked more or less securely to nontraditional devices that don't suffer human shortcomings. The immediate objective of Hales' *Flyspeck project*, "an undertaking . . . that has the potential to develop into one of historical proportions," is the *formalization* of the proof of the Kepler conjecture. Formalization is an ideal of mathematical argument acceptable to logicians and to like-minded philosophers of Mathematics: instead of writing a proof in some approximation to ordinary language and appealing to intuition in the transition from one step to the next, a formal proof is a series of propositions written in a formal language from which all ambiguity has been eliminated, with each proposition obtained from its predecessors by strict application of one of a short list of permitted transformations. The idea is that individual propositions written in a formal language, and the transitions between them, can be certified licit by a *proof-checker*, a computer program written for this purpose. The long-term objective is naturally more ambitious:

> We are looking for mathematicians . . . who are computer literate, and who are interested in transforming the way that mathematics is done.[13]

A substantial fraction of elementary mathematics has been successfully formalized in this way, and there's no reason to doubt that the proof of the Kepler conjecture will be converted into one that can be checked by machine. What happens next—the "transforming" part—is not so clear. As far as I can tell, no one is suggesting that mathematicians actually spend time reading the formal proofs, only that we take comfort in the knowledge that a specialized mechanical entity has done so and has given its seal of approval.

What's noteworthy about the comfort being offered is that it is *collective* and *metaphysical*, not merely personal and psychological. The search for such comfort has a long history, like the history of the attempts to impose a corporate model on research but here rooted in what historian Jeremy Gray calls "anxiety." "[O]nce the safe havens of traditional mathematical assumptions were found to be inadequate," Gray writes, "mathematicians began a journey that was not to end in security, but in exhaustion, and a new prudence about what mathematics is and can provide." Gray is writing about the turn of the twentieth century, just before the Foundations Crisis, when the "new discipline of physics could compete very favorably with mathematics on utilitarian grounds . . . more critical mathematicians were aware that they therefore had to base their claims for the quality and value of mathematics on more intrinsic grounds."[14] "Useful uselessness" in Flexner's sense no longer sufficed; a metaphysical anxiety now haunted our spare moments when we were not busy trying to convince the Birdseyes and Goodyears that we are not parasites.

I would argue that this particular anxiety is overstated, the ghost story mathematicians take down from the shelf when we want to share a morbid moment. But it remains a pressing concern for philosophers of both mathematics and Mathematics, as we'll see later; and it's one of the rare mathematical themes readily accessible to audiences outside our circle. Thus IAS professor Enrico Bombieri, in a public lecture in 2010, recalled the concerns about the consistency, reliability, and truthfulness of mathematics that surfaced during the Foundations Crisis and alluded to the ambiguous status of "computer proofs" (like the Appel-Haken proof) and "too-long proofs" (like the classification of finite simple groups)[15]— Bombieri's IAS colleague Pierre Deligne once remarked that "what interests him personally are results," unlike the two just mentioned "that he can, by himself and alone, *understand* in their entirety."[16]

Another IAS professor, the Russian-born Vladimir Voevodsky, is more actively anxious about the future of mathematics. "If one really thinks deeply about" the possibility that the foundations of mathematics are inconsistent, he said on the occasion of the IAS eightieth-anniversary celebration, "this is extremely unsettling for any rational mind."[17] Voevodsky obtained his prestigious position at the IAS and his Fields Medal[18] for his work in a field in which "too-long proofs" are common and in which the relatively small number of competent potential referees typically spend much of their time writing "too-long proofs" of their own, so he might understandably be concerned that proofs are not being read as carefully as they should. But Voevodsky is worried about something more central to mathematics. Consistency, in the technical sense, is what guarantees that the basic axioms don't allow you to prove a statement as well as its negation. Without consistency—if you could prove both **p** and **not p**, as logicians put it—then the distinction between mathematical truth and falsehood collapses and with it all the philosophy that seeks its grounding in mathematical logic.

Philosophy's reliance on the standards of mathematical reasoning is an old story. Descartes, for example, thought that "even in relation to nature there are some things that we regard as not merely morally but absolutely certain . . . Mathematical demonstrations have this kind of certainty. . . ."[19] "Mathematical knowledge," wrote Heidegger, "is regarded by Descartes as the one manner of apprehending entities which can always give assurance that their Being has been securely grasped."[20] In the language of politicians, "mathematical certainty" is the certainty that brooks no contradiction. Thus, in the middle of the continuing euro crisis in 2012, one could read in *Time* that budget adjustments being demanded of the Greek government (with catastrophic consequences, by the way, for scientific research, mathematics included) were "not politically, socially, or even mathematically possible." Under the influence of Frege and Russell, among others, twentieth-century philosophy focused on *symbolic logic* as the most secure branch of mathematics. The brief period of the Foundations Crisis, following Russell's discovery of paradoxes in Frege's interpretation of Cantor's set theory, marked the culmination of this concern with certainty. The interaction of mathematicians with philosophical logicians was so intense at the time that both groups were convinced that mathematical certainty was built on logical foundations. Hilbert's program went so far as to reinterpret mathematics as a kind of game, bound by a system of rules, inherited from the tradition (but metaphysically neutral), and rigid but flexible enough to provide a proof of their own consistency.

Unfortunately, the second incompleteness theorem of the Austrian logician Kurt Gödel asserts precisely that the consistency of a system of basic logical axioms cannot itself be guaranteed by a formal proof within the system.[21] Voevodsky's talk outlines his reasons for suspecting that inconsistency is more likely than consistency. Are you unsettled? Is your mind rational? To the anxious, Voevodsky offers a carrot and a stick. The carrot is that getting used to using (possibly) inconsistent systems of reasoning would be "liberating" in that "we could then use kinds of reasoning known to be inconsistent but closer to our intuitive thinking" to construct proofs that could then be verified to be reliable. His new project, to construct *univalent foundations* of mathematics, is a piece of the carrot.

The stick is hidden in this sentence: after finishing the project for which he won the Fields Medal, mathematics' highest honor, in 2002, Voevodsky "became convinced that the most interesting and important directions in current mathematics are the ones related to the *transition into a new era*, which will be characterized by the widespread use of automated tools for proof construction and verification"[22] [my emphasis]. In February 2011, at an IAS After Hours Conversation (a kind of before-dinner party), Voevodsky predicted it would soon be possible to design *proof checkers* based on univalent foundations that could effectively verify correctness of mathematical proofs written in the appropriate machine-readable language. In a few years, he added, journals will only accept articles accompanied by their machine-verifiable equivalents.

Bombieri ended his lecture with the surprisingly Kuhnian conclusion that "mathematics follows a kind of Darwinian evolution" and that "mathematical truth is not irrelevant, nor tautological; it is the glue that holds the fabric of mathematics together." Bombieri, like most mathematicians, has his eye on the fabric. He writes that "the working mathematician"—a character we'll meet repeatedly in the course of this book—"is guided by clear aesthetic considerations: Intuition, simplicity of arguments, linearity of patterns, and a mathematically undefinable aristotelian 'fitting with reality.' "[23]

Voevodsky worries about the glue. Bombieri's "intuition" or Deligne's "understanding" do not offer the "absolute certainty" at which Descartes was aiming. In developing his "conceptual notation" [*Begriffschrift*] for mathematics, Frege worked hard "to prevent anything intuitive [*Anschauliches*] from penetrating . . . unnoticed."[24] For "the working mathematician" understanding and intuition are *real*. The project of mechanization, which can be traced through Frege back to Leibniz,[25] proposes to substitute a model of *real* closer to Frege's, one that lacks the defects of subjectivity. "I never would have guessed," wrote Mike Shulman in 2011, "that *the* computerization of mathematics would be best carried out . . . by an enrichment of mathematics"[26] [my emphasis] along the lines of Voevodsky's univalent foundations project, to which Shulman is an active contributor.

I italicized the definite article in the Shulman quotation to emphasize how mechanization of mathematics—with or without human participation in the long term—is viewed in some circles as inevitable. To my mind, the fixation on mechanical proof checking is less interesting as a reminder that standards of proof evolve over time, which is how it's usually treated by philosophers and sociologists,[27] than as a chapter in the increasing qualification of machines as sources of validation, to the detriment of human rivals. In Yevgeny Zamyatin's novel *We*, whose protagonist is a mathematician named D-503, the scourge of human subjectivity—what virtual reality pioneer Jaron Lanier calls "the unfathomable penumbra of meaning that distinguishes a word in natural language from a command in a computer program"[28]—is ultimately eliminated by an X-ray operation that turns people into machines.

Who finds machines more appealing than humans, and for what ends? Most mathematicians have long since recovered from the exhausting journey of the Foundations Crisis, but certification of mathematical knowledge is still a preoccupation among the "mainstream" philosophers of Mathematics, for whom participation by human mathematicians is optional. Convinced, like Hume, that no *ought* can follow from an *is*, philosophers of Mathematics skip straight ahead to the *ought*, disdaining the *is* as unfit for philosophical consideration.[29] When I insist, like philosophers of mathematics (from my prephilosophical internal vantage point) on getting the *is* right—the "what" question—I am in fact addressing an *ought* that overlaps with the *ought* of the philosophers of Mathematics but is by no means identical, nor is the overlap necessarily very substantial.

• •

Some observers suggest the time might be ripe in mathematics for a paradigm shift, arguing either on *epistemic* or *stylistic* grounds.[30] The proposed paradigm shifts discussed above, in contrast, are based on conceptions of mathematics as serving competing *ethical* ideals: productivity for the international market in the first case, a standard of certain truth in the second. The two ideals coexist in contemporary mathematics departments, but they have been in tension since the time of Plato. The classic account in is Plutarch's *Life of Marcellus*:

> To [his mechanical engines Archimedes] had by no means devoted himself as work worthy of his serious effort, but most of them were mere accessories of a geometry practised for amusement, since in bygone days Hiero the king had . . . persuaded him to turn his art somewhat from abstract notions to material things, and by applying his philosophy somehow to the needs which make themselves felt, to render it more evident to the common mind [*hoi polloi*].
>
> For [this] mechanics*, now so celebrated and admired, was first originated by Eudoxus and Archytas, who embellished geometry with its subtleties, and gave to problems incapable of proof by word and diagram, a support derived from mechanical* illustrations that were patent to the senses. . . . But Plato was incensed at this, and inveighed against them as corrupters and destroyers of the pure excellence of geometry, which thus turned her back upon the incorporeal things of abstract thought and descended to the things of sense, making use, moreover, of objects which required much mean and manual labour. For this reason mechanics* was made entirely distinct from geometry, and being for a long time ignored by philosophers, came to be regarded as one of the military arts.[31]

Plutarch adds that Archimedes himself "regard[ed] mechanics and every art that ministers to the needs of life as ignoble and vulgar." Centuries later, Omar al-Khayyām, no less indignant than Plato, contrasted "the search for truth" with "vile material goals":

> [M]ost of those who in this epoch of ours compare themselves with philosophers drape truth with falsehood, and so they do not go beyond the borderline of deceit and the simulation of knowledge; and they do not use the amount of sciences they know except in vile material goals. And if they witness someone who is interested in the search for truth and has a predilection for veracity, endeavoring to reject falsehood and lie and to abandon hypocrisy and deceit, they consider him stupid and they laugh at him.[32]

"[A]s producers of cultural products the ultimate legitimacy of which is at the mercy of society at large, mathematical communities . . . at some point need to make a case for their own relevance, and take up a position within their social milieu."[33] But we have seen that "truth" is no longer as compellingly relevant as it was for Plato and Khayyām; and the priorities of Goodyear and Birdseye appeal as little to today's pure mathematicians as they did to Archimedes. Other ideals are available.[34] Mathematicians wishing to treat our discipline as an end rather than a means often fall back, as Bombieri did, on justification on *aesthetic* grounds. G. H. Hardy's *A Mathematician's Apology*, analyzed at length in chapter 10, provides the prototypical formulation: "real mathematics . . . must be justified as arts if it can be justified at all."

A curious art, not recognized as such except by mathematicians, without listings in the Sunday supplements and without *critics*, as the philosopher Thomas Tymoczko noted.[35] One obvious difficulty with this use of *art* is that the aestheticizing mathematician uses the word *art* as a synonym of *beauty*. Thus Cédric Villani, his freshly minted 2010 Fields Medal in hand, told a French parliamentary committee that "the artistic aspect of our discipline is [so] evident" that we don't see how anyone could miss it . . . immediately adding that "what generally makes a mathematician progress is the desire to produce something beautiful."[36] It has been a long time since one could unproblematically substitute artist for mathematician in that last sentence. Catalan conceptual artist Antoni Muntadas interviewed more than one hundred major players in the art world between 1983 and 1991 and published excerpts from the transcriptions; the word *beautiful* is used only three times: twice incidentally, once to invite museum-goers to "let go of your preconceived notions that art has to be beautiful."[37] But I think the confusion is deeper. When mathematicians refer to *beauty*, there are reasons to suspect they really mean *pleasure*. But it's a specific kind of pleasure they have in mind—*mathematical pleasure*—and if our goal is to explain how aesthetics can serve as an ideal for mathematics, we find we are practically back where we started.

• •

The utility of practical applications, the guarantee of absolute certainty, and the vision of mathematics as an art form—the *good*, the *true*, and the *beautiful*, for short,—have the advantage of being ready to hand with convenient associations, though we should keep in mind that what you are willing to see as *good* depends on your perspective, and on the other hand the *true* and the *beautiful* can themselves be understood as goods. Even when such justifications are given, it shouldn't be assumed mathematicians and the Department of Business, Innovation, and Skills understand *utility* in the same way, or that *truth* means the same to mathematicians as to philosophers of Mathematics, and I've already argued that mathematicians who talk about *beauty* really mean something else.

The *good* and the *beautiful* will be examined at length in subsequent chapters, especially in chapter 10. I want to focus for a moment on the *true*, the source of some of the most persistent misconceptions about the point of mathematics. In the first place, we should get out of the habit of assuming that mathematics is about being rational, at least as this is understood by philosophers of Mathematics. David Corfield's paper quoted in the previous chapter was originally entitled "How Mathematicians May Fail to be Fully Rational." Corfield, we have seen, is a philosopher interested in mathematics rather than Mathematics, and his paper was a sympathetic exploration of what it meant to be faithful to the specific kind of rationality that implicitly guides mathematical practice. Whether or not one agrees with Corfield's solution—namely, that mathematics is a "tradition-constituted enquiry" in the sense of Alasdair MacIntyre, to whom we return below[38]—there's no reason to assume the rationality proper to pure mathematics is exhausted by the attempt to ground proof on unshakeable foundations. Quite the contrary. Mathematicians frequently write that they see a proof not as a goal in itself, but rather as a confirmation of intuition. Thus G. H. Hardy once claimed that "proofs are what [collaborator J. E.] Littlewood and I call *gas*, rhetorical flourishes designed to affect psychology, pictures on the board in the lecture, devices to stimulate the imagination of pupils." More recently, the eminent British mathematician Sir Michael Atiyah could write:

> I may think that I understand, but the proof is the check that I have understood, that's all. It is the last stage in the operation—an ultimate check—but it isn't the primary thing at all."[39]

The answers are less important than how they change the way we look at the questions. So the proof confirms that we have found the right way.[40]

This fits into a general debate among philosophers on the epistemic significance of heuristics. In his classic *Homo Ludens*, Johan Huizinga quotes a Dutch proverb: "It is not the marbles that matter, it's the game."[41] To those of my colleagues who predict that computers will soon replace human mathematicians by virtue of their superior skill and reliability in proving theorems, I am inclined to respond that the goal of mathematics is to convert rigorous proofs to heuristics. The latter are, in turn, used to produce new rigorous proofs, a necessary input (but not the only one) for new heuristics.[42]

What passes for paradigm shifts in mathematics can plausibly be traced to heuristic rather than axiomatic innovation. Voevodsky's univalent foundations project looks like,[43] and in fact is, a sophisticated and highly technical research bridge between ways of thinking about *topology*, the kind of mathematics that has been a constant inspiration in Voevodsky's work, and ways of thinking about what are usually called the *foundations of mathematics*, by which I mean how one goes about axiomatizing formal mathematical reasoning, a topic one would expect to be of interest to all mathematicians. The bridge is astonishing in many ways—I will try to do it justice in chapter 7—and it is already stimulating thinking in the two domains it joins. If and when univalent foundations is adopted as a replacement for today's (largely informal) standard foundations, it will probably be on roughly Kuhnian grounds, but in a positive sense: rather than being provoked by an insoluble crisis, any change is more likely to be triggered by a demonstration of the new method's superiority in addressing old problems or in accommodating *neue Erscheinungen* in Kronecker's sense.[44] This is why Grothendieck's approach to algebraic geometry replaced André Weil's "foundations" in the late 1950s and why Langlands' unifying vision brought the somewhat arcane theory of automorphic forms into the center of mathematical life about ten years later. An account of paradigm shifts in contemporary mathematics might focus on such reworkings of "foundations," understood not as the bedrock of certainty but as a common language. Yuri Ivanovich Manin, one of the most philosophically sensitive Russian mathematicians of his generation, points to the ambivalence of the word as used by mathematicians:

> I will understand 'foundations' neither as the para-philosophical preoccupation with the nature, accessibility, and reliability of mathematical truth, nor as a set of normative prescriptions like those advocated by finitists or for-

malists. I will use this word in a loose sense as a general term for the historically variable conglomerate of rules and principles used to organize the already existing and always being created anew body of mathematical
knowledge of the relevant epoch. . . . [F]oundations in this wide sense is
something which is relevant to a working mathematician, which refers to
some basic principles of his/her trade, but which does not constitute the essence of his/her work.[45]

Hermann Weyl hints that the quest for reliable foundations has spiritual roots: "Science is not engaged in erecting a sublime, truly objective
world . . . above the Slough of Despond in which our daily life takes
place." Mathematicians will judge univalent foundations on pragmatic
mathematical grounds—as we did when we adopted Grothendieck's or
Langlands' new foundations for their respective subjects—rather than
on the logical grounds of its conformity to norms of absolute certainty;
as *foundations* rather than *Foundations*. But even if certainty were mathematicians' primary target, there's no reason to think this would satisfy
philosophers of Mathematics. Paul Benacerraf's 1973 article *Mathematical Truth* has landmark status in the mainstream tradition for its claim
that philosophers cannot simultaneously rationally account for truth and
knowledge in mathematics, whatever Foundations one chooses. Other
philosophers of mathematics retort that logical Foundations are not the
point:

> When Paul Benacerraf limits the articulation of mathematical truth to logic
> and then complains that the ability of mathematicians to refer has been lost,
> it is no wonder; it is also no wonder that number theorists and geometers
> have not borrowed the language of logic to do their work.[46]

Faced with the task of identifying the point of mathematics, sympathetic
authors fall back on terms like *experience* or *practice* or *intellect* or
human thought, all of them ambiguous and resistant to logical formalization. Referring to what he called the "antiphilosophical doctrines" of
philosophers of Mathematics, logician Georg Kreisel argued in 1965
that "*Though they raise perfectly legitimate doubts or possibilities*"—in
other words, they help mathematicians keep their riotous impulses in
check—"*they just do not respect the facts, at least the facts of actual
intellectual experience.*" In his account of the Foundations Crisis, Jer-

emy Gray argues that "The logicist enterprise, even if it had succeeded, would only have been an account of part of mathematics—its deductive skeleton, one might say. . . . mathematics, as it is actually done, would remain to be discussed." Philosophy of Mathematics remains encumbered by its "tendency to reduce [mathematics] to some essence that not only deprives it of purpose but is false to mathematical practice." The Italian geometer Federigo Enriques put it this way in his lecture at the 1912 International Congress in Cambridge:

> The act of free will that the mathematician embarks upon in the formulation of problems, in the definition of concepts, or in the assumption of a hypothesis, is not arbitrary. It is the opportunity to get closer—from more than one perspective, and by continuous approximations—to some ideal of human thought, i.e. an order and a harmony that reflect its intimate laws. [47]

Capital-F Foundations may be needed to protect mathematicians from the abyss of structureless reasoning, but they are not the source of mathematical legitimacy. Herbert Mehrtens views the Foundations Crisis as the culmination of a conflict between *moderns* and *countermoderns* [*gegenmoderne*]. Mechanization of mathematics is modern in Mehrtens's sense; the countermoderns "wanted both to preserve in signs a residue of reference and to continue to find the certainty of mathematics "in the human intellect" and not "on paper" (today's *Papier* is digital).[48] Practically all mathematicians working today are countermodern in that sense. Sociologist David Bloor imagines that he is reinventing the countermoderns when he writes that "[e]ven in mathematics, that most cerebral of all subjects, it is people who govern ideas not ideas which control people."[49] Handing over mathematical communication to mechanical proof-checkers would eventually fix that, in an ultimate triumph for the moderns, but it would also withdraw mathematics from the purview of sociology altogether, while paradoxically vindicating Bloor's questionable decision to turn his sociological eye to philosophy of Mathematics rather than to human mathematical practice.

So what is the point of human mathematical practice? In line with Hans Reichenbach's classic separation of the context of justification, considered to be the proper domain of philosophy of science, from the (contingent and, therefore, philosophically irrelevant) context of discovery, Jeremy Avigad has argued that

[w]e have all shared such 'Aha!' moments and the deep sense of satisfaction that comes with them. But surely the philosophy of mathematics is not supposed to explain this sense of satisfaction, any more than economics is supposed to explain the feeling of elation that comes when we find a $20 bill lying on the sidewalk . . . we should expect a philosophical theory to provide a characterization of mathematical understanding that is consistent with, but independent of, our subjective experience.[50]

The obvious problem with this paragraph is that if the border between subjective experience and mathematical understanding were clearly delineated, there would be less need for philosophy of mathematics. But maybe that "deep sense of satisfaction" is a key to an understanding deeper than philosophers of Mathematics might encounter in their Foundational dreams. Embedded in Hilbert's list of twenty-three problems is one answer to the "why" question. Explaining in 1900 how he compiled that list, Hilbert wrote that "[a] mathematical problem should be difficult in order to entice us, yet not completely inaccessible, lest it mock at our efforts. It should be to us a guide post on the convoluted paths to hidden truths, ultimately rewarding us with the pleasure [*Freude*] of the successful solution."

The short answer to the "why" question is going to be that mathematicians engage in mathematics because it gives us *pleasure*, Hilbert's *Freude*. "What," ask David Pimm and Nathalie Sinclair, "is mathematics's debt to pleasure? Specifically, what is the pleasure of the mathematical text? Might there even be an aesthetics of mathematical pleasure to be developed?"[51] Already in the fourth century BC, the Greek mathematician Eudoxus was disapprovingly singled out in the first and last books of Aristotle's *Nicomachean Ethics* for identifying "the good" with pleasure. The word *plaisir* [pleasure] occurs on 106 of the 900+ pages of *Récoltes et sémailles*, Grothendieck's unpublished (and unpublishable) memoirs. Grothendieck is off the scale in many ways, but Bernard Zarca's recent sociological survey of attitudes of mathematicians (see chapter 2, note 56) shows that in his pursuit of pleasure, Grothendieck fits squarely within the mathematical norm. Zarca's subjects were given a list of sixteen "dimensions" of their work and asked to rate their importance. "The pleasure of doing mathematics" was rated "absolutely important" or "very important" by 91% of pure mathematicians, a significantly higher proportion than among applied mathematicians (88%) or scientists in related fields (76%) and much higher than the second choice—pleasure, again, this time in "sharing, by exchange and communication, the results one obtains"—given top rating by 66%.[52] Some of the other dimensions referred to the *good* in the form of applications, while the true and the beautiful were curiously absent; but all three reappeared, slightly disguised, on a list of four "virtues" of mathematics— "artisanal pleasure" was the fourth—that respondents were asked to place on a scale from 0 to 4. No surprise: for pure mathematicians, beauty comes first, followed closely by "artisanal pleasure," then "understanding of the world," with "utility" dead last. (For applied mathematicians, utility comes far ahead of the other three.) And since Zarca specifically inserted the word *artisanal* to forestall confusion with other kinds of pleasure (notably aesthetic, intellectual, or moral), we should assume that today's pure researchers still see pleasure as mathematics' primary virtue.[53]

I fervently hope this book makes a modest contribution to some future comprehensive and unexpurgated guide to mathematical pleasure. It deserves to be investigated, for example, why no one has ever put much energy into arguing, as Aristotle might have done, that *moderation* in mathematical pleasure is a virtue. In the meantime, we have not yet explained why we should be paid for the time we spend working on mathematics. If it's so enjoyable, wouldn't we do it for free? Well, maybe we wouldn't be so efficient or creative if we had to work for a living—in our spare time, so to speak—at something we presumably don't enjoy so much.[54] It might help to put this question in perspective to remember that mathematicians are not the only scholars (or academics, to use a term better adapted to contemporary reality) under pressure to defend their pursuit of a pleasurable activity, at public expense, in the context of generalized austerity. In a letter to David Willetts, the minister responsible for universities, science, innovation, and space (within that Department for Business, Innovation and Skills), the British Philosophical Association (BPA) attempted to rise to the challenge in 2010, shortly after the government eliminated the Middlesex University philosophy department. While agreeing "that the money provided by the taxpayer for philosophical research should be put to good use" and that "it would be entirely appropriate for the government to expect us to be able to justify our view that that money is indeed being well spent," the philosophers denied that "impact" can "meaningfully be measured over short periods of time or at the level of individual researchers or groups of researchers within a particular institution." Instead, they offered an eclectic menu of some of the "enormous benefits that philosophy has been responsible for" (justice, democracy, the scientific method, computers, secular ethics, animal welfare . . .), in the process making painfully obvious the gulf between the disciplinary ethos of philosophers and today's world of "business, innovation and skills."[55]

The philosophers chose not to answer another question: why is it a matter of general interest, independently of the uncertain prospects of short- or long-term benefits to human welfare, to have a small group of people working at the limit of their creative powers on something they enjoy? If a government minister asked me that question, I could claim that mathematicians, like other academics, are needed in the universities to teach a specific population of students the skills needed for the development of a technological society and to keep a somewhat broader population of students occupied with courses that serve to crush the dreams of superfluous applicants to particularly desirable professions (as freshman calculus used to be a formal requirement to enter medical school in the United States). The prospect of taking pleasure in the research we freely determine would then be the lure that gets us into the classroom (the assumption that not all of us are eager to be there is a gross misreading of the true state of affairs). Or I could revert to the Golden Goose argument. But if the question is taken at face value, it answers itself. Indeed, if the notion of general (or public) interest means anything at all, it should be a matter of general interest that work be a source of pleasure for as many people as possible.

Play, as analyzed in *Homo Ludens*, shares many characteristic features with the practice of pure mathematics. Zarca asked his subjects to choose two from a list of eight descriptions of mathematical activity; 62% of pure mathematicians chose "a very pleasurable sort of game, with the appeal of the unknown," nearly twice as many as the next most popular choice ("an inner necessity"). Huizinga defines play as "an activity which proceeds within certain limits of time and space, in a visible order, according to rules freely accepted, and outside the sphere of necessity or material utility."[56] The condition on time and space may be irrelevant, but the other three items on Huizinga's list are unmistakable characteristics of pure mathematics.

On ethical grounds, Aristotle's account of music can serve as a model (if debatable) justification for mathematics: "Nowadays most people practice music for pleasure, but the ancients gave it a place in education, because Nature requires us not only to be able to work well *but also to idle well*" (my emphasis). Huizinga comments that "for Aristotle [idleness] is preferable to work; indeed, it is the aim (*telos*) of all work." Aristotle's "demarcation between play and seriousness is very different from ours. . . ." "Idleness" for Aristotle means the freedom to pursue "such intellectual and aesthetic preoccupations as are becoming to free men . . . . The enjoyment of music comes near to being . . . a final aim of action because it is sought not for the sake of future good but for itself. This conception of music sets it midway between a noble game and 'art for art's sake.' " Bellah adds that, for Aristotle "*Theoria*"—contemplation, the search for knowledge—"is useless. It is a good internal to itself, but it has no consequences for the world."

In trying to distinguish science from play, Huizinga argues that the two differ in that the rules of science are subject to paradigm shifts, "whereas the rules of a game cannot be altered without spoiling the game itself."[57] What does this say about mathematics? We have seen that the rules of mathematics—the "foundations" in Manin's epistemologically neutral sense—are regularly modified in response to changed priorities; Voevodsky's univalent foundations may be adopted as new axioms. But the perception on the part of those who adopt new rules is that they represent an improvement over the earlier rules, and I think Huizinga would have been wrong to argue that a proposed improvement of the rules would spoil a game.

Freud claimed that "the opposite of play is not seriousness, but—reality."[58] I am tempted to say that, whereas Mathematics aspires to be Real as well as Serious, mathematics may well be serious but is, at best, virtually real: not at all opposed to play. But play, like Aristotle's idleness, is not only hard to sell to the Gradgrinds of Britain's Department of Business, Innovation, and Skills; it also has awkward associations for mathematicians. Most mathematicians agree with Voevodsky that Hilbert's *formalism*, his reading of mathematics as a rule-based game, failed on technical grounds to achieve its primary goals. But it was scarcely more successful as an ideal. Long before Gödel announced his incompleteness theorems, Weyl wrote that mathematics "is not the sort of arbitrary game in the void proposed by some of the more extreme branches of modern art." More recently, Neil Chriss, who chose to forgo a promising future in the Langlands program to work for a series of wildly successful hedge funds, explained that he gave up mathematics in part because "it is so disconnected with the rest of the world."

> "[Hermann Hesse's] *The Glass Bead Game* [*Magister Ludi*—literally "master of the game"] is a favorite novel among my mathematician friends." In it "a monastic order of intellectuals has split apart from [a future] society. They engage in a complicated, all consuming game—the Glass Bead Game—and strive to achieve mastery of it . . . to members of the order it represents everything that is important in the world. . . . it is hard to believe that Hesse wasn't describing research mathematics."[59]

"The master of the moderns defines himself as a 'free mathematician,' as a 'creator,'" writes Mehrtens, "and the countermoderns accuse him of creator's arbitrariness."[60] Even when we grant that mathematical play is not sterile, in the way that the Glass Bead Game strikes most readers as sterile, it's risky to identify mathematics as a form of creative play. Computerization may have stripped chess of much of its pathos, of the sense that it engages the players on the deepest strata of their existence (imagine Bergman's *Seventh Seal* with HAL the computer in the role of Death); and a mathematics that sees itself as a game played on an endless chessboard leaves itself open to a similar devaluation.

So let's leave play to the side for the moment and retain from *Homo Ludens* a theoretical principle—"We may well call play a 'totality' in the modern sense of the word, and it is as a totality that we must try to understand and evaluate it" and an ethical stance: "Most [hypotheses] only deal incidentally with the question of what play is in itself and what it means for the player. . . . 'So far so good, but what exactly is the fun of playing?'" Reviel Netz echoes this stance in a book entitled *Ludic Proof*: "people do the things they enjoy doing."[61] What Netz calls a "modest" assumption seems unremarkable; but we have seen that the dominant justifications of mathematics all take for granted that mathematics is a "thing" people do with some other goal—goodness, truth, or beauty—in mind. That it should be a matter of general interest that work be a source of pleasure for as many people as possible, as I wrote before, ought to go without saying. But apparently it doesn't where pure mathematics is concerned; otherwise there's no explaining why mathematical research is consistently, and implausibly, presented as a means rather than an end.

When the German mathematician Heinrich Liebmann characterized mathematics as a "free, creative art" [*freie, schöpferische Kunst*] in his 1905 Leipzig inaugural address, the stress was on the adjectives, the word *art* being no more than the noun required by syntax. The advantage of using art—rather than that-which-is-free-and-creative, which is what is really meant here—isn't just that "art" is shorter but that it conjures up a picture. It's a misleading picture: when mathematicians say art, we've seen that they generally mean beauty. Other pictures are available.[62] Ethologist Gordon Burghardt proposes five criteria to distinguish play from other forms of animal behavior; the fifth is that play takes place "in a 'relaxed field.' " "One of the prime characteristics of play" for Burghardt "is that the animal is not strongly motivated to perform other behaviors. The animal is not starving (or even very hungry), it is not at the moment preoccupied with mating, setting up territories, or otherwise competing for essential resources or escaping predators."[63]

Burghardt's relaxed field is Eden (or Flexner's paradise) before the fall. The deer and the antelope play in their relaxed field, but they don't apply for National Science Foundation research grants. The theory of the relaxed field posits play as the opposite of *stress*. Play, for John Dewey, is *not* the opposite of work: "Play remains as an attitude of freedom from subordination to an end imposed by external necessity, that is, to labor; but it is transformed into work in that activity is subordinated to *production* of an objective result."[64] Reviewers asked[65] to assign objective grades to mathematics grant proposals are also mathematicians and interpret their instructions according to the values impressed upon them by their socialization; they seek objectivity in the locus of Liebmann's free creativity, which is another name for the relaxed field.

If we take the values themselves as the starting point, rather than the relaxed field in which they are free to express themselves, then we find ourselves in the sphere of *ethics*. Crispin Wright, a prominent philosopher of Mathematics, was more candid and more credible than the British Philosophical Association when asked "Should philosophy be funded, even if funding it holds forth almost no prospect of improving the lives of ordinary people? If so, why?"

> [C]ontrary to what one might first assume, philosophical process is a very large part of the value of philosophy. Suppose it became possible to program computers to take over most of the projects of philosophical research currently being pursued in academia . . . Few would feel, "Well good, we can leave all that to the machines now, and get on with other things." . . . it is crucial that [the products of philosophical research] be attained by human beings, and strongly preferable that they be attained by a shared process in which there is conversation and mutual understanding of why what results results, of the conceptual pressures and constraints that shape it (Wright 2011).

In other words, philosophy is what Alasdair MacIntyre calls a *tradition-based practice*—though Wright might not use this vocabulary to describe it — and this is responsible for much of its value.

MacIntyre uses the term *practice* to mean

> any coherent and complex form of socially established cooperative human activity through which goods internal to that form of activity are realized in the course of trying to achieve those standards of excellence which are appropriate to, and partially definitive of, that form of activity, with the result that human powers to achieve excellence, and human conceptions of the ends and goods involved, are systematically extended.[66]

This isn't really meant as a definition; the part from "standards of excellence" to "form of activity" looks temptingly circular, for example, and I wouldn't want to pin down what excellence or complex means to MacIntyre any more than what important or ambitious or major might mean to the authors of the ERC guidelines (or what visual perfection, quality, and aesthetic mean in the interviews recorded in Muntadas's book[67]). But I don't see that as a problem. The tradition-based approach is circular: like the Mehrtens quotation with which I ended the first chapter, it asserts that the point of doing mathematics is to do mathematics. And the tradition-based approach is not circular: it invites us to attend to *description* rather than *prescription*—to address the "why" question by returning to the "what" question.

Or to put it another way: there is no need to seek the meaning of mathematics elsewhere than in the practice constituted by tradition; and the *telos* of mathematics is to develop this meaning as a way of expanding the relaxed field. MacIntyre is at pains to distinguish his notion of tradition from the conservative version associated with Edmund Burke. "[A]n adequate sense of tradition" for MacIntyre "manifests itself in a grasp of those future possibilities which the past has made available to the present." "[T]he history of a practice in our time is generally and characteristically embedded in and made intelligible in terms of the larger and longer history of the tradition through which the practice in its present form was conveyed to us."[68] MacIntyre's nondefinition of practice draws attention to just what it is that matters about the "socially established cooperative human activity" of pure mathematics; and this focused attention is what makes mathematics *work* in Dewey's sense, "production of an objective result," while remaining relaxed, free "from subordination to an end imposed by external necessity."

• •

Crispin Wright had this to say about "improving the lives of ordinary people":

> A good philosophical process will be one which, necessarily, is appreciated as such by its participants, as interesting, eye-opening, inspiring, and perhaps importantly revisionary. . . . If it is a good for those who participate, then in a pluralist and civilized society, participation should be encouraged, and its scope should be widened as far as possible.[69]

This book does not pretend to justify mathematics; no one needs another *Mathematician's Apology*. It's goal is not so much to explain mathematics as to convey to Wright's "ordinary people"—presumably those whose lives are not necessarily directly bound up with mathematics— what it is like to be a mathematician, freely choosing a tradition to which to adapt, not to serve the Powerful Beings of market rationality nor the metaphysical Powerful Beings of our own creation. The pleasure of the mathematical tradition is inseparable from the pathos of its relaxed field, where the poles of some familiar antinomies—mind versus body, finite versus infinite, and necessity versus contingency, just to mention some of the themes of this book's three central chapters—stage confrontations that evolve along with the preoccupations of participants in the tradition, who don't necessarily imagine or even care how they might eventually reach resolution.

My hope is that exposure to the peculiarity of this pathos may enrich the lives of "ordinary people." In this book I am trying to make the mathematical tradition just real enough to make the pathos palpable. Here the presumed, but largely unsubstantiated, parallel between mathematics and the arts offers unexpected clarity. Anyone who wants to include mathematics among the arts has to accept the ambiguity that comes with that status and with the different perspectives implicit in different ways of talking about art. Six of these perspectives are particularly relevant: the changing semantic fields the word *art* has historically designated; the attempts by philosophers to define art, for example, by subordinating it to the (largely outdated) notion of beauty or to ground ethics in aesthetics, as in G. E. Moore's *Principia Ethica*, which by way of Hardy's *Apology* continues to influence mathematicians (see chapter 10); the skeptical attitude of those, like Pierre Bourdieu, who read artistic taste as a stand-in for social distinction; the institutions of the art world, whose representatives reflect upon themselves in Muntadas's interviews; the artist's personal creative experience within the framework of the artistic tradition; and the irreducible and (usually) material existence of the art works themselves.

Conveniently, each of these six approaches to art has a mathematical counterpart: the cognates of the word *mathematics* itself, derived from the Greek *mathesis*, which just means "learning," and whose meaning has expanded and contracted repeatedly over the millenia and from one culture to another, including those that had no special affinity for the Greek root; the Mathematics of philosophers of "encyclopedist" schools; school mathematics in its role as social and vocational filter; the social institutions of mathematics with their internal complexity and their no-less-complex interactions with other social and political institutions; the mathematician's personal creative experience within the framework of the tradition (the endless dialogue with the Giants and Supergiants of the IBM and similar rosters); and the irreducible and (usually) immaterial existence of theorems, definitions, and other mathematical notions. This book touches on each of these ways of talking about mathematics, but the guiding orientation is to a point somewhere between the "relaxed field" and the "tradition-based practice" in the fifth item on the list, which is where the pathos dwells in mathematics as well as in the arts.[70]

I would not compare mathematics to religion, but in reading Robert Bellah's *Religion in Human Evolution*, I am struck by structural parallels at least as relevant to the analysis of the respective fields as the parallels with the arts. Each of the ways of talking about art and mathematics listed before has its equivalent in Bellah's historical sociology of religion; it was in Bellah's book that I first learned about the relaxed field; and Bellah's allusion to MacIntyre's notion of practice reinforced what Corfield's approach had already taught me, namely, that many of the most persistent questions about the nature of mathematics are rooted in ethics rather than epistemology. In Bellah's work there is also a tension, to which he draws attention only briefly, between the relaxed field of religious ritual, "a form of life not subject to the struggle for existence," and the need to perform ritual in order to propitiate "Powerful Beings." During the ritual dances of the Kalapalo of the Upper Xingu Basin of central Brazil, " 'common humanity' . . . takes over from the divisions of everyday life." Bellah quotes the anthropologist Ellen Basso: "In ritual performance . . . the body is an important musical instrument that helps to create . . . the experience—however transient—that one is indeed a Powerful Being."

Bellah insists that the " 'Powerful Beings' of the Kalapalo, who were there 'at the beginning,' " are not (yet?) gods. But like other cultures, the Kalapalo believe that "Human life derives ultimately from the Powerful Beings" who "could be terribly destructive when crossed . . . but with whom people could identify if they followed the proper ritual, and, through identification, their power could become, at least temporarily, benign." In more complex societies, according to Bellah, the Powerful Beings are worshipped as gods, and rituals are largely connected with maintaining the basic material conditions of life; the Hawaiian Lono, to whom the four-month carnival-like *Makahiki* festival is devoted, is "the nourishing god."[71]

It is understood that the arts, however they are defined, entail reflection or representation of the ambient society. Reflection on this reflection is the province of critical theory and cultural studies, academic-based institutions parallel to the institutions of the art world proper, whose methods are routinely appropriated by artists themselves (like Muntadas). Mathematics has given rise to no such institution. This conspicuous difference between mathematics and the arts is more consequential than the absence, already noted, of mathematics *critics*; it means mathematics lacks the institutional capacity to reflect critically on the material conditions of its own creation. In pursuing their interests, today's Powerful Beings manage to create the conditions for keeping as many people as possible in a state of stress rather than relaxation. A particularly insistent narrative of Powerful Beings in the next chapter shows how mathematicians have been innovating and applying their skills to find new ways to make this possible.

chapter 4

Megaloprepeia

*Il est difficile de vivre dans un monde dont le support pro-
ductif est l'argent.[1]*

*Mephisto: Yes it is I who have been deceived I the devil who
no one can deceive yes it is I who have been deceived.[2]*

Crimes against Humanity

Economics originated in and has never completely broken with moral
philosophy. A president of the American Economic Association recalled
in 1968 that "when I was a student, economics was still part of the moral
sciences tripos at Cambridge University." On the grounds of the discipline's history and its vocabulary (goods, utility, satisfaction) and
because

> the fundamental principle that we should count all costs . . . and evaluate all
> rewards . . . is one which emerges squarely out of economics and which is
> at least a preliminary guideline in the formation of the moral judgment,

"mathematical ethics" is a plausible name for the dismal science.[3] So the
explosion of finance mathematics, and its implication in the 2008 crash,
has had the welcome, but unintended, consequence of establishing a
common border between mathematics and morality. Somebody was to
blame for the crisis, and primitive retribution was on the minds of many.
Protestors against the September 2008 bank bailout invited Wall Street
bankers to "Jump!" "Uncle Sam wants your money," a *Forbes* headline
screamed, "and the crowd outside the gate wants your head." The newly
elected President Obama warned a group of businesspeople that "My
administration is the only thing between you and the pitchforks."[4]

I had spent the spring of 2008 visiting Columbia University, adding
monthly to my pension and watching its balance drift slowly back down

to its previous value before the next month's installment, in that spring's market of slow decline. Manhattan mathematical geography was, therefore, still vivid in my mind when the collapse of the Lehman Brothers triggered the crash later that year, and was conspicuous in my own visions of retribution. All that fall and through the next spring, I would wake up having dreamed of crucified engineers of quantitative finance—the so-called quants—lining the 7.7 miles of Broadway joining the Columbia mathematics department with Wall Street, with a brief rally to the left at 34th Street, stopping at the CUNY Graduate Center, and then a dip to the right at 8th Street to NYU's Courant Institute ("a top quant farm" since the 1990s), in an ironic class reversal of the suppression of the Spartacus slave revolt by the future triumvir Marcus Licinius Crassus.[5]

I'm not being completely candid: those weren't quants up on the lampposts, they were mathematics professors. Hadn't Warren Buffett, as early as 2002, referred to derivatives as "financial weapons of mass destruction"? Hadn't former French prime minister Michel Rocard gestured at the border between mathematics and morality in 2008 and accused the "math professors who teach their students how to make a killing on the market" of "crime[s] against humanity"?[6] It's now routine to refer to much of what goes on in finance as criminal—"prima facie criminal behavior," in the words of economist Jeffrey Sachs. Charles Ferguson wrote *Predator Nation*, subtitled *Corporate Criminals, Political Corruption, and the Hijacking of America*, in part "to lay out in painfully clear detail the case for criminal prosecutions"—"What is this," he writes, "if not organized crime?"—and his point of view is hardly marginal; his documentary *Inside Job* won an Oscar in 2011, while in 2013 Britain's Parliamentary Commission on Banking Standards "envisages a new approach to sanctions and enforcement against individuals," including the establishment of a "criminal offence" with "Senior Persons" liable to imprisonment.[7]

But the equation of finance with criminality is usually based on *acts*, not *ideas*. Though Rocard stopped short of suggesting that mathematics professors be hauled before the International Criminal Court,[8] the notion that mathematicians could do something reprehensible was novel enough to generate six months' worth of heated rejoinders. Three of my eminent Paris colleagues asked whether politicians like Rocard weren't "much more guilty than mathematicians by not imposing real restrictions on the risks" taken by financial institutions.[9] Even before *Le Monde* published Rocard's accusation and again in the months that followed, French specialists in finance mathematics wrote articles and gave lectures in which they acknowledged the need for "a minimum of ethics," but insisted that banking culture and lax public supervision were responsible for the use of mathematical techniques without regard to their manufacturers' guidelines. "It's clear," wrote Marc Yor, professor at the Université Pierre et Marie Curie, better known as Paris 6, "that mathematicians have got caught in an infernal spiral," but the creation of subprimes or the seventy-year mortgages in Spain "are not really the fault of mathematicians." Nicole el-Karoui, also of Paris 6 and the École Polytechnique, suggested that a mathematical model may be "reasonable" at 100,000 € but not when the sums involved are much greater.[10]

Others found these arguments disingenuous. Two mathematicians claimed to have heard "students freshly sharpened by our university programs boasting of having created speculative products on stocks of basic necessities." While Marc Rogalski argued that financial mathematics is "directed essentially toward acquiring techniques for *increasing the rate of financial profits*" and wondered whether "mathematics [should] be on the owners' and stockholders' side in the class struggle," the late Denis Guedj, author of *The Parrot's Theorem*, wrote in *Liberation* that he saw the mathematicians who chose to join "the camp of the powerful" as "enemies . . . working for the misfortunes of the majority. . . ."[11]

One crash later, not much has changed. The Oxford-Man Institute for Quantitative Finance has been "going from strength to strength"[12] since it opened its doors in late 2007, just down the road from Oxford's Mathematical Institute. In Paris, you can read about the "explosion of the financial risk industry" on the Web site for the "Probability and Finance" program run by Karoui, Yor, and their colleagues at Paris 6 and the École Polytechnique; any allusions there to professional ethics are in print too fine for my eyes. There is no shortage of candidates. At the École Polytechnique, the most élitist of French engineering schools, 70% of mathematics majors aim at a career in finance; the École Centrale, only slightly less prestigious, had "become a school of managers," much to the regret of an official responsible for mathematics education.[13] Even the venerable École des Ponts et Chaussées (School of Bridges and Roads), one of the grandest of the Grandes Écoles, offers a masters program that diverts a sizable proportion of its graduating class onto the virtual highway of quantitative finance.

$ € £ ¥ $ € £ ¥ $ € £ ¥ $ € £ ¥ $ € £ ¥ $ € £ ¥ $ € £ ¥

During the 2012 French presidential campaign, economics minister François Baroin dismissed the opposition's proposals on the grounds that "attacking finance is as idiotic as saying 'I'm against rain,' " (or cold, or fog), while Prime Minister François Fillon advised the Socialist Party candidate to run his electoral platform by Standard & Poor's.[14] Rosa Luxemburg, criticizing the Russian revolution, wrote that "Freedom is always the freedom of those who think otherwise." But what use is the freedom to be against lousy weather?

The equations of mathematical economics are peculiar in that their variables are already familiar as characters in a morality play, actors in a cosmic drama: *Investor* (Capital)—the prime mover, *Finance*, *Industry*, *State*, *Labor*, *Consumer*. Messieurs Baroin and Fillon distantly echo Margaret Thatcher's motto—*TINA*: There Is No Alternative—rooted in a worldview that places market forces and natural forces on the same metaphysical plane.[15] I have nothing to add to the rich literature on this topic and on the origins of the crisis of 2008 beyond a few reflections on the new mathematically trained member of the cast: *Quant*, the mathematical engineer whose job is to protect investments from the hazards of finance, just as *Ponts et Chaussées* graduates used to brace their bridges against the perils of gravity.[16]

The consensus is that, Rocard's allegations notwithstanding, the quant is not a crook. Ferguson's subtitle alludes to "corporate criminals" and "political corruption," and most of his text is a scathing indictment of bankers; mathematicians get off the hook—as they do in Scott Patterson's *The Quants*, a journalistic melodrama in four personalities and four overlapping mathematical models, which accuses financial engineers of hubris, greed, and misjudgment but not of criminal intention. Quant is merely one of the Evil Principle's minions, like Goethe's Mephistopheles, who defines himself in Euclidean fashion as "a part of the Power that *always* wills the Evil, and *always* creates the *Good.*"

How should finance mathematics define itself? Adam Smith anticipated the main source of instability of financial markets—leveraging—
with his customary clarity:

> The interest of money is always a derivative revenue, which, if it is not paid
> from the profit which is made by the use of the money, must be paid from
> some other source of revenue, unless perhaps the borrower is a spendthrift,
> who contracts a second debt in order to pay the interest of the first.[17]

Financial engineering makes life profitable for Smith's spendthrifts by
endlessly recombining their multiple debts into *financial derivatives*,
obligations denominated in virtual money, in order to continue to rack
up profits while postponing the day of repayment. The impossibility of
exponential growth in the real world is what brings down all Ponzi
schemes, and the real estate bubble whose bursting in 2007 led inexorably to the cascade of banking and insurance failures was no exception.
But wealth was already largely virtual when Smith sought to reveal its
guiding principles. Referring to Renaissance innovations in paper money
and credit ("bills of exchange"), Fernand Braudel wrote that "this type
of money that was not money at all, and this juggling of money and
book-keeping to a point where the two became confused, seemed not
only complicated but diabolical."[18] It's a recurrent image in Braudel's
work, and he's on to something. Here is Mephistopheles himself, in
Faust, Part II, explaining his invention of paper money to the emperor:

> Such paper's convenient, for rather than a lot
> Of gold and silver, you know what you've got,
> . . . Since the paper, in this way, pays for itself,
> It shames the doubters, and their acid wit,
> People want nothing else, they're used to it.

The emperor thanks Mephistopheles for his innovation:

> The Empire thanks you deeply for this bliss:
> We want the reward to match your service.
> We entrust you with the riches underground,
> You are the best custodians to be found.
> . . . your roles make the Underworld, and the Upper,
> Happy in their agreement, fit together.[19]

Once Mephistopheles has his hands on "the riches," he will not readily let go. But the Fiend's inspiration was superfluous when it came time to price financial derivatives; a Nobel Prize–winning team of economists had inspiration enough.

The natural world is not always predictable; human beings don't always keep their promises. Aristotle examined the first source of uncertainty in his *Physics* and *Metaphysics*, the second in his *Nicomachean Ethics*. The financial risk market provides instruments to protect against either form of uncertainty. Most readers know that an *option* is a kind of bet on the future value of an investment (such as a stock, a bond, or a commodity). They, and more complex derivatives, can be used to insure against unexpected changes in crop prices, or accidents, or the failure of borrowers to repay their debts. This sort of thing is at least as old as Greek mathematics itself. In his *Politics*, Aristotle recalls how Thales purchased options on the use of "all the olive-presses in Chios and Miletus" and "made a quantity of money" when the harvest was unusually rich. Aristotle's conclusion, which the reader is encouraged to keep in mind while reading this chapter, is "that philosophers can easily be rich if they like, but that their ambition is of another sort."[20]

Derivatives can also be used for purposes of pure speculation. "Thanks to mathematics, the financial risk market works, roughly speaking [*à peu près*]." That was Karoui, quoted in the French business press in 2005, explaining her fear that a "Tobin tax"—a small tax on financial transactions proposed (years before) by James Tobin to discourage parasitic speculation—would "collapse the fragile edifice patiently constructed over decades."[21] The foundation of the edifice was the *Black-Scholes equation*, invented (or discovered?) in 1973 by Fischer Black and Myron Scholes, soon joined by Robert C. Merton (son of the Merton who named the Matthew Effect). Patterson, with a flair for overstatement, describes what Black-Scholes wrought:

> Just as Einstein's discovery of relativity theory in 1905 would lead to a new way of understanding the universe, as well as the creation of the atomic bomb, the Black-Scholes formula dramatically altered the way people would view the vast world of money and investing. It would also give birth to its own destructive forces and pave the way to a series of fi-

nancial catastrophes, culminating in an earthshaking collapse that erupted in August 2007.

Ten years earlier, just a few months after Merton and Scholes shared the 1997 Nobel Prize in Economics, Long Term Capital Management, which managed hedge funds in large part on the basis of the Black-Scholes model, had collapsed, along with the finances of much of East Asia and Russia. The catastrophe seemed earthshaking at the time, but the losses were a thousand times greater in 2008 (and they continue to mount).[22]

It's not the equations that make it difficult for a mathematician like me to grasp quantitative finance. My problem is with adopting the psychology, the motivations, the *persona* of *Investor*, rational economic agent, and invariable protagonist of the allegory through which politicians, the mainstream press, and the finance math textbooks ram the moral lessons of public economic policy into popular consciousness. Someone who (that TIAA-CREF account notwithstanding) has never aspired to playing *Investor*, a figure whose cardinal virtue is maximizing returns, is at a distinct disadvantage. This is one reason I am grateful for sociologist Donald MacKenzie's lucid deconstruction of the allegory, in an impressive series of articles and books. In an analytic history of Black-Scholes, MacKenzie called option-pricing theory *performative*: "it did not simply describe a pre-existing world, but helped create a world of which the theory was a truer reflection." More precisely, "The availability of the Black-Scholes formula, and its associated hedging techniques, gave participants the confidence to write options at lower prices, . . . helping options exchanges to grow and to prosper, becoming more like the markets posited by the theory . . . the model [was] used to make one of its key assumptions a reality."[23] If we are tempted to view Ngô's proof of the fundamental lemma as "performative"—both because it brought a stage of the Langlands program to maturity and be-

$$\frac{\partial w}{\partial t} = rw - rx\frac{\partial w}{\partial x} - \frac{1}{2}\sigma^2 x^2 \frac{\partial^2 w}{\partial x^2}$$

Figure 4.1. The Black-Scholes partial differential equation. Here x is the price of the underlying stock or other security, w is the price of the derivative as a function of x and time t, r is the riskless rate of return, and σ is a measure of volatility.

cause by borrowing the *Hitchin fibration* from its home in a different part of geometry, it revised our understanding of the interrelations among central branches of mathematics—we can say that the Black-Scholes model was performative in the latter sense, with incomparably more profound material consequences.

Applied Stochastic Analysis

[A]s the mathematics of finance reaches higher levels so the level of common sense seems to drop.[24]

"You're in for it now" is what the Time Traveller told himself when he descended into the underground world of H. G. Wells's *Time Machine.* And that's what I told myself in 2000 at the European Congress of Mathematicians in Barcelona, when a speaker at a round table on the future of mathematics hailed the recent explosion of finance mathematics that providentially brought so many undergraduate and masters' students to our departments' lonely corridors.[25] A visit to Columbia in 2004 revealed the full scope of the phenomenon. A colleague boasted that Columbia's mathematical finance program was underwriting the lavish daily spreads of fresh fruit, cheese, and chocolate brownies, when other departments, including mine in Paris, were lucky to offer a few teabags and a handful of cookies to calorie-starved graduate students. I spent a month living on my own, and when I walked down the stairs after staying late at the office, even at 9 or 10 p.m., I could hear traders, sent up by their Wall Street offices, struggling with their late-night equations. It brought to mind *Morlocks*, the underground-workers Wells' Time Traveller encountered at the bottom of his descent, toiling in the darkened basement of applied stochastic analysis to provide fresh fruit for the daily teas of the sunny *Eloi* of pure mathematics. A crater still disfigured lower Manhattan when I returned for a longer visit four years later, but the city's economy had rebounded and traders were still staying overtime for basement classes. Then, in one of those ironic reversals that are much more entertaining when they take place in a novel set 800,000 years in the future than when they unfold in real time with us in the middle, the Morlock traders turned out to be too big to fail, which entailed not literally *eating* all the rest of us, but setting us to working for the foreseeable future to bail them out.

[Two colleagues, having read to this point, gently but firmly advised me to remove this chapter from the book, suggesting independently that this sort of talk is "more suitable for conversations over coffee" in the common room (with or without fresh fruit) "than for a serious book." Regarding the seriousness expected of an author with my degree of charisma, the reader is directed to chapter 8. More troubling is that both my colleagues seem to have missed my disclaimers. So I repeat them here, more explicitly. My purpose is not to assign responsibility for the 2008 crash and certainly not to imply that mathematics professors are specifically to blame. Nor, as I write shortly, does this chapter aim to change anyone's mind about fundamental questions of economic policy. Its primary purpose, rather, is to explain some of the context for a debate that is actually taking place, within and around mathematics, in connection with the growth of mathematical finance. The tensions between the *internal* and *external* goods, in MacIntyre's sense, involved in the creation of mathematics, are well illustrated by this debate; to expose these tensions is not to take sides. Readers will have no trouble guessing what I believe, but I don't necessarily expect them to agree with me.

The secondary purpose of this chapter is to provide a very brief introduction to the mathematical modeling of reality. Thus, we return to the main narrative.] Stochastic analysis is the marriage of probability theory with calculus, mathematics' two main contributions to modeling physical reality. A deterministic model is one that allows exact prediction of the future state of a system on the basis of knowledge of the present, combined with relevant physical laws. Deterministic laws typically take the mathematical form of *differential equations*.[26] Physicists represent the state of a physical system—like the global climate—by *functions* that encode the numerical values of relevant quantities. In the case of the global climate, these quantities would include (among many others) temperature, concentration of certain gases and particles in the atmosphere, surface reflectivity, and the earth's position relative to the sun. Most of these quantities vary, depending on time and position, and when we assign the temperature (say) to *every point* on the earth's surface at *every time*, we define a *function* of position and time. We can't really measure temperature at *every* point and *every* time, but it is regularly measured at certain positions and times, and this measure provides a model of the abstract temperature function as well as an input for the daily weather report.

Physical laws predict how quantities of interest influence one another mutually, and this prediction takes the form of a differential equation whose solutions include the future values of functions that represent the state of the system: solving the differential equations determines how physical quantities evolve in time. This is the kind of information we want to know: the prediction of global warming is the claim that, with the passage of time, the average temperature at any given place is likely to increase, eventually with catastrophic consequences. This is read from the output of the model, which is an expression for the temperature function—call it **T** for temperature—that includes not only past values, which can be measured, but also future values, which we can also measure—but only when it's too late to do anything about them. Solving the differential equation to find the function **T** is in every way analogous to solving a polynomial equation like $y^2 = x^3 - x$ (cf. chapter β) except that (1) we are finding not a number or pair of numbers but rather a way to assign a number (temperature) to every point on the earth's surface at a given time in the future and (2) the differential equation is not a simple algebraic expression but rather a means for determining how **T** changes in time and space as a result of interaction with the other factors.

Since these factors are changing simultaneously, in part in response to changes in **T**, there are multiple feedback effects, and the solution of differential equations is extremely difficult both conceptually and technically. A *difference* (rather than *differential*) *equation*, in which time is measured in discrete units, is simpler but works the same way. Here is *Stable Equilibrium*, a difference equation game you can play at your next dinner party. You are seated around a circular table; the *initial conditions* for the equation are the amount of money each of you has brought to dinner. At regular intervals, you give 10% of your money to each of your neighbors to the right and left, and the others do likewise. Since, at each move, each of you gives away a fixed proportion of what you have, you might expect that after enough time has passed, you will all have the same amount of money, and this is indeed what happens, though it

Start	1	33	3	4	5	70	300	3	2	5	11	3.4	2
1	4	27	6	4	11	87	247	33	2	5	10	4	2
2	6	22	8	5	18	95	210	51	6	5	9	4	2
3	8	19	9	7	25	99	182	62	10	6	8	5	3
4	8	17	10	9	30	100	162	69	15	6	7	5	4
5	9	16	10	11	35	99	147	73	20	7	7	5	4
6	9	14	11	13	39	97	134	75	24	9	7	5	5
7	9	14	12	16	42	95	125	76	27	10	7	5	5
8	9	13	12	18	45	93	117	76	30	11	7	5	6
9	9	12	13	20	47	91	110	75	33	13	7	6	6
10	9	12	14	22	49	88	105	75	35	14	8	6	6
100	29	31	33	35	37	39	39	38	36	34	32	30	29
1000	34	34	34	34	34	34	34	34	34	34	34	34	34

Figure 4.2. The evolution of Stable Equilibrium to a stable equilibrium after 1000 rounds.

may take many lifetimes (or play has to be exceedingly rapid) if the initial distribution is highly unequal.

Stable Equilibrium is a discrete model of the diffusion of heat in a ring. It's technically a difference equation on a graph (see chapter 7) and, although it's analogous to some models in economics in that it demonstrates evolution to an equilibrium position, it's clearly a silly example— experience shows that heat flows from warmer to colder bodies but that money trickles up from poorer to richer. Figure 4.3 illustrates the *Mat-*

Start	1	1	1	1.00	1.01	1.00
1	1	1	1	1.00	1.01	1.00
2	1	1	1	1.00	1.01	1.00
3	1	1	1	1.00	1.02	1.00
4	1	1	1	0.99	1.02	0.99
5	1	1	1	0.99	1.03	0.99
6	1	1	1	0.98	1.04	0.98
7	1	1	1	0.98	1.05	0.98
8	1	1	1	0.97	1.06	0.97
9	1	1	1	0.95	1.08	0.95
10	1	1	1	0.93	1.10	0.93
100	2×10^{11}	(2×10^{11})	3×10^{11}	(3×10^{11})	3×10^{11}	(3×10^{11})
127	1×10^{15}	(2×10^{15})	2×10^{15}	(2×10^{15})	2×10^{15}	(2×10^{15})

Figure 4.3. The Matthew Effect difference equation game illustrates in simplified form the danger that leveraging on the basis of exponential growth will lead to a vicious circle of rapidly expanding debt. At some point Standard and Poor's or some

thew Effect difference equation game (rules based on Matthew 25:29; see chapter 2, note 27): instead of receiving from each neighbor 10% of his or her wealth, you receive 10% of your own. At the start, one player has $1.01 and all the others have $1.00. After 127 iterations, half the players are quadrillionaires and the other half are quadrillions in debt. In this game, initial conditions are preserved indefinitely if they are *precisely* equal, but this is an *unstable* equilibrium: a slight change of the initial conditions leads to an enormous change in the outcome.

Both games illustrate the kind of information differential equations provide, and both also exhibit an obvious circular *symmetry*: if all the participants straggled simultaneously one seat to the right, moneybags

1	1	1	1	1	1	1
1	1	1	1	1	1	1
1	1	1	1	1	1	1
1	1	1	1	1	1	1
1	1	1	1	1	1	1
1	1	1	1	1	1	1
1	1	1	1	1	1	1
1	1	1	1	1	1	1
1	1	1	1	1	1	1
1	1	1	1	1	1	1
1	1	1	1	1	1	1
3×10^{11}	(2×10^{11})	2×10^{11}	(1×10^{11})	3×10^{11}	3×10^{11}	(1×10^{11})
2×10^{15}	(2×10^{15})	1×10^{15}	(8×10^{14})	3×10^{14}	3×10^{14}	(8×10^{14})

other rating agency casts doubt on the ability of some of the players to repay the quadrillions they owe, and the entire system unravels in short order.

in hand, the rules of the game would not change.[27] And both games are *deterministic* in the sense that every time they are played, the outcome is completely determined by the initial conditions; There Is No Alternative, as Thatcher would say. But how many times would you want to play a game whose outcome is known in advance, like tic-tac-toe?[28] That would be boring! The soldier in Stravinsky's *A Soldier's Tale* becomes fabulously wealthy when he hands over his violin to the devil in exchange for a book that—deterministically—predicts the future; but such a book is only useful to an owner who possesses the unique (insider) copy. The Black-Scholes model, in contrast, is based on a *stochastic differential equation*: it doesn't determine exact future values of a stock—that would be truly diabolical—but rather uses a probabilistic model of the range of possible values to calculate the "correct" price of an option.

We could incorporate *stochasticity*—randomness—into our deterministic party games. At each round, for example, each player might flip a coin: heads means hand over 10% of your neighbor's wealth; tails, 10% of your own wealth. Some rounds are played like *Stable Equilibrium*, some like *Matthew Effect*, and most are a mixture of the two. Like a board game or like casino gambling, the randomness the coin toss introduces means the outcome will be different each time the game is played. Probability theory calculates your *expected winnings*—your average take if you play the game thousands or millions of times—but it can't tell you whether you'll have enough money left for cab fare home when the party's over.

Generations of philosophers have racked their brains trying to reconcile the random aspects of life in human society with a deterministic worldview—or with free will, for that matter.[29] A rational *Ponts et Chaussées* engineer will use deterministic models to design bridges to withstand extreme stress—a once-in-a-thousand-year earthquake, for example. But once in a thousand years is a probabilistic prediction that means only that the number of earthquakes will get closer and closer to the number of 1000-year periods that elapse as we march toward eternity: what is the philosophical status of such a claim? A rational agent at the investment casino will eschew philosophers and will look instead for a broker with a credible model to maximize expected winnings while minimizing risk. Being rational, the agent may prefer not to bet the pension on which his or her survival depends. But Wall Street culture loves risk! Patterson's *The Quants* traces the dominant models in quantitative finance back to a 1962 book entitled *Beat the Dealer* and enlivens his narrative of investment house routine, not with dreary columns of figures (like figures 4.2 and 4.3) but with adrenaline-charged vignettes of high-risk poker games, extreme sports, and emotional volatility. "The greatest gambling game on earth is the one played daily through the brokerage houses across the country."[30]

"It is not the least of the paradoxes," writes David Steinsaltz in the *AMS Notices*, "that Patterson's protagonists eagerly seek risk in gambling, while their core mathematical models presume that investors pay a premium to dispose of risk." The Black-Scholes model is an equation expressing the insight of what Ivar Ekeland calls the "three fundamental theorems [of] mathematical finance," all of which say exactly the same thing: "If you take no risk, you get the riskless rate." The riskless rate is the marvelous feature of modern economies that lets money grow just by staying in the bank (not much these days). Since an option can be used to hedge against potential losses—for example, you can buy the option of selling a share of stock (or a ton of wheat) at a fixed *strike price* at a fixed date in the future—the price of the option should reflect just what is needed to compensate the risk of loss, compared to the risk-less rate, so that "as the asset price increases (or decreases) the option value increases (or decreases) and the sum of the two, suitably weighted, will remain constant . . . equating the return on the . . . [riskless] port-folio with the bank rate of interest results in . . . the Black-Scholes equation."[31]

Terence Tao devoted a lengthy blog entry to the reasoning that Black, Scholes, and Merton used to obtain the equation in figure 4.1, which calculates the price of an option as a function of the price of the underly-ing "instrument." The Black-Scholes-Merton model caused quite a sen-sation at the time (1973) because the equation expresses a deterministic relation between option prices and stock prices, while the prices them-selves vary stochastically in time. Tao emphasized that the model, while mathematically appealing, depends for its validity on a series of unreal-istic hypotheses.[32] Financial analyst Paul Wilmott agrees—"The world of markets doesn't exactly match the ideal circumstances Black-Scholes requires"—but argues that "the model is robust because it allows an intelligent trader to qualitatively adjust for those mismatches. You know what you are assuming when you use the model, and you know exactly what has been swept out of view." On the other hand, once the Black-Scholes paradigm made options pricing performative, common sense abandoned even the most intelligent traders. The modeling of CDOs—collateralized debt obligations, the derivatives most responsible for magnifying the mortgage crisis and nearly bringing the house down—is "confusedly elegant," according to Wilmott. "The CDO research papers apply abstract probability theory to the price co-movements of thou-sands of mortgages. . . . all uncertainty is reduced to a single parameter that, when entered into the model by a trader, produces a CDO value. . . .

over-reliance on probability and statistics is a severe limitation." Like Aristotle, Wilmott favors causal over chance relations: "Statistics is shallow description, quite unlike the deeper cause and effect of physics, and can't easily capture the complex dynamics of default."[33]

$ € £ ¥ $ € £ ¥ $ € £ ¥ $ € £ ¥ $ € £ ¥ $ € £ ¥ $ € £ ¥

Aristotle sounds uncharacteristically cynical when he writes about debt, in the section of the *Nicomachean Ethics* devoted to friendship:

> Benefactors are thought to love those they have benefited, more than those who have been well treated love those that have treated them well, and this is discussed as though it were paradoxical. Most people think it is because the latter are in the position of debtors and the former of creditors; and therefore as, in the case of loans, debtors wish their creditors did not exist, while creditors actually take care of the safety of their debtors . . . (Book IX, 7).

Proponents of finance math, in contrast to Aristotle, are upbeat when they insist on its utility and have no trouble justifying "mortgage-backed securities, the root of our present problems." For everyone who wants to buy a home, there has to be an investor willing to provide the capital for the loan. "Today, foreign institutions have the big money—and they would not make deposits in U.S. Savings and Loans even if such institutions were available." If Southwest Airlines has a long string of profitable quarters, it's "because it used derivative securities to hedge against price increases" of jet fuel. Derivatives, we are told, allow international firms "to hedge currency risk" and insurance firms to "hedge against increases in longevity. The quants," their defenders conclude, "did not create derivative securities . . . [they] help us understand them, price them, trade them and manage the risk associated with them."[34]

Opponents also invoke utilitarian arguments but come to different conclusions. "Despite what some academics (primarily in business schools) claimed, the vast sums of money channeled through Wall Street did not improve America's productive capacity by 'efficiently allocating capital to its best use.' " On the contrary, their "financial chicanery, outrageous compensation packages, and bubble-infected stock price valuations" have actually "diminished the country's productivity." Another opponent adds: "It has been claimed that the increasing use of borrowing is an inevitable part of economic growth, and that the rise of global finance is necessary to sustain a globalised world economy. A more sceptical view is that much of the recent economic growth in the West is illusory, since it is based on the higher and higher levels of debt made possible by ill-conceived deregulation and financial experimentation."[35]

That's a nice summary of the massive literature that has been filling best seller lists since the crash. Readers will have decided for themselves which of the preceding accounts is most credible; this chapter does not aim to change anyone's mind. What's not in dispute is that the derivatives markets have grown to account for roughly ten times the total annual product of the entire planet[36] and that this has been made possible by the mathematical innovations of the last few decades. How have future quants been prepared for the inevitable ethical challenges of the new situation?

Less than a year after they reacted to Rocard's outburst, the directors of France's leading financial mathematics program invoked the virtue of expertise favored by the markets. Quants engage in "two types of activity: developing "derivatives (options, warrants, swaps . . .)" where the quant "often participates along with . . . traders, structurers or . . . clients . . . essentially as an expert to assess amenability to mathematical treatment"; and "global risk management" for which "[h]e [*sic*] must . . . design (in part) and calculate an array of indicators of short- and medium-term risk. . . . " Growth is anticipated in areas "such as energy and climate derivatives" and "the creation of markets in polluting rights."[37] A popular American textbook (now in its ninth edition at a list price of $333) explains no less technocratically that "derivative markets provide a means of managing risk, discovering prices, reducing costs, improving liquidity, selling short, and making the market more efficient" but does not shrink from addressing the underlying ethical questions, however briefly:

> An important distinction between derivative markets and gambling is in the benefits provided to society. Gambling benefits only the participants and perhaps a few others who profit indirectly. The benefits of derivatives, however, extend far beyond the market participants. Derivatives help financial markets become more efficient and provide better opportunities for managing risk.[38]

After warning the student not to succumb to "the temptation to speculate when one should be hedging," pointing to the "many individuals [who] have led their firms down the path of danger and destruction" as a result of "excessive confidence in one's ability to forecast," the text concludes with the reassuring words:

> Fortunately, derivatives are normally used by knowledgeable persons in situations where they serve an appropriate purpose. We hear far too little about the firms and investors who saved money, avoided losses, and restructured their risks successfully.

To reinforce the lesson, students are asked to answer test questions:

> Why is speculation controversial? How does it differ from gambling?
> What are the three ways in which derivatives can be misused?
> Assume that you have an opportunity to visit a civilization in outer space. Its society is at roughly the same stage of development as U.S. society is now. Its economic system is virtually identical to that of the United States, but derivative trading is illegal. Compare and contrast this economy with the U.S. economy.

Storm clouds were already gathering over the mortgage markets in 2007, but publishing, with its built-in time lag, remained cheerful about quant prospects. In a book of accounts of personal success published under the alluring title *How I Became a Quant*, Neil Chriss, whom we already met in chapter 3, gave his view of how the profession became an option for mathematicians:

> In the early 1990s, most of the people I know who went into finance started off going to graduate school making an earnest attempt to become academics. The ones who move to finance did so either because they discovered finance and found it an exciting alternative or because they discovered academia was not for them.

What I remember from that time is that even the most earnest new PhDs were having trouble becoming academics, in part because support for Russian mathematics evaporated when the USSR collapsed and the spectacular Russian mathematical school was emigrating en masse and taking many of the available jobs. The joke at the time was that those who failed to find university jobs could console themselves with starting salaries in finance significantly higher than those of their thesis advisers. Chriss goes on to describe how things had changed:

> First, because of the rise in popularity of mathematical finance programs. . . many would-be math and physics PhDs are getting masters degrees in financial mathematics and going directly to finance. Second, many PhDs and academics looking to leave academia are going directly into quantitative portfolio management and not quantitative research. . . . It's a change in perspective from "I want to do research for the firm" to "I want to trade and make money for the firm."
>
> I think that this change in perspective . . . is extremely good for financial markets.[39]

Cathy O'Neil, a mathematician who had recently left academia, published a first-hand account that same year in the *AMS Notices*. Offering guidance to those considering following in her path, she enumerated a few of the attractions of finance:

> Finance is a huge and rapidly growing, sexy new field which combines the newest technology with the invention of mathematics to deal with ever more abundant data . . . finance has provided me with the opportunity to come up with good, new ideas that will be put into effect, be profitable, and for which I will be directly rewarded.

And its professional virtues:

> For me and for many of my colleagues it is intrinsically satisfying to be in a collaborative atmosphere as part of a functional, productive, and hard-working team with clear goals.

She did not minimize the ethical challenges of being a quant, which she conceived in purely personal terms, without reference to "the path of danger and destruction" for the political process or the global economy:

> Many mathematicians who talk to me about moving to finance are genuinely worried about the potentially corruptive power of money. . . . I think one can resist being corrupted by money by keeping a perspective and maintaining personal boundaries.[40]

O'Neil began to have doubts when the whirlwind hit:

> Three months into her job, the stock market dipped, a hedge fund failed, and her firm lost a ton of money. She started to question many of the assumptions underpinning her work.[41]

Readers of the earlier texts had no inkling that those who followed the path would soon be accused of unspeakable crimes and much worse. "People assume that if they use higher mathematics and computer models they're doing the Lord's work." (These are the words of Warren Buffett's "longtime partner, the cerebral Charlie Munger," quoted in *The Quants*.) He continues: "They're usually doing the devil's work"[42] . . . THERE HE IS AGAIN!!! Readers who think I'm overworking the Faust motif should check out post-crash quant allegories—for example, this outline history sketched by a French colleague: faced with an "alarming" drop in enrollments, French mathematicians "implicitly" accepted a "Faustian pact: international finance would save our . . . departments from closure and in exchange we would offer the lion's share of our best brains." Or consider this quant, anonymously quoted in a 2009 *New York Times* article, speaking of "a thousand physicists on Wall Street"—physicists are implicated as deeply as mathematicians—who "talk nostalgically about science":

> "They sold their souls to the devil," she said, adding, "I haven't met many quants who said they were in finance because they were in love with finance." "They [i.e., quants] get paid, a Faustian bargain everybody makes," [added former trader] Satyajit Das.[43]

David Steinsaltz managed to review Patterson's book for the *AMS Notices* without once invoking the devil's name. He did say, with circumspection, just where mathematicians needed to acknowledge responsibility:

> Economic historians teach us that one indispensable ingredient in a financial crisis is an excuse for ignoring the lessons of the past, . . . for believing that 'this time is different.' The most recent round of excuses was provided, if not directly by mathematicians, then under the banner of mathematics, and the crisis that ensued was of terrifying proportions.

To summarize moralistically: before the crash, *How I Became a Quant* was published in New York and the *AMS Notices* ran O'Neil's article on *the Transition from Academia to Finance*; after the crash, *The Quants* was published in New York (as were many other titles in the same vein), and its *AMS Notices* review was the occasion for a scathing denunciation of a culture of finance that flourished "under the banner of mathematics." Cathy O'Neil, meanwhile, was identified in a 2011 *Mother Jones* profile as organizer of "a branch of Occupy Wall Street known as the Alternative Banking Group." Here's what she wrote on her *mathbabe* blog on New Year's Day 2013:

> It's our ideas that threaten, not our violence. We *ignore* the rules, when they oppress and when they make no sense and when they serve to entrench an already entrenched elite. And ignoring rules is sometimes more threatening than breaking them.
>
> Is mathbabe a terrorist? Is the Alternative Banking group a threat to national security because we discuss breaking up the big banks without worrying about pissing off major campaign contributors?
>
> I hope we *are* a threat, but not to national security, and not by bombs or guns, but by making logical and moral sense and consistently challenging a rigged system.

$ € £ ¥ $ € £ ¥ $ € £ ¥ $ € £ ¥ $ € £ ¥ $ € £ ¥ $ € £ ¥

> [A] science is said to be useful if its development tends to accentuate the existing inequalities in the distribution of wealth, or more directly promotes the destruction of human life.*

Writing before the crash, Cliff Asness, one of Patterson's protagonists, had attributed his decision to leave Goldman Sachs ten years earlier (in part) to "simple ambition and greed," adding "[A]nyone who starts a hedge fund and doesn't admit this motivation probably shouldn't be trusted with your money."[44] It's fascinating to unpackage this aside, starting with the "you" to whom it is addressed: if "you" are an individual who wants to hedge in the United States, "you" had better be a High-net-worth Individual (HNWI)—"your" net worth should exceed

---

* G. H.Hardy in 1915, quoted with an apology in *A Mathematician's Apology*.

one million dollars, enough, one hopes, to have given you a feeling for "simple ambition and greed."[45]

Gordon Gekko's "Greed is good" is an explicitly post-Aristotelian ethical stance. Greed, in the *Nicomachean Ethics*, is called *pleonexia* and is the vice responsible for injustice. Jesus, in Luke 12:13–21, also had harsh words for *pleonexia*. But Jesus and Aristotle didn't know about Adam Smith's Invisible Hand. Ayn Rand, founder of "Objectivism," whose acolyte Alan Greenspan helped usher in the age of Gekko, would have explained to them that

> [s]ince time immemorial and pre-industrial, "greed" has been the accusation hurled at the rich by the concrete-bound illiterates who were unable to conceive of the source of wealth or of the motivation of those who produce it.[46]

Were it not for greed, the argument goes, no one with access to capital would risk losing it on an uncertain investment, and there would be no growth; so the greedy are the motors of general prosperity. Hand in hand with Greenspan and Rand, the Hand has been working flextime, by day apportioning goods, capital, labor, and raw material where they are most needed in gentle equilibrium, by night rewriting Aristotle to retool the vice of *pleonexia* as the virtue of *megaloprepeia*, variously translated "magnificence" or "munificence." Aristotle writes, "The magnificent man is like an artist [*epistemon*] in expenditure: he can discern what is suitable, and spend great sums with good taste."[47] For an Objectivist, investing to make oneself even richer is already a sign of taste, a work of art in the medium of wealth creation.

It's often pointed out that the historic Adam Smith was not at all sympathetic to Gekko's ethics. The Invisible Hand was first spotted in his *Theory of Moral Sentiments* dealing a version of the egalitarian *Stable Equilibrium* game of figure 4.2, making "nearly the same distribution of the necessaries of life, which would have been made, had the earth been divided into equal portions among all its inhabitants, and thus without intending it, without knowing it, advance the interest of the society." Smith's opening sentence is cited in a recent effort to prove that we are "a cooperative species":

How selfish soever man may be supposed, there are evidently some principles in his nature, which interest him in the fortune of others, and render their happiness necessary to him, though he derives nothing from it, except the pleasure of seeing it.[48]

Studies of cooperative behavior may or may not be consistent with Objectivism. One study found most experimental subjects "generally cooperative or public spirited," with the notable exception of "a group of first-year graduate economics students: the latter were less cooperative, contributed much less to the group, and found the concept of fairness alien . . . 'Learning economics, it seems, may make people more selfish' . . . [other experiments found that] students of economics, unlike others, tended to act according to the model of rational self-interest to which they are exposed in economics." The conclusions: "exposure to the [economists'] self-interest model does in fact encourage self-interested behavior" and "differences in cooperativeness are caused in part by training in economics."[49]

"When you find an innately hedged environment," said a character in Neal Stephenson's *Cryptonomicon*—a novel, one of whose premises is that number theory can make you rich—"you lunge into it like a rabid ferret going into a pipe full of raw meat."[50] This looks like an allusion to Eugene Fama's Nobel Prize–winning *Efficient Market Hypothesis* (EMH), in the vivid imagery of Robert C. Higgins: "The arrival of new information to a competitive market can be likened to the arrival of a lamb chop to a school of flesh-eating piranha. . . . Very soon the meat is gone, leaving only the worthless bone behind. . . ." Modern market theory has transmuted the Invisible Hand to efficient rows of Visible Teeth. EMH was celebrated as long as the party lasted—the quants in Patterson's book saw themselves as piranhas rewarding themselves for keeping markets efficient—and widely criticized when it ended, some going so far as to blame EMH for "chronic underestimation of the dangers of asset bubbles breaking" that made the crash so severe.[51] I wouldn't presume to judge how well EMH or any other academic hypothesis fares as a model of "objective" economic reality. As a mathematician I can point out, however, that the information-devouring ferrets and piranhas restoring the market to equilibrium are in exactly the same line of work as the impersonal forces in the deterministic difference equations of figures 4.2 and 4.3.

The syllabi and textbooks don't need to remind students that a good way for a quant to become an HNWI is to manage the fortunes of HNWI clients.[52] For Cliff Asness, the "mandate" of the Quantitative Research Group he created in 1994 at Goldman Sachs came down to this: "let's see if we can use these academic findings to make clients money."[53] I have seen the "best quantitative financial minds" of a generation drinking designer cocktails in theme bars in the Lower East Side neighborhood where my father's parents met 100 years ago[54] and saying things like

> [w]e look for places where the math is right . . . [where] you get the kind of exponential growth that should get us all into fuck-you money before we turn forty.[55]

My colleagues around the world refrained from showcasing Asness's "mandate" in the 1990s when, one after another, they welcomed financial mathematics programs into their departments, but it would have been a salutary challenge to the conventions of academic decorum to adopt something along the same lines as the new track's motto: "let's see if we can teach our graduate students to use what they learn here to make clients fuck-you money!"

$ € £ ¥ $ € £ ¥ $ € £ ¥ $ € £ ¥ $ € £ ¥ $ € £ ¥ $ € £ ¥

Practically anyone can strive to perfect most of the Aristotelian virtues, like courage [*andreia*], truthfulness [*alētheia*], liberality/generosity [*eleutheriotēs*], and wittiness [*eutrapelos*], but "only the affluent and those of high status can achieve certain key virtues,"[56] notably *megaloprepeia*, which is generosity on an order conceivable only to the richest 1% or 0.01%, or as we would say now, HNWI or UHNWI. "There is no virtue more suitable to a great man than liberality and magnificence," wrote Sigismondo Sigismondi, "but magnificence is the greater of the two, because liberality can also be exercised by a poor man."[57] Marcus Licinius Crassus, who crushed the Spartacus rebellion, was definitely one of the late Roman Republic's UHNWIs. Although Plutarch writes that as consul, he "made a great sacrifice to Hercules, and feasted the people at ten thousand tables, and measured them out corn for three months," Plutarch saw Crassus as a model of *pleonexia* rather than *megaloprepeia*.[58] In our time *megaloprepeia* has been transmuted into the cardinal capitalist virtue, the prerogative of the investor, the entrepreneur, the job creator, the HNWI to whose creative destruction TINA.

Magnificent display is still associated with *megaloprepeia*, which was already a motor of the arts when Lorenzo the Magnificent reigned in Florence.[59] It's no less true today. Muntadas' interviews with art world movers give collectors a special prominence. For gallery owner Daniel Templon, "There's no other way to evaluate a work of art than to put a price tag on it. And that's how it's been ever since art has existed." Artist Hans Haacke, in contrast, stresses the calculations of corporate collectors: they "got involved in the arts because the arts were something with which they could polish their image, and through the arts, they became more forceful in their lobbying for interests very close to the bottom line."

Some individual collectors allude to speculation ("that is a different type of collector") but insist on their own spiritual link to the art and the artist. "My motivations," writes one, "have very much to do with my own spirit and my own soul," just like "the clothes you buy, the shoes you wear, the furniture you buy . . . the woman you love. . . . You love it or you don't. And after you have loved it and you want it, you want to live with it.[60]

MacIntyre reminds us that "no practices can survive for any length of time unsustained by institutions. . . ." which are "characteristically and necessarily concerned with" what MacIntyre calls "external goods" like "money, power and status." These are contrasted with what MacIntyre identifies as the "internal goods," the virtues on which a tradition-based practice is founded. "[C]haracteristically," MacIntyre's external goods "are such that the more someone has of them, the less there is for other people."[61] This is not the case for internal goods. Muntadas's interviews point to the ambivalent relations between external and internal goods:

> If you become successful in terms of the economic system that art is part of, you become automatically bound up with the value systems . . . of those who have the money to have leisure time and to purchase artwork. That's a problem for a lot of us, because we don't all share the values of those who have the power.[62]

The collector seems to have no mathematical counterpart in chapter 3's table of parallels with the arts. Our practice's relations of ambivalence are mainly with governments and university administrations. But times are changing. James Simons was one of the world's most respected pure mathematicians, winner of the prestigious AMS Veblen Prize in geometry, before he became the man Patterson calls "the reclusive, highly secretive billionaire manager of Renaissance Technologies, the most successful hedge fund in history." "I went into the investing business without any thought of applying mathematics at all," explains Simons. "I had some ideas and they worked out well." Ferguson devotes a page to his version of how Simons obtained "extraordinary returns by using powerful computers and proprietary trading algorithms to exploit tiny market trends invisible to humans, and which often last only a fraction of a second, . . . Normal investors who don't own gigantic computer systems end up paying slightly more for stock trades and have slightly lower investment returns, while Simons and his imitators pile up micropennies by the billions."

Ferguson concludes that such high-frequency trading has "no social benefit at all . . . no economic utility." Simons disagrees:

> [A]s the markets have become electronic and computers have been applied to generating prices and accepting trades and all the rest, the markets have grown tremendously more liquid. Spreads have come down. . . . So, is [high-frequency trading] socially useful? Well, if you think highly liquid markets are socially useful then I think so.[63]

Hedge funds, remember, are exclusively for HNWIs, and I leave it to them to debate the utility of "highly liquid markets" in a world of ferrets and piranhas; as this chapter's title suggests, I'm still waiting for someone to convince me of the social utility of HNWIs. But Simons has devoted much of his personal fortune to supporting research in mathematics as well as physics and biology. The Simons Foundation sponsors fellowships and conferences and has made substantial donations to scientific institutions on every continent, including tens of millions of dollars to the IAS alone; the dining hall there now bears his name, as does the Simons Center for Systems Biology. Berkeley's MSRI has a Simons Auditorium and the IHES outside Paris has a Marilyn and James Simons Conference Center; there is a Simons Center for Geometry and Physics on Long Island and a Simons Institute for the Theory of Computing in Berkeley. As one country after another has entered into crisis, as universities lose positions and libraries see their subscription budgets cut, I've lost track of the number of times colleagues looked for salvation to the "white-bearded wizard": think Gandalf, not Mephisto.[64]

When the trader Satyajit Das disclosed to the *New York Times* that financial mathematics devises models for "[m]aking money" and that "[t]hat's not what science is about,"[65] he was saying that the work of quants does not conform to the internal goods, in MacIntyre's sense, of the practice of science. But Simons is clearly as concerned as any mathematician, physicist, or biologist about the internal goods of their respective fields. His foundation bases its decisions on the advice of those within the fields who share these concerns. Other foundations—the Clay Mathematics Institute, the American Institute of Mathematics (AIM), and (on a much smaller scale) the charitable *Fondations de l'Institut de France*, are run along similar lines. And thus finance mathematics enters as part of a novel triangular trade, an exponentially virtuous circle: academic mathematics departments host finance mathematics programs that generate the UHNWI within financial institutions and they, in turn, provide the "external goods" necessary to maintain the practice of pure mathematics, a kind of perpetual *megaloprepeia* machine from which the Columbia math department even manages to extract a limitless cornucopia of fresh fruit.[66]

The philanthropy of Powerful Beings is not new in mathematics. Princes, counts, and kings financed prize competitions in mathematics starting in the eighteenth century. Mathematics was a state priority in Napoleonic France, and the German school grew to dominate nineteenth-century mathematics under the patronage of Prussian ministries.[67] In a section entitled "the general mathematical market," Mehrtens analyzes relations between science and industry in Germany at the turn of the twentieth century, focusing on Felix Klein's work with the *Göttinger Vereinigung*.[68] Even then, it was difficult to force mathematicians into the industrial mold. The ambiguities of mathematics were a handicap: "it was considerably harder to find funds" for mathematics than for chemistry with their "unequivocal and firmly established link to industry." Natural scientists, unlike mathematicians, may have had closer personal relations to industrialists and bankers. Professionally speaking,

Mehrtens thought, "mathematicians are neither economically nor technically especially interesting, nor are they particularly enterprising." But funds could still be found for mathematics—"special encouragement of applied mathematics" and especially "prizes for outstanding work." The IHP building in Paris was built in 1928, in large part with Rockefeller and Rothschild money.

National governments now provide the lion's share of funding for mathematical research, even at private universities in the United States. Cutbacks to research budgets under current conditions of austerity are a source of anxiety, but no one complains about the ups and downs of private philanthropy—it's "their" money, after all, and we are fortunate to be seeing any of it. The deeper irony is that the (ostensibly) democratically based social institutions of government are perceived as less sympathetic to the "internal goods" of mathematical practice than the structures of *megaloprepeia* endowed by Powerful Beings like Clay, AIM, or Simons. Whatever can be said about the sources of their generosity, the immediate effects of this kind of philanthropy on mathematics have been uniformly positive. Mathematicians fill the seats on foundations' advisory boards and operate these structures as relaxed fields; in practice, their scientific decisions are answerable to mathematicians alone, within the budget the benefactors provide. There is no set format for proposals, but they should be short—one page for a Clay workshop, two to three pages for AIM—and the foundations are not expecting to hear about potentially useful applications. This is usually presented as a narrative about the generosity of certain Powerful Beings; but it could also be seen as a commentary on the state of democracy that is no less depressing for its homology to what has long been taken for granted in the art world.[69]

"Responsibility Seems to Dissolve"

[L]ike all the phenomena of our societies, the current financial crisis is only seen as a technical problem. . . . We reason solely in mechanical terms, as if living individuals were elementary particles subject to laws . . . conceived on the model of Galilean physics.[70]

In his lecture at the 2010 International Congress of Mathematicians in Hyderabad, Ole Skovsmose, a specialist in mathematics education, analyzed how the "symbolic power" of mathematics is exercised through "the grammatical format of a mechanical and formal world view." Skovsmose puts his finger on the ethical danger: "Through mathematics in action, we are in fact bringing our social, political, and economic environment deeper into a mechanical format. . . . mathematics-based actions often appear to be missing an acting subject . . . to be conducted in an ethical vacuum." Aspects of the ongoing crisis can be related to "mathematics-based avalanches of decisions" in financial markets and elsewhere. "But who could be held responsible? Somehow responsibility seems to dissolve."[71]

In addition to our technical qualifications, mathematicians bring to finance a scorn for mere empiricism that under appropriate conditions can blossom into full-scale denial; as Diogenes the Cynic is said to have marveled that "mathematicians should gaze at the sun and the moon, but overlook matters close at hand." Whether or not the data corroborate mathematical models, mathematics has the merit for decision makers of being perceived as "incorrigible"—"it provides the kind of truth with which there is no possible argument."[72] Economics is only the most politically influential of the fields that call upon mathematics to adjudicate potential conflict between alternative realisms: in this case, the TINA realism of the market and the ethical realism that provides the basis for democracy as an expression of human values.

By playing the role of provider of the appearance of scientific objectivity, mathematics indentures itself in turn to the model of scientificity that underpins the philosophy of Mathematics. Our truth flows outward, clutching its warrant of incorrigibility, into the larger society, where it is deployed by Powerful Beings in what is demonstrably the only way possible. As far as finance is concerned, Steinsaltz calls this process "legitimacy exchange" in his review of *The Quants* for the *Notices of the AMS*:

> University mathematicians have served the finance industry with a steady stream of students trained . . . to accept models axiomatically, without skepticism. At least as important, we have participated in what G. Bowker . . . has termed "legitimacy exchange" (though with a view to the huge blow suffered by the reputations of both groups, perhaps it might be called a "credibility default swap"). . . . In return for lending the reputation of their

subject, academic mathematicians are compensated by sharing the financiers' reputation as important people doing important work, and the enviable status as a conduit to high-paying careers."[73]

"The observable economic facts," thought Albert Einstein, correspond to "what Thorstein Veblen called 'the predatory phase' of human development." Einstein saw socialism as a way "to overcome and advance beyond the predatory phase."[74] To believe TINA—There Is No Advancement—is to deny any positive role for ethics in politics; to believe TINA on the grounds that it can be proved mathematically is to hold mathematics responsible for the "ethical vacuum" of contemporary politics. A few more years of this and an Exterminating Angel will surely arise to cast out the Fiend. Remember that in the final pages of Goethe's Faust, heaven reneged on the bargain Mephistopheles contracted in the opening scenes. Will this be good for mathematics? Don't count on it.

chapter β

How to Explain Number Theory at a Dinner Party

SECOND SESSION: EQUATIONS

Everyone who attends high school learns to solve simple algebraic equations. A *linear equation* with a single variable x has exactly one solution;[1] for example,

$$3x + 1 = 7$$

has the unique solution $x = 2$, by which we mean that (a) $3 \times 2 + 1 = 7$ and (b) if you put any other number in the place of 2, the two sides will no longer be equal.

A *quadratic equation* can have two solutions, no solutions, or (more rarely) a single solution:

$$x^2 - 4x + 3 = 0$$

has the two solutions $x = 3$ and $x = 1$, whereas

$$x^2 - 4x + 16 = 0$$

has no solutions unless you allow *imaginary* numbers, such as $i = \sqrt{-1}$; more about them later. But there are also cases like

$$x^2 = 2,$$

whose solution, $\sqrt{2}$, as we saw in the previous section, is an irrational number. (The distinct but no less valid negative solution, $-\sqrt{2}$, is irrational as well.)

The *quadratic formula* taught in high school algebra solves the problem in all cases: if the equation is

$$Ax^2 + Bx + C = 0$$

with A a number different from 0, then the solutions are[2]

$$\frac{-B+\sqrt{D}}{2A}, \frac{-B-\sqrt{D}}{2A}, D = B^2 - 4AC,$$

whose nature (rational, imaginary, or occasionally equal) can be determined by inspection: If $D > 0$, we don't worry about taking its square root; if $D = 0$, the two roots are equal; and if $D < 0$, then its square root is *imaginary* (a paradoxical name if there ever was one, because the square root of D is the side of a square whose area is D, and who has ever imagined a square with negative area?) and the solutions are *complex numbers*.

Thus, if we are willing to allow square roots into our arithmetic, we can consider the quadratic equation a problem whose solution has been long understood (in some cases by the ancient Babylonians), though it might be argued that it was not until imaginary numbers were generally accepted that the preceding formula was established. Equations of degree 3 and 4, such as

$$x^3 - 2x^2 + 14x + 9 \text{ and } x^4 + 5x^3 + 11x^2 + 17x - 29,$$

were first solved in Renaissance Italy to great acclaim; the solutions are given by formulas involving cube roots and fourth roots.

Nils Henrik Abel and Evariste Galois both died tragically young in the early nineteenth century, contributing to the romantic image of mathematics still operative in certain quarters, after showing independently that there is *no formula* for the roots of general polynomial equations of degree 5 and up. Before he was killed in a duel at age twenty-one (see chapter 6), Galois did much more: he provided a method for understanding the roots even though they cannot be written down.

Galois' work represents the starting point of modern *algebraic number theory*, which is probably the most accurate name for the kind of mathematics in which I specialize. In general, a polynomial equation of Nth degree in one unknown x has at most N solutions, or exactly N if some of them are counted more than once, as it is often useful to do. Galois theory does not provide a formula for these solutions but rather a way of looking at the solutions that is completely general and is more useful for answering most questions than a formula would be. In this sense the problem of equations in one unknown can be said to be com-

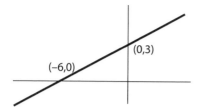

(0,3)

(−6,0)

Figure β.1. The line $y = x/2 + 3$ in the coordinate plane.

pletely settled, thanks to Galois theory, although many questions remain. Equations in more than one unknown are another matter. For example, in high school algebra one encounters linear equations in two variables, such as

$$y = \tfrac{1}{2}x + 3.$$

This equation has infinitely many solutions: for any number x, the equation provides a formula for y, and the choice of possible x is infinite. This is hardly surprising, because the solutions of the equation form a line in the coordinate plane, as shown in figure β.1, and the line is understood to have infinitely many points. The collection of rational solutions is just the set of pairs $(x, 1/2x + 3)$ whenever x is rational.

The coordinate plane represents pairs of numbers as points on the page (assumed infinite in all directions). To reach the point (x, y), one moves x units to the right and y units up (which means down if y is negative and to the left if x is negative), starting from the origin $(0, 0)$, which is the point where the horizontal and vertical lines meet in figure β.1. The Pythagorean theorem then calculates the distance of the point with coordinate (x, y) to the origin $(0, 0)$ as

$$\sqrt{(x-0)^2 + (y-0)^2} = \sqrt{x^2 + y^2}\,.$$

Thus the solutions to the equation of degree 2,

$$x^2 + y^2 = 1,$$

are given by the points (x, y) at distance 1 from the origin $(0, 0)$—in other words, by the points on the circle of radius 1 around the origin. We can temporarily forget about the picture and look for solutions in whole numbers, but we quickly realize that there are only four, namely, $(1, 0)$, $(0, 1)$, $(-1, 0)$, $(0, -1)$—the four cardinal points of the circle.

Solutions in *rational* numbers are more interesting.[3] Write x and y as fractions with a common denominator:

$$x = \frac{a}{h}; y = \frac{b}{h},$$

where a, b, h are whole numbers, $h \neq 0$. Then

$$x^2 + y^2 = 1$$

is equivalent to

$$a^2 + b^2 = h^2.$$

By the Pythagorean theorem, this is the equation for the lengths of the sides of a right triangle: the hypotenuse is h, and a and b are the two sides adjacent to the right angle.

For example, $a = 3, b = 4, h = 5$ is a solution everyone has seen; so $x = 3/5, y = 4/5$ is a solution to the original problem. If your high school teacher was a bit more energetic, you will have seen $a = 5, b = 12, h = 13$ or even $a = 9, b = 40, h = 41$.

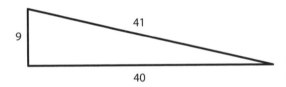

Figure β.2. A right triangle with integer sides.

Thus $x = 5/13, y = 12/13$ is another solution using rational numbers. It turns out that there are *infinitely many rational solutions* of this kind, and there is even a formula for writing them all down. We start with two whole numbers r and s without common factors; then

$$a = r^2 - s^2; b = 2rs; h = r^2 + s^2.$$

We can also multiply all three terms by a common factor d:

$$a = dr^2 - ds^2; b = 2drs; h = dr^2 + ds^2.$$

Up to interchanging a and b, this gives all possible solutions in whole numbers, a fact proved in Book X of Euclid's *Elements* in the third century BC and (in the preceding geometric interpretation) about five hundred years later by Diophantus of Alexandria. Anyone with sufficient patience can establish this result using elementary properties of even and odd numbers and the unique factorization of whole numbers as products of prime numbers, as mentioned in the previous session.

This problem is special because it is attached to two stories—the story of right triangles with integer sides, on the one hand, and the story of Diophantus' formula for the solutions. One can consider equations of *any* degree in *any* number of variables. A theorem in logic[4] tells us that there is no general way to find integer solutions or even to determine whether or not there are solutions. That is also a story—but not one with a happy ending. One suspects that a problem in the field of *Diophantine equations*—the problem of finding integer or (at least) rational solutions to equations with integer coefficients—is interesting to the extent that it forms part of a story. The story is naturally internal to mathematics but, like any story, is most interesting when the problem does not appeal only to specialists.

An equation like $x^2 + y^2 = 1$ or $x^2 - 3xy + 3y^2 = 1$ is called *quadratic*[5] because the highest power of either variable is the second power and any monomial including both variables is a constant multiple of xy. Quadratic equations in two variables can be considered completely understood. There is the example $x^2 + y^2 = 1$ discussed earlier, but there is also the very similar example

$$x^2 + y^2 + 1 = 0.$$

This one has no solutions at all in which x and y are rational numbers or even real numbers. The square of a real number is positive, so for any real numbers x and y, the left-hand side of the equation is always at least 1 and, therefore, can never be equal to zero. This is analogous to the quadratic equations in one variable, discussed previously, with no roots that don't involve imaginary numbers. It is to solve such equations that one introduces imaginary numbers. From the standpoint of imagination imaginary numbers pose special challenges:

We have never seen any curve or solid corresponding to my square root of minus one. The horrifying part of the situation is that there *exist* such curves or solids. Unseen by us they do exist, they must, inevitably; for in mathematics, as on a screen, strange, sharp shadows appear before us. One must remember that mathematics, like death, never makes mistakes. If we are unable to see those irrational curves or solids, it means only that they inevitably possess a whole immense world somewhere beneath the surface of our life (Ye. Zamyatin, *We*, Record 18).

... in that kind of calculation [involving imaginary numbers] you have very solid figures at the beginning, which can represent metres or weights or something similarly tangible, and which are at least real numbers. And there are real numbers at the end of the calculation as well. But they're connected to one another by something that doesn't exist. Isn't that like a bridge consisting only of the first and last pillars, and yet you walk over it as securely as though it was all there? (R. Musil, *The Confusions of Young Törless*, trans. S. Whiteside, p. 82)

It takes a fictional character to restore all its pathos to the mystery of *existence*. As far as pure algebra is concerned, imaginary numbers are no more imaginary, and their existence no more problematic, than irrational numbers such as $\sqrt{2}$.

Be that as it may, the quadratic equations in two variables fall into two classes: those with infinitely many rational solutions, like the Pythagorean equation, and those with no solutions at all. Given a quadratic equation in two variables, for example,

$$x^2 + y^2 = 3,$$

there is a simple procedure for determining whether it has infinitely many or finitely many rational solutions.

Look first for integer solutions, and start by testing small integers. It's obvious that most of these won't work because the left-hand side quickly becomes bigger than the right-hand side, but let's carry out the calculation in the hope that this will help us with rational solutions.

$$(x, y) = (0, 0): \quad 0^2 + 0^2 = 0 \neq 3,$$
$$(x, y) = (0, 1): \quad 0^2 + 1^2 = 1 \neq 3,$$
$$(x, y) = (1, 1): \quad 1^2 + 1^2 = 2 \neq 3,$$
$$(x, y) = (0, 2): \quad 0^2 + 2^2 = 4 \neq 3,$$
$$(x, y) = (1, 2): \quad 1^2 + 2^2 = 5 \neq 3,$$
$$(x, y) = (2, 2): \quad 2^2 + 2^2 = 8 \neq 3,$$
$$(x, y) = (0, 3): \quad 0^2 + 3^2 = 9 \neq 3,$$

$$(x, y) = (1, 3): \quad 1^2 + 3^2 = 10 \neq 3,$$

$$(x, y) = (2, 3): \quad 2^2 + 3^2 = 13 \neq 3.$$

We don't need to check pairs like $(x, y) = (1, 0)$ because x and y can be exchanged and the result on the left is the same. We notice that the sum is often an even number (0, 2, 4, 8, 10, not 6 and not 12 either). With more attention, one observes that the odd numbers that do occur have something in common:

$$1 = 4 \times 0 + 1; 5 = 4 \times 1 + 1; 9 = 4 \times 2 + 1; 13 = 4 \times 3 + 1.$$

The missing numbers on the list are

$$3 = 4 \times 3 + 3, 6 = 2 \times 3; 7 = 4 \times 1 + 3, 11 = 4 \times 3 + 3, 12 = 4 \times 3.$$

This suggests a pattern. Say a positive integer n is *congruent to* 1 *modulo* 4 if it is of the form $n = 4k + 1$, where k is a positive number; in other words, if $n - 1$ is evenly divisible by 4. Likewise, say n is *congruent to* 3 *modulo* 4 if it is of the form $n = 4k + 3$, in other words, if $n - 3$ is evenly divisible by 4. The terminology "congruent to" and "modulo" will be explained in chapter γ.

Our calculations suggest the following hypothesis.

Two Square Theorem: An odd number that can be written as the sum of two squares must be congruent to 1 (and not to 3) modulo 4. An odd *prime* number can be written as the sum of two squares if and only if it is congruent to 1 modulo 4.

This hypothesis is correct and was proved by a number of mathematicians, notably by Gauss, and for that reason, it was promoted to the status of a **theorem**.[6] With additional work one can show that this theorem implies that the equation $x^2 + y^2 = 3$ has no solution in rational numbers either.

We have chosen for our example the equation for the circle with radius $\sqrt{3}$, which (as we have already seen) is an irrational number; if the radius were rational, then Diophantus' method would again give all the solutions. The two square theorem determines which circles with irrational radii have points with rational coordinates. This is the beginning of a long story. From the point of view of number theory, there is no essential difference between a circle and an *ellipse*, which you can think of geometrically as the collection of points in the plane, the *sum* of whose distances between *two fixed points*, the *foci*, is a constant. For example, the quadratic equation

$$x^2 + 2y^2 = 5$$

is the equation of the ellipse with total distance $2\sqrt{5}$ from the two foci $(\sqrt{\tfrac{5}{2}}, 0)$ and $(-\sqrt{\tfrac{5}{2}}, 0)$.

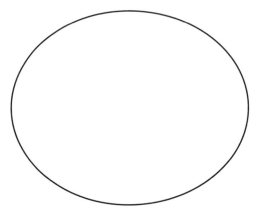

Figure β.3. The ellipse $x^2 + 2y^2 = 5$.

The quadratic equation

$$x^2 + xy + y^2 = 11$$

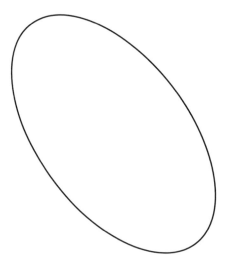

Figure β.4. The ellipse $x^2 + xy + y^2 = 11$.

is the equation of a tilted ellipse, with foci at $(\sqrt{\frac{22}{3}}, -\sqrt{\frac{22}{3}})$ and $(-\sqrt{\frac{22}{3}}, \sqrt{\frac{22}{3}})$. Neither of these equations has any rational solutions. On the other hand, each of the equations $x^2 + 2y^2 = 3$ and $x^2 + xy + y^2 = 7$ has an integer solution—$(x, y) = (1, 1)$ in the first case and $(x, y) = (1, 2)$ in the second case—and, therefore, the general theory implies it has infinitely many rational solutions.

It's easy to write down a complete list of rational solutions, using a procedure similar to the one invented by Diophantus to find Pythagorean right triangles. But it is even easier to write down a complete list of *integer* solutions, because squares of integers are positive, and after you have gone through the first few options, the sums on the left get bigger than the target on the right. For example,

$$0^2 + 2 \times 0^2 = 0,\ 1^2 + 2 \times 0^2 = 1,\ 0^2 + 2 \times 1^2 = 2,\ 1^2 + 2 \times 1^2 = 3,$$
$$2^2 + 2 \times 0^2 = 4;$$

after that, the sum $x^2 + 2y^2$ is bigger than 3. So $(1, 1)$ is the only solution. One checks similarly that $x^2 + xy + y^2 = 7$ has exactly two integer solutions: $(x, y) = (1, 2)$ and $(x, y) = (2, 1)$.

On the other hand, if you look at quadratic equations involving *differences* rather than *sums* of squares, you can't use this kind of argument to terminate the search for solutions. In fact, an equation like

$$y^2 - 7x^2 = 1$$

has infinitely many solutions with x and y both integers. The word *like* has to be understood appropriately. The general equation of the form $y^2 - dx^2 = 1$, where d is a fixed integer, is called *Pell's equation*,[7] and it has been known since the eighteenth century that it has infinitely many integer solutions provided d is not a square. If by mistake you choose $d = c^2$, with c an integer, then you can write

$$y^2 - dx^2 = (y + cx)(y - cx) = 1.$$

But then both $y + cx$ and $y - cx$ are integers, both of which divide 1, and the list of integers dividing 1 is very short. For example, the only solution to $y^2 - x^2 = 1$ in nonnegative integers is given by $x = 0, y = 1$.

But when *d* is not a square, there is a method for finding all the solutions. I will not describe this method,[8] but instead I will explain how from one solution you can derive infinitely many new solutions. The smallest solution to $y^2 - 7x^2 = 1$ is given by $x = 3, y = 8$. Take the number $8 + 3\sqrt{7}$ and multiply it by itself: you get

$$(8 + 3\sqrt{7})(8 + 3\sqrt{7}) = 64 + 48\sqrt{7} + 63 = 127 + 48\sqrt{7},$$

and you can check that $y = 127, x = 48$ is also a solution to $y^2 - 7x^2 = 1$. Multiply by $8 + 3\sqrt{7}$ again:

$$(8 + 3\sqrt{7})(127 + 48\sqrt{7}) = 2024 + 765\sqrt{7}.$$

and $2024^2 - 7 \times 765^2 = 1$.

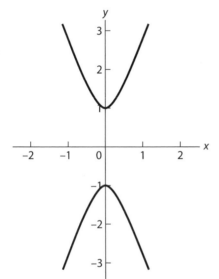

Figure β.5. The hyperbola $y^2 - 7x^2 = 1$.

So the quadratic curves with infinitely many rational points are further subdivided into *hyperbolas* (as in figure β.5) with infinitely many *integer* points and *ellipses* (as in figures β.3 and β.4) with only finitely many integer points. There is a simple explanation for the difference between the two kinds of equations, but you will have to take a course in algebraic number theory to see it.

Four more kinds of curves round out the list of quadratic equations in two variables (not counting the equations with no solutions, such as

$x^2 + y^2 + 1 = 0$). Most curvelike is the *parabola*, for example $y = 4 - x^2$, shown in figure β.6.

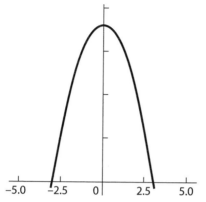

Figure β.6. The parabola $y = 4 - x^2$.

The parabola adds nothing especially interesting to what we have already said about rational solutions. Next comes an equation that looks like the equation of a hyperbola but with 0 replacing 1:

$$y^2 - dx^2 = 0.$$

If d is not a square, then there are infinitely many points, but the only rational solution is $(x, y) = (0, 0)$; if d is a square, then there are infinitely many rational solutions. In either case, the graph of this equation is a *pair of lines* crossing at the point $(0, 0)$

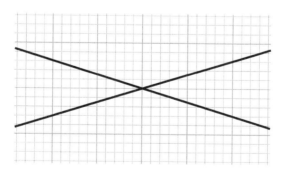

Figure β.7. Graph of $y^2 - 0.09x^2 = 0$.

unless $d = 0$, in which case the two lines coincide and you get just the line $y = 0$, but counted twice, the *double line*.

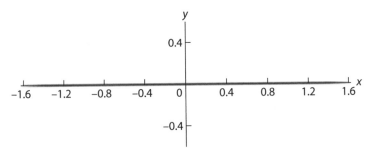

Figure β.8. Graph of the double line $y^2 = 0$.

The equation $x^2 + y^2 = 0$ has only the one solution $(0, 0)$, all of the other solutions being *imaginary*; its graph looks like figure β.9.

Figure β.9. Graph of the degenerate conic $x^2 + y^2 = 0$.

Finally, we have seen that the equation $x^2 + y^2 + 1 = 0$ has *zero* solutions. Its graph, the *imaginary circle* of radius -1, is therefore omitted because it looks like nothing at all, or alternatively, like the picture of *zero* objects, which is where logicians started constructing numbers and where we started our previous dialogue, and so the imaginary circle circles us back to our starting point.

PERFORMING ARTIST: Before you and I get started, I need to lodge a complaint about my homework assignment. You claimed that both lengths and quantities are numbers, but Aristotle insisted that these are the objects of two distinct sciences:

We cannot . . . prove geometrical truths by arithmetic. . . . in the case of two different genera such as arithmetic and geometry you cannot apply arithmetical demonstration to the properties of magnitudes unless the magnitudes in question are numbers.[9]

NUMBER THEORIST: You're well versed in the mysteries of the *Organon*!

P. A.: Once you've read the *Poetics*, and we performing artists all have, you really don't want to put Aristotle down.

N. T.: I'm not sure how to break the news to you, but Aristotle is not exactly the last word in philosophy of mathematics. The point I was trying to make was that the science of algebra is defined not by its *objects* but by its *methods*, as al-Khwārizmī argued in the ninth century, and the methods are the same whether applied to lengths or to quantities.

P. A.: I wonder what you think of William Rowan Hamilton's more recent complaints about algebra:

> [I]t requires no peculiar scepticism to doubt . . . that numbers, called imaginary, can be found or conceived or determined, and operated on by all the rules of positive and negative numbers, as if they were subject to those rules, although they have negative squares, and must therefore be supposed to be themselves neither positive nor negative, nor yet null numbers, so that the magnitudes which they are supposed to denote can neither be greater than nothing, nor less than nothing, nor even equal to nothing.[10]

N. T.: Extending al-Khwārizmī's reasoning beyond anything he could have imagined, a number is an answer to a question to which his methods can be applied. For example, the question answered by i, to which Hamilton in his quaternions did not hesitate to add a j and a k, with peculiar rules of arithmetic indeed. By the way, Omar al-Khayyām, in the twelfth century, was reportedly the first mathematician to admit irrationals as numbers.[11]

P. A.: Let's move on to this session, then. I'm afraid I don't see how your pictures of two ellipses shed any light whatsoever on what appears to be your main point, namely, whether there are rational solutions or not.

N. T.: It's hard to draw a picture of nothing, and I concede that it might be misleading to illustrate nothing by a picture of something. Nevertheless—and setting aside the fact that I'm going to need those pictures in the next chapter—when you want to understand rational solutions of equations, the pictures like the quadratic curves you've just seen, which show the solutions in *real numbers*, are indispensable, as are the *congruences* we'll be seeing in our next session.

P. A.: I hope you're not about to repeat what the mathematics teacher told young Törless: ". . . if you can do ten times as much mathematics as you do at the moment, you will understand, but for the time being: believe!"

N. T.: Not at all! That would completely defeat the purpose of our meetings, and of the whole book, for that matter. Anyway, I'm flattered that you would take time off from work to—

P. A.: What makes you think this is not part of my work? As Alasdair MacIntyre said, and I know you're fond of quoting him, "A conversation is a dramatic work, even if a very short one."[12]

N. T.: Conversations fall into many different genres . . .

P. A.: . . . and this one belongs to the genre of mathematics lesson. So to get straight to work, I gather that there are two kinds of equations . . .

N. T.: Many more than that, in fact, but go on.

P. A.: The ones whose solutions can be given by *formulas* and the rest.

N. T.: .. . . the latter being far more numerous but much less well understood.

P. A.: Certainly not understood by me, since you haven't said a word about them beyond claiming that there is no formula for the roots of equations of degree 5 and up and that you expect me to find this romantic.

N. T.: What is considered romantic is that the two mathematicians who made this discovery both died in their twenties. I don't think they died of absence of a formula, or at least I haven't seen any claims to that effect.

P. A.: But you will tell me how a solution can possibly answer a question if you can't write it down in a formula.

N. T.: An excellent question. You mentioned you were working on *Three Sisters*?

P. A.: Actually, I mentioned no such thing, but what if I were?

N. T.: If you were, which sister would you be?

P. A.: If I had to choose, I would certainly prefer to be Masha.

N. T.: Just as I thought. Now just as the quadratic equation $x^2 - 4x + 3 = 0$ has two solutions, the *cubic equation*

$$x^3 + px + q = 0$$

has three solutions.

P. A.: What do you mean by p and q?

N. T.: It's notation for unspecified rational numbers, just as A, B, and C were unspecified rational numbers in the quadratic equation. If $p = 1$ and $q = 2$, then we are talking about the equation $x^3 + x + 2 = 0$. But I'm going to write down a formula with p's and q's, not 1s and 2s.

So suppose that Chekhov's three sisters, Olga, Masha, and Irina, are the three solutions to this equation. How can we tell which is which?

P. A.: "Mathematicians . . . may look like the rest of us, but they are not the same."[13] What, pray tell, are you talking about?

N. T.: The formula of Scipione del Ferro tells us that the three roots are

$$\sqrt[3]{-\frac{q}{2}+\sqrt{\frac{q^2}{4}+\frac{p^3}{27}}} + \sqrt[3]{-\frac{q}{2}-\sqrt{\frac{q^2}{4}+\frac{p^3}{27}}}, \; \omega\sqrt[3]{-\frac{q}{2}+\sqrt{\frac{q^2}{4}+\frac{p^3}{27}}}$$

$$+ \omega^2\sqrt[3]{-\frac{q}{2}-\sqrt{\frac{q^2}{4}+\frac{p^3}{27}}},$$

and $\omega^2\sqrt[3]{-\frac{q}{2}+\sqrt{\frac{q^2}{4}+\frac{p^3}{27}}} + \omega\sqrt[3]{-\frac{q}{2}-\sqrt{\frac{q^2}{4}+\frac{p^3}{27}}}$, where $\omega = \frac{-1+i\sqrt{3}}{2}$,

and I'm asking you which of the three is Olga, which is Masha, and which is Irina. . . . Stop looking at me that way; you're the one who suggested I try to connect complex numbers with literature. You seemed to appreciate my allusion to Musil.

P. A.: The crucial word was *connect*. Only connect . . .

N. T.: Let me back up a moment. Suppose we think of the characters in *Hamlet* as solutions of some huge equation and try to imagine what would happen if a few of them exchanged their lines. Most of the time, for example, if Claudius and Gertrude changed places, you would get a vastly different play.

P. A.: Women have been playing Hamlet since before Sarah Bernhardt; so how could you tell the difference?

N. T.: That's exactly where I'm heading. But let's be conservative for the moment and think about the characters whose lines could be exchanged without making any substantial difference to the play.

P. A.: You want me to say that Rosencrantz could switch with Guildenstern.

N. T.: Little is lost if you give Voltemand's lines to Cornelius.

P. A.: But then the laconic Cornelius would become the voluble Voltemand, and vice versa.

N. T.: That's the point: it's only by choosing where to direct your attention—by making a choice—that you can decide which is which. Any other examples?

P. A.: How about Bernardo, Francisco, and Marcellus?

N. T.: Indeed: any one of the three could give voice to the vague sense that something is rotten in the state of Denmark. So let's say there are thirty-four characters, and seven of them can be permuted. For example in only one of the three groups:

RG	CV	BFM
GR	CV	BFM (Rosencrantz and Guildenstern change places, the others stay put);
RG	CV	BFM
RG	CV	MFB (Marcellus has switched with Bernardo)

Or, in all three at once:

RG	CV	BFM
GR	VC	MBF

It's easy to check that there are twenty-four possible ways to switch within each of these three groups, leaving the other characters in place. So I'll say the *Galois group* of Hamlet has twenty-four elements, the possible ways to permute the characters without affecting the play.

P. A.: When we last met, you "permuted" the four solutions to an equation. Do I detect foreshadowing?

N. T.: The four solutions to $x^4 + 14x^2 + 121 = 0$ had two parts, $\sqrt{2}$ and $3i$, each preceded by a + sign or a − sign. If you permute the signs, you exchange one solution with another. So the *Galois group* of the equation $x^4 + 14x^2 + 121 = 0$ has four elements, the possible ways to permute the solutions without changing the equation.[14]

P. A.: And now you're going to see what happens when Chekhov's three sisters change places. But just a minute ago you asked which of the three I would prefer to play. In other words, you've acknowledged that each one has her own uniquely recognizable personality. So, they're not interchangeable.

N. T.: Imagine you're watching a scene from a production in a language you don't understand and you try to determine which sister is which.

P. A.: If the production is any good it should be obvious.

N. T.: Maybe it's not such a good production . . .

P. A.: Then why would I keep watching?

N. T.: So you switch it off after one of the sisters speaks a line to a second sister, and then you wonder: what scene was that? Was it Olga speaking to Irina, with Masha biding her time in the background? Or Irina speaking to Masha? There are six possibilities, with the speaking sister listed first, followed by the listening sister, and the third in parentheses:

$$O\,I\,(M) \quad I\,M\,(O) \quad M\,O\,(I) \quad O\,M\,(I) \quad I\,O\,(M) \quad M\,I\,(O)$$

Each of these is represented in the play, and a single line is not enough to determine which you've seen. So, the Galois group of *Three Sisters*, seen from the standpoint of a single line, is the collection of these six permutations.

But maybe the play could be rewritten so that Olga never speaks to Masha, Irina never speaks to Olga, and Masha never speaks to Irina. In that case, only the first three configurations would be possible, and then the Galois group would have only three permutations.

Now Chekhov could have gone up to four sisters and Ferrari's formula (figure β10) would still give us a way of writing them down. Andrew Wiles used just this fact in a crucial step of his proof of Fermat's Last Theorem.

$$x_{1,2} = -\frac{b}{4a} - S \pm \frac{1}{2}\sqrt{-4S^2 - 2p + \frac{q}{S}}$$

$$x_{3,4} = -\frac{b}{4a} + S \pm \frac{1}{2}\sqrt{-4S^2 - 2p - \frac{q}{S}}$$

Figure β.10. Ferrari's formula for the solution to the general quartic.

P. A.: Four sisters wouldn't have fit Chekhov's purpose. They would inevitably have split into two factions in any of three different ways.

N. T.: And twenty-two sisters could have split up into two football teams in 705,432 different ways.

P. A.: That's why it's so important for there to have been a *prime* number of sisters; otherwise they would constantly be in danger of breaking up into rival gangs of equal size.

N. T.: But as soon as there are five sisters or more, then Abel and Galois have told us there is no formula like those of Scipione del Ferro or Ferrari to label them.

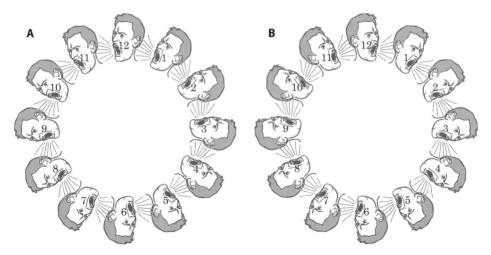

Figure β.11. (a) Twelve angry men yelling at each other counterclockwise. (b) Twelve angry men yelling at each other clockwise.

P. A.: I suspect Chekhov had compelling reasons to stop at three sisters even without consulting Abel and Galois. You might have better luck with *Six Characters in Search of an Author*.

N. T.: Or with *Twelve Angry Men* . . . try to imagine twelve angry men sitting around a round table, with the first angry man yelling at the second angry man, who in turn yells at the third angry man, and so on, until we get to the twelfth angry man, who is yelling at the first one. Then there are twenty-four possible seating arrangements, as illustrated in figure β.11(a) or (b); in either case we are allowed to rotate the table around its center, like a ferris wheel, with the angry men following their place settings. Just as in the last chapter's *Stable Equilibrium* game.

P. A.: But if the angry men are all deaf in their right ears—

N. T.: —then only the arrangements in figure β11(a) would be allowed. In the former case the Galois group has twenty-four arrangements and in the latter case, only twelve.

P. A.: Why was it so important for us to imagine this scenario?

N. T.: Because that's practically all I can do with groups of sisters, or angry men, to illustrate Galois theory. As things now stand, if you want to see some order in the general theory, you'll have to look into Grothendieck's theory of motives[15] and the Langlands program. You can find pointers in the book if you're willing to look beyond the confines of this dialogue. I can give you some page numbers.[16]

P. A.: Why do you think I would bother reading the book when you can't even explain how a number can answer a question if it can't be written down in a formula?

N. T.: How do you know I haven't already explained that? But I, too, can play the "spirit who always says 'no.' " Can you tell me what a character is without explaining what would happen if you permuted Claudius and Gertrude?

P. A.: Nothing you've said convinces me that permuting characters is in any way relevant to what makes a play a play.

N. T .: In other words, we agree that whatever kind of art mathematics is, it's not primarily a *dramatic* art.

An Automorphic Reading of Thomas Pynchon's *Against the Day* (Interrupted by Elliptical Reflections on *Mason & Dixon*)

All mathematics leads, doesn't it, sooner or later, to some kind of human suffering.

—*Against the Day,* p. 541

Thomas Pynchon, postmodern author, is commonly said to have a non-linear narrative style. Inger H. Dalsgaard suggests that "a novel like *Against the Day* may be read in non-linear fashion, in keeping with the operations of a time machine."[1] No critic, however—not even the "seventeen of the foremost heavyweights from over forty years of Pynchon criticism"[2] who contributed to the *Cambridge Companion to Thomas Pynchon*—seems to have taken seriously the possibility, to be explored in this chapter, that his narrative style might in fact be *quadratic*.

Google gives no matches whatsoever for "quadratic narrative style,"[3] and this hypothesis—more precisely, that Pynchon's major novels are structured by *conic sections*, at a rate of roughly one per book—sheds no light on the deeper import of his writing. Just as with an "obvious" proof in mathematics, however, once you've seen the thesis, it's impossible to stop seeing it.

This "insight" came to me instead of a solution to another puzzle that had been troubling me since I started reading *Against the Day* in the spring of 2008. The prominence of mathematics in this book is exceptional even for Pynchon. Two of the main characters are at least part-time mathematicians; Hilbert, Minkowski, and Gibbs make cameo appearances; several chapters are set in the Göttingen mathematics department; and among Pynchon's signature silly ditties there is this romantic number:

Her idea of banter
Likely isn't Cantor
Nor is she apt to murmur low
Axioms of Zermelo,
She's been kissed by geniuses,
Amateur Frobeniuses
One by one in swank array,
Bright as any Poincaré . . .

and so on in that vein.

It was when I came upon the word *automorphic* on page 409,

Earth making its automorphic way round the sun again and yet again . . . ,

that I began to wonder what was going on—and then on pages 452 and 453, there it was again:

periodic functions, and their generalized form, automorphic functions

as a prelude to a scholarly discussion of time travel:

Time no longer 'passes,' with a linear velocity, but 'returns,' with an angular one. All is ruled by the Automorphic Dispensation. We are returned to ourselves eternally, or, if you like, timelessly.

Between the two mentions of automorphic is a scene reminiscent of Odysseus' voyage to the river Styx, in which one of the Chums of Chance crosses a recognizably non-Euclidean landscape:

. . . the more "respectable" parts of town . . . at each step were receding, it strangely seemed, disproportionately farther as the young men went on.

The puzzle was: did all this technical and, for the most part, legitimate mathematics serve as mere atmospheric accompaniment to the turn of the century's hesitant exploration of the relations between time, space, and light? So that reviewers could write, with Luc Sante in the *New York Review of Books*:

My own eyelids drooped when the subject was mathematics, for example, but that is something I am profoundly ignorant about . . .

and still feel qualified to shower the book with unstinting praise? Or, for that matter, to write, like Louis Menand in *The New Yorker*:

I can't do the math, but I think that the idea behind "Against the Day" is something like this . . . [continuing with an elaboration of what I just wrote above],

but concluding in disappointment that "Pynchon must have set out to make his readers dizzy and, in the process, become a little dizzy himself."[4]

Or was all this mathematics there for a reason integral to the structure of the book, inaccessible, perhaps by design, to "profoundly ignorant" reviewers who "can't do the math"? Perhaps I was missing a cryptic message intended expressly for people on familiar terms with the word "automorphic," for participants in the Langlands program—for people exactly like me, in other words. . . .

I'd better stop here to reassure the reader that I have not lost my marbles. Paranoia is one of Pynchon's favorite topoi, and although this is very much a self-referential paranoia—as when. in *The Crying of Lot 49*, the secret network of communication Oedipa Maas believes may have been created for her benefit was indeed created by the author himself—Pynchon's reclusiveness combined with his choice of theme must make the interpretation of his books a magnet for all sorts of cranks as well as genuine paranoids. So I repeat that what impels me to write is not the belief that I have somehow penetrated "the bright, flowerlike heart of a perfect hyper-hyperboloid," to quote a passage from the last page of *Against the Day*, to which I will soon return but that here can stand for the book itself or for the entirety of Pynchon's work. Whatever I've seen is just something I can't help noticing, meaningless or not.

Returning to my narrative, it was not reassuring to discover that Christophe Claro, who had recently translated *Against the Day* into French to considerable acclaim in the literate press, chose to render automorphic as "automorphique," as in "fonctions automorphiques," instead of the correct term. "automorphe," although he was supposedly indirectly in communication with Pynchon himself.[5] If Pynchon uses mathematics as background music, an imprecise translation makes no difference; but if it serves a structural purpose, the choice of word may be very important.

I'm afraid I can't solve the puzzle. But I can say that when you start to look, you find an awful lot of hyperbolas in *Against the Day*. For example: the hyperbolic geometry to which I alluded in connection with automorphic functions; the "Automorphic Dispensation," which seems to be a "function . . . by which, almost as a by-product, ordinary Euclidean space is transformed to Lobachevskian" (p. 453); and that "perfect hyper-hyperboloid" that "only Miles" Blundell, the one character to have apprehended the meaning of space-time, "can see in its entirety." There are (hyperbolic) wave equations (and a whole family of Vibes) and the "noted Quaternionist V. Ganesh Rao of Calcutta University" who, by rotating himself in an imaginary direction, performs something "like reincarnation on a budget, without the element of karma to worry about" (pp. 130, 539).

I remembered having understood Pynchon's *V.* as the convergence, *V*-like, of two narrative lines.[6] *Gravity's Rainbow*, obviously dominated by the image of the parabola and full of explicit references to the shape, such as:

> He had noted this parabola shape around on Autobahn overpasses, sports stadiums u.s.w. and thought it was the most contemporary thing he'd ever seen. Imagine his astonishment on finding that the parabola was also the shape of the path intended for the rocket through space (p. 298).

has also been compared to a parabola in its narrative structure, not least by Salman Rushdie.[7] The obvious guess was that *Mason & Dixon*, which I hadn't read, would turn out to have been written under the sign of the ellipse. And sure enough, here's what I found on page 555 of *M&D*:

> In a slowly rotating Loop, or if you like, Vortex, of eleven days, tangent to the Linear Path of what we imagine as Ordinary Time, but repeating itself,—without end.

M&D's main characters are Astronomers; there are orreries and orbits "as elegant as Kepler's," and then this vision, near the end of the book, when the explorers are at the point of being turned back:

> "In the Forest . . . ev'ryone comes 'round in a Circle sooner or later. One day, your foot comes down in your own shit. There, as the Indians say, is the first Step upon the Trail to Wisdom."

• •

> *The infinite boredom of conic sections . . . their calm and tantalizing respectability . . .*
>
> —F. Scott Fitzgerald, *This Side of Paradise*

But does *M&D* have an elliptical or even merely circular structure? You can find plenty of ellipses, not to mention circles, in all Pynchon's novels, as well as the ellipses that look like this. . . . For that matter, you can find circles in practically every story ever recorded, starting with Gilgamesh. Besides, what better literary representation of a double line (figure β.8) than the border between Pennsylvania and Maryland, mapped by Mason *AND* Dixon?[8]

Of course I decided I had no choice but to read *M&D*, bitterly regretting I had not done so ten years sooner and, of course, unlike the book's eponymous heroes, finding exactly what I was seeking, specifically ellipses of all shapes and sizes: the word *Ellipse* to start with, or rather to end with, since it occurs twice and rather superfluously in close succession in chapters 75 and 76.

M&D comes closest of all Pynchon's novels to the traditional sense of closure. The account of the Mason-Dixon expedition is neatly sandwiched between introductory and concluding sections, as it is enclosed comfortably within the story told by the Rev[d] Cherrycoke to his family circle, itself reflected in the thematically overdetermined "Mirror in an inscrib'd Frame" that appears on the novel's second page. The main narrative is, in turn, studded like a Fruit-Cake with digressions: side trips to New York and Virginia, a tale of a Tub, and a chapter devoted to the mechanickal Duck's love story and another one on the fairy tale of the Court Astronomers Hsi and Ho. . . . The *Ghastly Fop* episode that interrupts the novel without warning in chapters 53 and 54 looks like an exception: far from being self-contained, its unexpected fusion with Cherrycoke's narrative leads directly to this dialogue between Captain Zhang and Dixon:

> "*We* happen to be the principal Personae here, not you two! Nor has your
>   Line any Primacy in this, being rather a Stage-Setting . . . "
> "And Mason and I,—"
> "Bystanders. Background. Stage-Managers of that perilous Flux,—little
>   more."
> "Eeh." Dixon thinks about it. "Well it's no worse than Copernicus,
>   is it . . . ?"

This blurring of multiple levels of fiction may be a pinnacle of novelistic self-referentiality, but it may also be a sly reminder of the simple geometric fact that an ellipse has not one but *two* foci. . . . [9] And perhaps we should understand the double Mason-Dixon line as what remains of an ellipse when its two foci, in this case the two main characters, are sent to infinity in opposite directions.

In my initial sketch for this chapter, written in 2008, I set out to write down unquiet thoughts about automorphy in *Against the Day* that came to me only some time after I had put the book aside. Having turned my attention to *Mason & Dixon* with a Thesis to defend or discard, it's not surprising that my reading turned up a great deal more in the way of material. I just mention in passing

—the "Geometry more permissive than Euclid," apparently spherical rather than hyperbolic, in chapter 33,
—the Möbius smoke ring in chapter 34,
—and the ruminations on Time, the "Space that may not be seen," the "true River than runs 'round Hell" in chapters 32–34.

Time is undoubtedly of thematic import in *M&D*, and this particular sequence of attempts to circumscribe Time is initiated by the swallowing, by a character known only as R.C., of a Watch that operates by Perpetual-Motion. Having completed his reflections on the consequences of swallowing Time, Pynchon opens chapter 35 with the Cherrycoke's family's most heated, and most quoted, argument, on the relation between History, Truth, and 'Novel.' Philadelphia lawyer Ives LeSpark puts it this way:

Time on Earth is too precious. No one has time, for more than one Version of the Truth.

For the author, swallowing Time amounts to enclosing it in a novelistic structure. And Time, in the form of the watch inside R. C. who is buried in a tomb in the middle of a story-within-Cherrycoke's-story-within-Pynchon's-story at the heart (or should I say focus?) of the book, is the cosmic principle around which all the plot's planets revolve.

Most suggestive of the *Mason & Dixon*'s cyclical structure is the prediction in its very last line:

We'll fish there. And you too.

Addressed to Mason by his two older sons, after (or as?) they "ensign their Father into his Death," the last three words lose none of their poignancy if they are taken literally as a prediction, which becomes true only on the condition that the narrative is meant to recommence at that point from the beginning. As it is a convenient way to reconcile Mason's abrupt decision near the end of his life to return with his family to Pennsylvania with this summary of his life, made shortly before Dixon's death:

> To leave home, to dare the global waters strange and deep, consort with the highest Men of Science, and at the end return to exactly the same place, us'd,—broken. . . .

Gluing the novel's front and back together in this way is also the only way to validate what the talking "British Dog" promises Mason and Dixon at the end of the penultimate chapter:

> The next time you are together, so shall I be, with you.

The words Mason speaks "the next time," back at the front of the book in chapter 3, could well serve as my motto for this chapter:

> Isn't it worth looking ridiculous, at least to investigate this English Dog, for its obvious bearing upon Metempsychosis. . . .

• •

Determining how, if at all, *Against the Day*'s narrative structure is hyperbolic is more challenging, but here are some thoughts. As a hyperbola has two connected components—*bilocation?*—one would expect *Against the Day* to have *two* nonoverlapping narrative arcs. So it is significant that the Chums of Chance and the main characters of the Traverse family narrative never meet. The Chums open the novel with a landing of the airship *Inconvenience* and close it with the same airship returning to the sky, to "fly toward grace." The Traverses naturally spend much of their time underground in mines or tunnels or underwater in a submarine or, in one case, a torpedo. The two arcs do come very close in three successive chapters set in the Low Countries—in Oostende, to be precise, exactly in the middle of the book.[10]

Legitimate Pynchon scholars have also noted the presence of two narrative arcs. Nina Engelhardt, one of the rare professional readers to take Pynchon's mathematics seriously, has made the ingenious suggestion that the two narratives echo *Against the Day*'s frequent references to complex numbers and quaternions and their real and imaginary parts. The Chums, recurring heroes of a series of adventure books, live—like the square root of –1—on an imaginary axis, while the Traverse family and their companions traverse the all-too-real landscape of war and class struggle. Webb Traverse is "murdered by men whose allegiance [. . .] was to that real axis and nothing beyond it" (p. 759), while his son Reef encounters the Chums in his imagination, reading one of their books; and his future companion Yashmeen meets them in her dreams. This is appealing and also makes sense—not least because it's so easy to extend the name (0, 0) of the meeting point of the real and imaginary coordinate axes to spell out Oostende. The hyperbolic and real/imaginary readings are no less mutually exclusive than two strands in one of *Finnegan's Wake*'s multilingual puns—and remember that Joyce boasted that he had "put in [*Ulysses*] so many enigmas and puzzles that it will keep the professors busy for centuries arguing over what I meant. . . ."[11]

Against the Day is filled out by a host of secondary characters who bounce or vibrate from one narrative strand to the other—mostly more or less mad scientists (Heino Vanderjuice, Merle Rideout, V. Ganesh Rao, Roswell Bounce) obsessed with time travel and quaternions, but also including the detective Lew Basnight and (at least tangentially) members of the Vibe clan. The plot is beginning to look like a hyperbola whose two arcs are joined by a sinusoidal curve whose graph one might hope to find in a mathematical analysis of double refraction in calcite (*Iceland spar*, the title of the second part of the book). But this is too simplistic. Calcite is merely *uniaxial birefringent*; for hyperbolic interference patterns you have to look at *biaxial birefringent* minerals, as in figure 5.1.

I leave these speculations to a later version of this text that no one may ever need to write. But let me insist again that the hypothesis would be no less frivolous if it turned out to be in some sense correct. Pynchon is such a relentlessly playful writer—I would have said relentlessly silly, except that Freud made the point that play (unlike silliness, perhaps) is not opposed to seriousness—that frivolousness must often be part of his

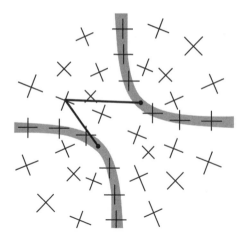

Figure 5.1. http://www.uwgb.edu
/dutchs/Petrology/intfig1.htm.

design. Choosing to structure his successive novels by conic sections, if that is in fact what Pynchon did, may have been a private joke, just like the recurring trope of entropy in his work may be an extended allusion to C. P. Snow's comparison of ignorance of the second law of thermodynamics to not having read Shakespeare. Thanks largely to Pynchon,[12] it's hard to find a critic ignorant of the second law of thermodynamics, though if Sante and Menand are typical, "don't know much trigonometry" is still a popular refrain among the literary Elect. Or, entropy may have been an arbitrary (dis)organizing principle, a disciplinary constraint to guarantee the nonlinearity of his narratives—or maybe a little of both.

Even if you're willing to grant this interpretation, you may well ask, so what? Is Pynchon deliberately broadcasting conspicuously futile signals across the Two-Culture divide? Is his goal to delimit a bounded zone of reference from which the habitual insiders are excluded but which, in any case, can be mined for depthless interpretation, at best? Does it matter, for example, that the initials R.C. in the story of the perpetual motion watch clearly allude to the "Ticking Stone" of George Alfred Townsend's *Tales of the Chesapeake*, which features Mason and Dixon as minor characters—also known as the Ticking Tomb, a tourist attraction in Landenberg, Pennsylvania? Are Pynchon's conic sections like the *Figure in the Carpet* in Henry James's novella?

[T]he thing without the effort to achieve which [the writer] wouldn't write at all, the very passion of his passion, the part of the business in which, for him, the flame of art burns most intensely? . . . It stretches, this little trick of mine, from book to book, and everything else, comparatively, plays over the surface of it. . . . I live almost to see if [the secret] will ever be detected.

To answer questions like these you would have to consult a licensed Pynchonian—someone like Inger Dalsgaard, for example, who wrote:

[L]iterary scholars naturally tend to look less at what natural sciences described in fiction say themselves about the real world or reality than at what fictional texts or literary strategies which include or mirror scientific fields, theories or methods say about the world.[13]

Or like Michael Harris, professor of English at Central College, Iowa, whose article "Pynchon's Postcoloniality" sees *M&D* as a "meditation on lines, boundaries, borders, margins [but not ellipses, MH[14]], and their consequences," quoting Captain Zhang to the effect that drawing a straight "Line upon the Earth is to inflict . . . a long, perfect scar . . . hateful Assault." Though his essay aimed to "reinscribe Pynchon as, at least in part, a political writer"—the mapping of the Mason-Dixon line is a clear allegory for the colonial conquest of America and prefigures the subsequent drawing of the closed curves of reservations—he adds that, with Pynchon, "almost nothing can be taken literally."[15] What does quadratic narrative form say about the world? Does a postcolonial border run between the Two Cultures?

I don't know how to answer those questions. Nor have I figured out how or whether Pynchon's other novels fit in this tableau of conic sections. By a process of elimination, it would make sense to associate *Vineland* with the circle, which is what you get when the two foci of an ellipse coincide. But apart from the name of the main character (Zoyd Wheeler) and a throwaway allusion to the Arctic Circle Drive-In, I've found nothing more conclusively circular in *Vineland* than in any of the other novels. It's true that the circle is the most visibly symmetric of the conic sections; and *group theory*, the branch of mathematics concerned with symmetry, figures prominently in *Vineland*, through the saga of Weed Atman, "a certain mathematics professor, neither charismatic nor even personable" at College of the Surf in Trasero County, California. The magic of marijuana transformed this absent-minded group theorist to the center of a "moving circle of focused attention" of a cluster of groupies and the leader by consensus of "The People's Republic of Rock and Roll," a short-lived utopian community—a self-governing "relaxed field"—smashed by the violence of the state. Gunned down by a comrade as a result of the machinations of the supremely sinister FBI agent Brock Vond, Atman becomes a symbol of nostalgia for an unfinished revolution—just like Galois, popularly considered the inventor of group theory.

Shortly after *Vineland* appeared, Joseph W. Slade published an analysis in *Critique* in which he lists parallels —too many to be coincidental—between the details of Atman's story and popular accounts of Galois' life. "What probably attracts Pynchon to group theory," he speculates, "is that it can support contradictory schemes of reality."[16] Slade has produced another mathematical reading of a Pynchon novel, one that again elicits the reply so what?—from Slade himself, who warns that "almost any work of fiction is subject to analysis by group theory." One more thing, though: in Pynchon's major novels, much attention is paid to the movement of objects through space. Newton's law of universal gravitation, one of the oldest of all differential equations, tells us that the motion of an object once launched from the earth is determined forever—predestined, one might say, in keeping with the Calvinistic spirit of *Gravity's Rainbow*—by its *initial conditions*, in this case the starting point and its initial velocity. If a projectile is launched at a speed below escape velocity, it will fall to the earth, like a V-2 rocket, its path tracing a parabolic arc. If launched precisely at escape velocity, earth's gravity will restrain its motion to an elliptic orbit. But if it attains and exceeds escape velocity, it will recede endlessly into space along a hyperbolic trajectory. At the end of the Chums of Chance's journey in *Against the Day*,

> it is no longer a matter of gravity [for the Inconvenience]—it is an acceptance of sky.

There aren't any more conic sections, except for the most degenerate of all, the single point, a possible plan for a novel in which absolutely nothing happens. On the other hand, there's literally no limit to what Pynchon might do with cubic narrative structures. . . .

PART II

chapter 6

Further Investigations of the Mind-Body Problem

The Kabbalist: Thinking too much can lead to madness.
Don Pedro Velasquez: Reason can accept anything, if it is used knowingly.

—from *The Saragossa Manuscript* (1965),
film by Wojciech Has

*The real lover of learning [*philomathes*] . . . will not . . . desist from eros until he lays hold of the nature of each thing in itself with [the rational] part of the soul . . . drawing near to and having intercourse with the really real.*

—Plato, *The Republic*, 490b.[1]

If, as French film critic Jean Mitry claimed, "a film is a mirror in which we recognize only what we present to it through what it reflects back to us,"[2] do we mathematicians recognize ourselves in the increasingly frequent cinematic images of our profession? When we look in our private mirrors, do we see Carlo Cecchi (*Death of a Neapolitan Mathematician*), Tilda Swinton (*Conceiving Ada*), Matt Damon (*Good Will Hunting*), Gwyneth Paltrow (*Proof*), Russell Crowe (*A Beautiful Mind*), Rachel Weisz (*Agora*), Sean Gullette and Mark Margolis (*Pi*), David Wenham (*The Bank*), or Béatrice Dalle (*Domaine*)? Or perhaps the autistic savants played by Dustin Hoffman (*Rain Man*) and Andrew Miller (*Cube*)? Do we agree that we are misfits (Damon, Hoffman, Miller, Dalle, Wenham) if not martyrs (Weisz, Margolis, Cecchi, Swinton), on the verge of madness (Paltrow, Crowe, Gullette, Cecchi) if not over the edge, and stunningly attractive (nearly everyone)?

Now joining the cast is Edward Frenkel, a Berkeley professor and one of the rare mathematicians who would not look out of place on the red carpet at Cannes, as author, codirector, and star of a twenty-six-minute independent film entitled *Rites of Love and Math*. The action is well summarized by the synopsis included with the film's press packet:

> Mathematics is first and foremost pursuit of Absolute Truth and Beauty. This is the story of a Mathematician who has found, after many years of hard work, the ultimate Formula of Love. At first, he was thrilled that his formula would benefit people, bringing them eternal love, youth and happiness. But later he discovered the flip side of the formula: it could become, if used in the wrong way, a weapon against Humanity. And so forces of Evil are now after the Mathematician. They want to take possession of the magic powers of his formula and misuse them in order to achieve their sinister goals. The Mathematician knows that there is no escape for him, and he is ready to die. But he wants his formula to live.
>
> The Mathematician has a secret love affair with a beautiful Japanese woman, Mariko. At midnight, he comes to her place and tells her about his predicament. Having realized that they are seeing each other for the last time, they make love more passionately than ever before. And then the Mathematician tattoos the magic formula on Mariko's body. They both know that this is the last time they are seeing each other. Afterward their love will live in this formula engraved on her beautiful body.

Whenever I hear of a mathematician pursuing extraprofessional interests, I am reminded of Claude Levi-Strauss's impressions of São Paulo high society in the 1930s: "Society, being limited in extent, had allocated various roles to its different members. All the occupations, tastes and interests appropriate to contemporary civilization could be found in it, but each was represented by only a single individual. . . . There was the Catholic, the Liberal, the Legitimist and the Communist; or, on another level, the gourmet, the book-collector, the pedigree-dog (or -horse) lover. . . ."[3] Among contemporary mathematicians, one finds the novelist, the ballroom dancer, the dandy, the cultural critic, the theologian. In recent years Paris has seen a procession of mathematicians from America in novel roles: the installation artist (Benedict Gross), the fashion consultant (the late Bill Thurston), the *auteur* (Ed Frenkel), and the sex symbol (Frenkel again).

The more dynamic sectors of the French mathematical establishment—represented by the Fondation de Mathématiques de Paris-Centre, which contributed financially to Frenkel's film, and the Institut Henri Poincaré (IHP), under the direction of the media-friendly Fields Medalist Cédric Villani—have conspicuously welcomed these developments that hint at a kind of glamour to which we are mostly unaccustomed. Although his project suffered neither from corporate sponsorship nor from the support of the mainstream media, turning filmmaker has nevertheless brought Frenkel into some scandalous company, Japanese novelist and ultranationalist Yukio Mishima, for starters. Frenkel's *Rites* is consciously modeled on Mishima's film *Yûkoku*, also known as *Patriotism* or *The Rite of Love and Death*, "suppressed" by his widow after the author performed *seppuku* in 1970 (according to Wikipedia) until it resurfaced accidentally earlier this century. Through his codirector, the experimental filmmaker Reine Graves, Frenkel was exposed to the psychoanalytic theories of Jacques Lacan—Graves confided to me that Frenkel had read all of Lacan in a single day—as well as to Graves' former collaborator Jacques Henric, literary contributor to the magazine *Art Press*, which is directed by his companion, the sexual Stakhanovite Catherine Millet, whose *The Sexual Life of Catherine M.* is one of the founding documents of what French weekly magazines a few years ago were calling the "new libertinism." Henric supplied a sort of interpretation for the *Rites* press kit:

> There is also a lesser tradition in literature, philosophy and morals, which strives to ease and even bluntly cut the link between Eros and Thanatos. In the course of the XVIII century, French Libertines followed it, but in the XIX century, with the advent of Romantism [*sic*], "love to death" came back into fashion, and the past century. . . did nothing to liberate itself from the influence of this ideology and this moral philosophy. In Japan, on the other hand, a deep-rooted tradition, close in spirit to that of French Libertines, has nourished a grand and steady literary current as well as an essential trend in painting striving to produce the most beautiful images.

Both Mishima's film and its "homage" by Frenkel and Graves are highly stylized and take place on the Noh stage with Wagner's *Tristan*[4] playing in the background. Frenkel's Mathematician, played by Frenkel himself, replaces Mishima's army officer, who, passionately attached to the emperor as well as to his wife, performs *seppuku* to avoid being forced to attack his equally patriotic comrades-in-arms. We have seen in the synopsis that the Mathematician impales himself on a blade as well. The ritualized suicides take place after no less ritualized scenes of love-making—"love to death" is alive and well in Japan as well as in *Rites*— in front of a scroll painting. In Mishima's film, the scroll reads "Sincerity" in calligraphic Chinese characters; in Frenkel's it reads Истина [*istina*], Russian for "truth"—and though the screen informs us that that "In the face of death, the Mathematician and Mariko bid final farewell to every little detail of each other's body," what Lacan would have called the Mathematician's "signifier" remains at all times concealed from the spectator. Here's Henric again:

> What has motivated Edward Frenkel and Reine Graves to make their film *Rites of Love and Math*? Is it to drive not just the nail, but the knife, if one may say so, between spirit and flesh, or is it to finally reconcile them?

Relatively few professions are practiced even intermittently in the nude, and while *Rites* is likely to reopen the long overdue debate on whether mathematics, like the fieldwork for Catherine M.'s memoirs, should be one of them, I find the film most explosively scandalous in its confusion of genres—practically a category mistake—focused precisely on the reconciliation Henric evokes of "spirit and flesh," more classically known as the *mind-body problem*. Archimedes deserved a best-supporting-role nomination for dramatizing the problem in Plutarch's *Life of Marcellus*:

> He neglected to eat and drink and took no care of his person; ... he was often carried by force to the baths, and when there he would trace geometrical figures in the ashes of the fire, and *with his finger draws lines upon his body* when it was anointed with oil, being in a state of great ecstasy and divinely possessed by his science [my emphasis].

The Archimedes of classical literature embodies a metaphysical paradox. On the one hand, in the Plutarch quotation, as well as in his *Eureka!* scene—the most persuasive argument to date in favor of mathematical nudity—he created the classic figure of the mathematician distracted to the point of total withdrawal from the material world, reduced to mind alone. The archetype of the absent-minded mathematician was revived during the Enlightenment, as we see shortly, but is not conspicuous in recent cinematic representations of the profession, and this is perhaps surprising, given that the absent-minded professor is certainly a stock character in popular films. On the other hand, in the familiar anecdotes just recalled, Archimedes's body is literally visible and uncovered; in a third anecdote, also from Plutarch, a Roman soldier's sword severed the spirit from the flesh of the Greek mathematician found in a "transport of study and contemplation" on the beach near Syracuse. Seen from the outside, the mathematician's body is an object of ridicule, inappropriately displayed and in the way. But from the inside the body is irrelevant, at best serving as a convenient surface for the drawing of geometric diagrams, as Mariko's body in Frenkel's film is in the end only a surface for preserving Frenkel's "magic formula"[5] or as the bodies in Catherine M.'s narrative, not least her own, are little more than machines performing repetitive and largely predictable motions in a variety of natural and artificial settings.

The scandal of Catherine M. is the "*Je*" of the first-person narration, told from the viewpoint of one of these machines; the scandal of Archimedes, and of western metaphysics as a whole, is that the mind forever ceases its inventions and discoveries when the body is left in a heap on the sand. An unwritten rule of cinema, especially of erotic cinema, is that not all reminders of materiality are equally painful to behold; in this respect *Rites* is faithful to the tradition. But attempts to transcend our material limitations and to encompass the infinite within our finite bodies lead invariably to swift retribution and martyrdom: expulsion from the Garden of Eden, the fall of Icarus, crucifixion, or the insanity of the mathematicians represented in popular films. The hero of *Pi*, having computed the 216-digit number from which all patterns in nature arise, escapes martyrdom only by voluntarily ridding himself, with the help of a power drill, of the substance responsible for his mathematical understanding, located on the border between spirit and flesh in his right temporal lobe.

Members of the nonspecialist public more comfortable with words than images may base their impressions of mathematicians on the biographies, written by distinguished writers and published in Norton's Great Discoveries series, of Cantor, Gödel, and Alan Turing: two madmen and a martyr, all damaged by encounters with infinity. Those who prefer a balance of words and pictures can turn to the graphic novel *Logicomix*, authored by Apostolos Doxiadis and Christos Papadimitriou with a team of professional artists: they will learn that the creators of the logical foundations of mathematics, not excluding the consummate rationalist Bertrand Russell, were haunted by madness. Even those who know only what they see on TV have heard about Grigori Perelman, deemed crazy for turning down a million-dollar prize after having solved "one of the most difficult problems of the last ten centuries" and reportedly slated to be the subject of a James Cameron film.[6]

Frenkel has explained in a series of press interviews that he wants to "set the record straight" and offer an alternative to the stereotypic image of the mathematician as "the mad scientist" of *Pi* and *A Beautiful Mind*:

> My purpose is precisely to counter these stereotypes. I wanted the mathematician in our film to be seen as a human being with whom the public can relate: he tries to do his best in difficult times, he is someone who can love and be courageous, who fights for his ideals.[7]

Once in New York City, near *Pi*'s neighborhood, a friend's neighbors, to whom I had been introduced as a mathematician, told me how fortunate they felt to meet someone from such a sensitive profession. In those days, before the dark forces that finance mathematics serves had transformed lower Manhattan's sociocultural landscape, it was still possible to be moved by a sentiment expressed so unaffectedly, even by a couple marginally integrated into society who, my friend later told me, practiced domestic violence almost on a daily basis. It was the last time artists acknowledged me without prompting as one of their own, and I thought of them at the champagne reception following the film's first screening, when three of the codirector's friends, sharing a cigarette, insisted in response to my question that they were not at all shocked to see eros and mathematics treated in film as reflections of one another. On the contrary, although their professionally informed comments were sometimes sharply critical of the lighting, colors, sound, acting—of everything that makes *Rites* a film rather than an idea, in fact—they were in total agreement that Frenkel and Graves had found a *très beau choix de sujet*, a *beau theme*, in conceiving a film about a mathematician who is simultaneously a lover and a martyr to truth.

Reading the blogs, one learns that Frenkel's complaint about the treatment of mathematicians in popular culture is widely shared. But in an important sense it's beside the point. Cinema doesn't need to look to mathematics for unstable or unhappy character types. No one complains that films about Sid Vicious, Kurt Cobain, or Jim Morrison reinforce negative stereotypes about rock musicians, and who can keep track of all the cinematic representations of Van Gogh? The interesting question is not whether mathematicians are portrayed as deranged or tormented, but in what sense their torment or madness is characteristically mathematical.[8]

Indeed, the madman is only one of the stereotypes on the mathematician's storyboard. According to historian Amir Alexander,

> Among modern mathematicians, it seems, extreme eccentricity, mental illness, and even solitary death are not a matter of random misfortune . . . in the popular imagination . . . mathematicians feature prominently as loners and misfits who never find their place in the world.[9]

Frenkel's Mathematician is neither a loner nor a misfit, much less a madman, but his death is almost—though not quite—solitary. His film thus reproduces the image of mathematician as *romantic hero*, the stereotype that, for Alexander, has represented mathematics "in the popular imagination" since Galois was elevated to iconic status several decades after his death. Revisiting the lives of Chopin and the poets Byron, Keats, Shelley, and Novalis—he could have added Pushkin and Lermontov—Alexander sees the romantic hero as "a doomed soul whose quest for the sublime leads to loneliness, alienation, and all too often an early death. But in the few years allotted to him, the romantic hero burns more fiercely and shines brighter than any of his fellows." Familiar picture, to be sure. More surprising is Alexander's report that the romantic details of Galois's biography that have inspired generations of mathematicians—including Frenkel, whose important work on the Langlands program straddles the boundary between Galois theory and mathematical physics—were in large part fabricated in the decades after his death in order to fit Galois into a preexisting romantic mold.[10] Moreover, according to Alexander, the "troubled mathematical martyr," exemplified not only by Galois but also by Abel, János Bolyai, Riemann, Cantor, Gödel,

Turing, John Nash, Grothendieck, Perelman, and even, in a certain sense, Cauchy, remains to this day the dominant image of the "ideal mathematician," long after the romantic paradigm was exhausted in the arts.

Perhaps paradoxically, the well-known account of Galois' life is the more significant for being false. One sees more clearly that Galois' invented romantic persona fills and thereby reveals a cultural need; unadorned truth is not always up to the task. Alexander sees this persona as constructed in opposition to the Enlightenment ideal of the mathematician as a "natural man." The prototype of this ideal, with its obvious echoes of Rousseau, was the mathematician and encyclopedist Jean le Rond d'Alembert, especially as represented by the Marquis de Condorcet in his eulogy of d'Alembert at the Académie Française: another fabrication, as Alexander makes clear. A semifictional d'Alembert was made to play the natural man as "disconnected dreamer and hopeless bumbler" in Denis Diderot's *D'Alembert's Dream*. The "geometer" Don Pedro Velasquez of Jan Potocki's late enlightenment novel *The Saragossa Manuscript* offers a more complete and well-rounded fictional interpretation of the mathematician as natural man. Distracted, like Archimedes,[11] to the point of walking into a stream while engrossed in his calculations, Velasquez is nonetheless noble and elegant, an engaging storyteller, an enlightened philosopher, and, most importantly for our purposes, an object of romantic interest.

In this respect, the fictional Velasquez was a mathematician of his time. The Oxford English Dictionary informs us that in 1750 "The Wranglers"—top-ranked candidates in the mathematical tripos at Cambridge University—"usually expected, that all the young Ladies of their Acquaintance . . . should wish them Joy of their Honours."[12] In their obituary dedicated to the analyst and geometer Alexis Clairaut, Diderot and F. M. Grimm recalled the mid-eighteenth-century mathematical craze:

> *Clairaut avait vu ce règne brillant de la géométrie, où toutes nos femmes brillantes de la cour et de la ville voulaient avoir un géomètre à leur suite.*
> [Clairaut had witnessed the illustrious reign of geometry, when all our brilliant women of the court and the city wanted to have a geometer at their disposal.][13]

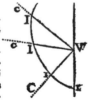

Figure 6.1. (*Left*) Clairaut (courtesy of the Smithsonian Institution Libraries, Washington, D.C.) and (*right*) his love equation, circa 1745. *V* is the fixed fulcrum and the *i*'s and *I*'s moving clockwise represent the motion of its endpoint.

Clairaut, who, according to Diderot and Grimm, *aimait éperdument le plaisir et les femmes*, may have been the first mathematician to propose a love equation [figure 6.1(b)] in the form of the Archimedean spiral, "on which the Geometers have so greatly exerted themselves, without having discovered its true nature."

> *On demande la Courbe iiI décrite par l'extrêmité d'un corps Vi, qui étant d'abord dans une situation verticale renversée Vi, change ensuite de grandeur & de position en devenant successivement Vi VI, &c.* [We seek the curve described by the endpoint of a body, initially vertical and pointing downward, that subsequently changes in length and position.][14]

Enlightenment mathematicians were more likely than not to be erotically curious. In his speculation on life on other worlds, Fontenelle claimed that "mathematical reasoning is made like love" and that "these two sorts of people [mathematicians and lovers] always take more than they are given." Maupertuis' thoughts[15] on evolution begin with a vivid depiction of the role of pleasure in the preservation of species not unlike the "rites of love" that occupy the middle of Frenkel's film:

> *Celle qui l'a charmé s'enflamme du même feu dont il brûle: elle se rend, elle se livre à ses transports; et l'amant heureux parcourt avec rapidité toutes les beautés qui l'ont ébloui: il est déjà parvenu à l'endroit le plus délicieux.*

And Potocki, a soldier and adventurer in his early life, has the beautiful and cultivated Rebecca of his novel, in the course of her extended flirtation with Velasquez, ask the geometer where love fits in his system:

> *Mais, dit Rébecca, ce mouvement que l'on appelle amour, peut-il être soumis au calcul?* [But, said Rebecca, this movement we call love, can it be calculated?]

Here on the twentieth day of the narrative,[16] Rebecca is concerned with the tendency of a man's love to diminish with intimacy while the woman's increases; Velasquez replies that there must, therefore, be an instant when the two love equally and adds, "I have found a very elegant proof for all problems of this kind: let X. . . ." (See figure 6.2.)

On the thirty-third day, Rebecca and Velasquez return to the question with the latter recasting the question in terms of positive and negative numbers:

> *Si je hais la haine de la haine, je rentre dans les sentiments opposés à l'amour, c'est-à-dire dans les valeurs négatives, tout de même que les cube de moins est moins.* [(If I hate the hatred of hatred, I return to the feelings opposed to love, that is to negative values, just as the cube of minus is minus.]

After several paragraphs of this, Rebecca, now called Laure, interrupts Velasquez:

> [S]*i je vous ai bien compris, l'amour ne saurait être mieux représenté que par le développement des puissances de X moins A beaucoup moindre que X.*
>
> *Aimable Laure, dit Velasquez, vous avez lu dans ma pensée. Oui, charmante personne, la formule du binôme inventée par le chevalier don Newton doit être notre guide dans l'étude du coeur humain comme dans tous les calculs.* [[I]f I have understood you, love is best represented as the development of powers of *X* minus an *A* that is much smaller than *X*.

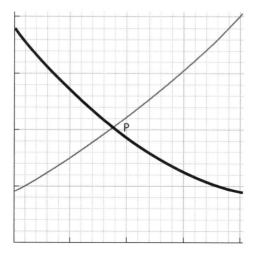

Figure 6.2. Illustration of Don Pedro Velasquez' Intermediate Love Theorem. The black curve measures the decline of the man's love and the gray curve, the increase of the woman's. The point *P* is the intersection predicted by Velasquez. Unfortunately, Potocki does not allow his character to present the proof. A more general proof of a similar theorem was first published by Cauchy in his course at the École Polytechnique in 1821, seventeen years after the first edition of Potocki's book.

Dear Laure, said Velasquez, you have read my mind. Yes, charming person, the binomial formula invented by the knight Don Newton must be our guide in the study of the human heart as in all calculations.]

• •

I agree with you quite upon Mathematics too—and must be content to admire them at an incomprehensible distance—always adding them to the catalogue of my regrets—I know that two and two make four—and should be glad to prove it too if I could—though I must say if by any sort of process I could convert two and two into five it would give me much greater pleasure.

—Lord Byron to his wife

Leading mathematicians grew increasingly intimate with power as the Enlightenment waned in the last decades of the eighteenth and first of

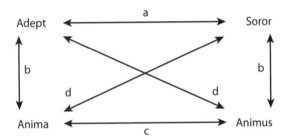

Figure 6.3. Jung's "marriage quaternio" (Jung 1969, p. 59).

the nineteenth centuries, and mathematics was prized as a source of charisma. The artillery specialist Choderlos de Laclos, for example, was trained in ballistics—as a mathematician, in other words—before he became known as author of the classic libertine novel *Les Liaisons dangéreuses*. "Even more than their [Enlightenment] predecessors," writes Alexander, the mathematicians Gaspard Monge, Pierre-Simon de Laplace, Lazare Carnot, and Joseph Fourier were "men of action, high officials of state under the revolutionary regime and Napoleon, and personal friends of the greatest men of the realm." And "for [the young Stendhal], at least, mathematics was . . . the royal road to Paris, glory, high society, women."[17]

Readers are nevertheless advised to contain their excitement a bit longer. Since the French revolution literary attitudes toward mathematics have been marked by rejection or indifference, as indicated by the Byron quotation given earlier.[18] When Zamyatin's protagonist D-503 in *We* asks, "*Can I find a formula to express that whirlwind which sweeps out of my soul everything save* [his opposite number, the sensual I-330] . . . *when her lips touch mine*," the author clearly wants us to answer: No, there is no formula. Even the economist Francis Ysidro Edgeworth, inventor of the indifference curve, acknowledged in his ambitious attempt to develop a "Hedonical Calculus" of "*Feeling*, of Pleasure and Pain" that "we cannot number the 'innumerable smile' of seas of love."[19]

Psychoanalysts did occasionally resort to diagrammatic representations of love, as in Jung's representation (see figure 6.3) of the archetype of the "marriage quaternio."[20]

There have also been bursts of enthusiasm in France. The writer Isidore Isou, who founded Lettrism shortly after World War II and wrote

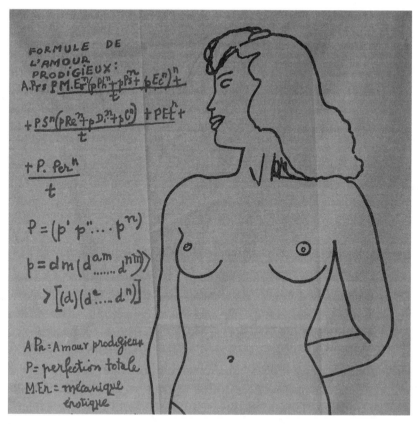

Figure 6.4. *Formule de l'amour prodigieux*, copyright Cathérine Goldstein. I thank her for bringing this image to my attention and for authorizing its reproduction. The formula is accompanied by a legend: A.Pr = *amour prodigieux*, P = total perfection, p = partial perfection, Per = perversion, D = discussion, M.Er = erotic mechanics, t = time, and so on.

a theoretical and practical study of the "mechanics of women," as well a treatise entitled *Traité d'érotologie mathématique et infinitésimale*, devoted a number of texts to formulas for love, including the one illustrated in figure 6.4. Existentialists were more ambivalent, but I have heard of an incident in which one of them *caressed* the monographs of a well-known mathematician, "the only kind of books they could not understand at all." Undeterred by their occasional lack of understanding, but deeply influenced by the Bourbaki group's insistence on the centrality of *structures* in mathematics, the structuralist generation in France had meanwhile placed mathematical models at the center of their philosophy, with Lévi-Strauss going so far as to include a chapter on group theory by André Weil in his *Elementary Structures of Kinship*.[21] Weil, by the way, is the author of what may be the single most quotable, erotically charged passage on mathematical creation. Attempting to describe how it felt to discover his topological approach to counting solutions to equations, Weil wrote that

> around 1820, mathematicians (Gauss, Abel, Galois, Jacobi) permitted themselves, with anxiety and delight, to be guided by the analogy [between an algebraic and a geometric theory] . . . [Now] gone are the two theories, their conflicts and their delicious reciprocal reflections, their furtive caresses, their inexplicable quarrels; alas, all is just one theory, whose majestic beauty can no longer excite us. Nothing is more fecund than these slightly adulterous relationships; nothing gives greater pleasure to the connoisseur, whether he participates in it, or even if he is an historian contemplating it retrospectively.[22]

Among the structuralists, responsibility for love, by way of psychoanalysis, fell to Jacques Lacan, and a flourishing collaboration in Lacanian knot theory remains active at the margins of French mathematics and psychoanalysis. In most of Lacan's seminars mathematical symbolism made only a fleeting appearance, as in this excerpt, which I will not attempt to translate, from his seminar in Vincennes December 3, 1969.[23]

> JACQUES LACAN—(*se tournant vers le tableau*). *Ça c'est une suite, une suite algébrique* . . .
> INTERVENTION—*L'homme ne peut pas se résoudre en équation.*
> JACQUES LACAN—. . . *qui se tient à constituer une chaîne dont le départ est dans cette formule*:

$$S2\ a\ s\ S \qquad S1\ S\ S\ a$$

Some of the terms in the "algebraic" formula may refer to the poles in the diagram (figure 6.5) of the "mirror stage," a version of the all-purpose structuralist double binary opposition scheme and the starting point for Lacan's approach to the mind-body problem, in which "the child antici-

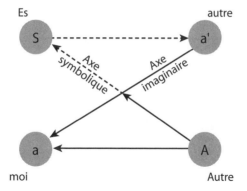

Figure 6.5. Lacan's Schéma L: http://fr.wikipedia.org/wiki/Sch%C3%A9ma_L.

pates the mastery of his bodily unity by an identification to the image of his likeness and by the perception of his image in a mirror."[24]

Interviewed by *Daily Californian* reporter Samantha Strimling, Frenkel explained "he wanted to challenge the notion of what mathematicians can do" by following Mishima and including a sex scene in *Rites*:

> "How about this: We have a mathematician . . . who is fighting for his ideas, and he is in love and there is actually a nude scene. He is actually making love to a beautiful woman," [Frenkel] said. "How about that compared to the stereotypes people are used to?"

I'm not sure I've ever met a professor other than Frenkel, in any specialty at any university, who has submitted the surface of his or her body to public inspection in quite this way.[25] But we have seen that the fashionable public of Diderot's time had no doubts about "what mathematicians can do." It is ironic that the rise of the romantic image of the mathematician coincides with the end of the image of the mathematician as a model romantic partner. George Sand had a full register of romantic men *à sa suite*, but none of them was a mathematician. Lord Byron's daughter Ada, in contrast, was encouraged by her mother to study mathematics expressly in order to protect her from her father's romantic madness.[26] Every mathematician knows that Galois was killed in a duel over a woman—sometimes identified as "Stéphanie"—but Alexander argues that "the affair, such as it was, did not go well, and [. . .] Stéphanie was trying to distance herself from the intense young mathematician."

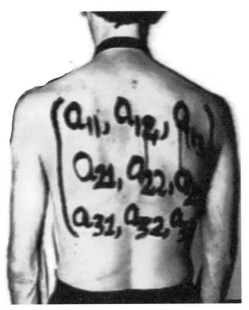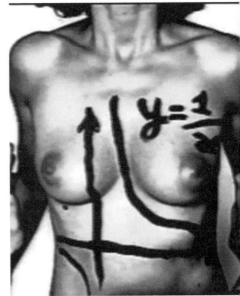

Figure 6.6. His and hers matching love formulas: from the short film *Riducimi in forma canonica* [Reduce me to canonical form] by Monica Petracci, 2000.

Galois' death, for Alexander, was the culmination of a pattern of arrogant and self-destructive behavior rationalized as uncompromising devotion to the truth. In the letter he addressed "to all republicans" on the eve of his death, Galois can only "repent having told a baneful truth to men who were so little able to listen to it calmly . . . I take with me to the grave a conscience clear of lies."[27] Alexander argues that the transition from the Enlightenment to the romantic ideal in the lives of mathematicians parallels a transformation of the subject matter. ". . . [T]he new mathematicians turned away from the Enlightenment focus on analyzing the natural world to create their own higher reality—a land of truth and beauty governed solely by the purest mathematical laws."[28] In his own mind and in the legend that grew up around his life, Galois died for the sake of what *Rites'* synopsis called "Absolute Truth and Beauty," just as Frenkel's Mathematician died for *istina*. But where, as Alexander continued, "[the romantic], whether poet, artist, or mathematician . . . was . . . an otherworldly creature who belonged in a better and truer world than ours," the Mathematician of *Rites* devotes his life's work to discovering a "Formula of Love" that "would benefit people, bringing them eternal love, youth and happiness."[29] Is Frenkel's Mathematician's romantic martyrdom for truth in contradiction with his materialist objectives, like the Enlightenment practitioners of mathematics as a "science of the world that had its roots planted firmly in material reality"?[30] Or do Frenkel and Graves agree with the Russian mathematician A. N. Parshin, who wrote, referring to the mathematics of elementary particles, that *"The deeper we plunge into the material world, the further we move from it in the direction of the ideal world"*? [31]

• •

Hal: [. . .] Mathematicians are insane. I went to this conference [. . .] last fall. I'm young, right? I'm in shape. I thought I could hang out with the big boys. Wrong. I've never been so exhausted in my life. Forty-eight straight hours of partying, drinking, drugs, papers, lectures . . .

Catherine: Drugs?

Hal: Yeah. Amphetamines mostly. [. . .] Some of the older guys are really hooked. [. . .] They think math's a young man's game. Speed keeps them racing, makes them feel sharp. There's this fear that your creativity peaks around twenty-three and it's all downhill from there. Once you hit fifty it's over, you might as well teach high school.[32]

A totally unfounded fantasy, and if you were expecting any serious gossip at this point, I'm afraid you're going to be disappointed. On the contrary, there seems to be a growing consensus that, in spite of the persistent public fascination with mathematics, it's not among mathematicians that you'll find the best parties and that life is more fun in the company of dancers, philosophers (Anglo-American or continental), hedge-fund managers, fashion designers, biomedical engineers, theater critics and/or performers, historians, industrialists, Russian Orthodox theologians, or anyone involved with the movies. In the new world of public-private partnerships, leading European research institutions are also encouraged to follow the dark side into its realm of dream and magic. Consorting with the men and women at the summit of the socioeconomic firmament is consistent with the injunction raining down on us from all quarters to tie our work more closely to the needs of production, especially the production of wealth. The culture industry, meanwhile, deals primarily in the production of novelty, and indications are that the added value of mathematics, as a cultural *signifier* of indeterminate scope, has yet to be exhausted. Aronovsky's *Pi*, whose plot is driven by Wall Street's search for a (deterministic) formula to predict the future course of stock prices, stands as a reminder that the "forces of Evil" will not invest in our work without expecting a substantial return. When a mathematical community still largely faithful to the romantic ideal crosses paths with the masterminds of enlightened materialist ruthlessness, will the result be a second Enlightenment mind-body synthesis, a Faustian bargain signaling a renewed appreciation of one another's desirability, with the "brilliant women of the court" replaced in the updated version by shapers of taste and prophets of economic competition, and with the reflection on human freedom of a Rousseau or a Condorcet replaced by obsessive attention to the bottom line?

In recent years, mathematics has made its mark on the inexhaustible witches' sabbath that is Paris nightlife. Harvard mathematician Benedict Gross passed through Paris in the fall of 2008 for the opening of his collaborative installation with multimedia artist Ryoji Ikeda at *Le Laboratoire*, a "creative space" near the French Ministry of Culture "dedicated to experimental collaboration between artists and scientists," directed by the author and biomedical engineer David Edwards. After exchanging messages for a year on "mathematics, infinity, the sublime,"[33] Gross and Ikeda settled on an installation of two horizontal monoliths, each covered with more than seven million tiny digits representing in one case a prime number, in the other a "random" number.

Biomedical engineers have the edge on parties, if the opening reception at *Le Laboratoire* is any indication. Wandering across the dimly lit exhibition space—the two numbers were illuminated from above and their digits could be inspected only with the help of a magnifying glass— the few mathematicians in attendance, Ed Frenkel and I among them, gradually found one another in a crowd of hundreds of the experimental collaboration's boisterous and exuberantly stylish celebrants. An office door slid open before us, and we were ushered in to join the collaborators. While Edwards fiddled with a magnum of VIP champagne, Ikeda dropped to his knees at the feet of Jean-Pierre Serre and announced, "For me you are a rock star!"

Figure 6.7. Thurston and assistant wrapping Dai Fujiwara in a knot (my photo); still from http://www.isseymiyake.com/isseymiyake_women/ (photo Frédérique Dumoulin/Issey Miyake).

Parisian mathematics had no official representation at *Le Laboratoire* but was in evidence at the March 2010 opening of another Japanese-American collaboration, this time between the late topologist Bill Thurston and Dai Fujiwara, creative director for the Issey Miyake fashion house.

"[Y]ou did not need a top grade in math to understand the fundamentals of this thought-provoking Issey Miyake show. . . ."[34] Fujiwara had contacted Thurston after learning about Thurston's Geometrization Conjecture and its connection with the Poincaré Conjecture. Fashion designer and mathematician, it turned out, both used the peeling of an orange to help their students understand geometry. "We are both trying to grasp the world in three dimensions," Thurston told the AP. "Under the surface, we struggle with the same issue."[35]

Alerted by a message from IHP director Villani (who starred as himself in the 2013 film *Comment j'ai detesté les maths*—we've seen his own fashion statements favor a nineteenth-century romanticism), my colleagues and I arrived in time to sample the hors d'oeuvres (American and topological: doughnuts, pretzels, bagels) and to register the shocked expressions of the insiders, too spontaneous to be concealed, as they witnessed the breaching of a fortress of Parisian fashion by the hopelessly unfashionable. Thurston (the "coolest math whiz on the planet," according to an admirer of his YouTube appearance[36] with Fujiwara) modeled an original Miyake blazer created for the occasion at the show and again at the reception. Playing neither the natural man nor the romantic hero ("I can't believe this mathematics guy. He's so . . . not like what I expected."), Thurston told his interviewer that "Mathematics and design are both expressions of the human creative spirit."[37]

Like the authors of *Rites*, Thurston invoked truth and beauty in his essay for the show. "The best mathematics uses the whole mind," he insisted, "embraces human sensibility, and is not at all limited to the small portion of our brains that calculates and manipulates with symbols." Thurston was back in Paris in June 2010 for the ceremony organized by the Clay Mathematics Institute to honor Grigori Perelman for his solution to the Poincaré and Thurston conjectures. A grandson of Poincaré was on hand, and the pantheon of the last fifty years of geometry, with only a few notable exceptions, had been assembled for the occasion, which was extensively covered by the French media (though not more than the *Laboratoire* show). One by one, the distinguished senior geometers stood up to praise the absent Perelman, who had not yet decided to refuse the Clay Institute's million dollars. Only Thurston took the opportunity to express sympathy for Perelman's defense of the romantic ideal against the onslaughts of the good intentions of *megaloprepeia*:

> Perelman's aversion to public spectacle and to riches is mystifying to many. . . . I want to say I have complete empathy and admiration for his inner strength and clarity, to be able to know and hold true to himself. Our true needs are deeper—yet in our modern society most of us reflexively and relentlessly pursue wealth, consumer goods and admiration. We have learned from Perelman's mathematics. Perhaps we should also pause to reflect on ourselves and learn from Perelman's attitude toward life.[38]

Most of the spectators at *Rites*' first screening were artists of some sort, rather than mathematicians, and were apparently drawn from Reine

Graves' extensive list of Facebook friends. Once again, it was not hard to identify the mathematicians in the crowd at the postscreening reception, but the contrast was not as jarring as at Thurston's fashion show. On the contrary, the champagne was a democratic vintage and everyone in the *Rites* audience seemed to share a rejection of the *couture* mindset—the artists by design, the mathematicians by indifference. Communication across the cultural divide was cryptic but unstrained. One of Graves' three friends speculated that *les maths sont là pour exprimer l'essence de la nature* [the math is (in the film) to express the essence of nature]; another saw *une beauté calligraphique* [a calligraphic beauty], analogous to the calligraphy at the center of the Noh stage, in the tattooed Frenkel-Losev-Nekrasov formula. Number theorist Loïc Merel, on the other hand, thought the film was an exploration of "how to preserve knowledge" but that the question was not taken seriously; the film's language was that of a *conte de fées* [fairy tale].

To mark the conclusion of a year spent in Paris as the occupant of the *Chaire d'Excellence de la Fondation Sciences Mathématiques de Paris*, Frenkel organized a mathematical conference entitled *Symmetry, Duality, and Cinema* at the Institut Henri Poincaré. Four lectures on mathematical topics of interest to Frenkel were followed by another projection of *Rites d'Amour et de Math*. At the champagne reception that followed,[39] I took notes while Gaël Octavia, the *Fondation*'s public relations specialist, asked Graves why she decided to make a film about mathematics.[40] Without hesitating, Graves, whose motto is *ne jamais avouer* [never confess], gave the very best possible answer. Mathematics, she began, is *un des derniers domaines où il y a une vraie passion* [one of the last areas where there is a genuine passion]. Cinema, according to Graves, is dominated by economics; so is contemporary art. Mathematics, like a very few other activities—she mentioned physics and sculpture—is practiced without complacency [*sans autosatisfaction*]; instead there is a true *exigence au travail* [demanding work ethic]. Mathematicians seek to *percer le mystère*. You can see it at once in *l'oeil qui brille* [the eye that gleams].

Let us gaze back a moment into the gleaming mathematical eye that Graves finds so compelling. Amir Alexander describes a portrait of Abel by the Norwegian painter Johan Gørbitz:

[I]t is the young man's eyes that grab our attention and draw us irresistibly toward them. Dark and intense, . . . [t]hey burn with a fire that suggests deep passions of the soul and profound insights of the mind. Their gaze shoots out from the painting's surface . . . focused not on us but on a distant point on the horizon . . . the portrait is of a man . . . absorbed by his own inner flame and a vision he perceives far beyond.[41]

And looking back at us from romanticism's troubled borderlands, the eyes of Pechorin, Lermontov's *Hero of Our Time*, "shone with a kind of phosphorescent gleam . . . which was not the reflection of a fervid soul or of a playful fancy, but a glitter like to that of smooth steel, blinding but cold."

Alexander's final chapter compares the portraits of Abel and Galois to the self-portraits of early romantic painters A. Abel de Pujol and O. Runge, as well as to contemporary portraits of Keats and Byron—always the same *"oeil qui brille."* In contrast, enlightenment mathematicians like D'Alembert and Johann Bernoulli resembled nineteenth-century physicists like Helmholtz and Lord Kelvin, "successful men of the world, showing no hint of the morose sentimentality that became a hallmark of the [romantic] mathematical persona" (pp. 260–61).

"In the film that [Frenkel and Graves] envisioned." writes Henric,

the central character . . . fights not for honor [in contrast to Mishima's hero, M.H.], but, like his ancestors in science and philosophy, for truth. So here is the question, philosophical, religious, political, moral: should one sacrifice himself and die for the truth?

Yes, said Socrates, Giordano Bruno, Michel Servet . . . , and all scholars and thinkers who did not compromise with the truth and preferred death to disowning it. No, said the philosopher Kierkegaard.

In his own mind, Galois had no doubt that he belonged to the first group. Stripped and doomed like a gladiator, Frenkel's Mathematician is a martyr following in the footsteps of Galois or Socrates . . . or of Hypatia of Alexandria, the third mathematician, with Archimedes and Frenkel, whose biography features a prominent nude scene. Along with another entirely gratuitous nude scene, this one, culminating in her martyrdom, is duly included in Alejandro Amenabar's film *Agora*, starring Rachel Weisz as the scientist and philospher of late antiquity. Amenabar's depiction of Hypatia's murder is lighthearted in comparison with Gibbon's version, in which the crowd literally inserts the knife between the victim's spirit and mortal flesh as if to punish the latter for its presumptuous affirmation of its primacy:

> On a fatal day, in the holy season of Lent, Hypatia was torn from her chariot, stripped naked, dragged to the church, and inhumanly butchered by the hands of Peter the Reader and a troop of savage and merciless fanatics: her flesh was scraped from her bones with sharp oyster-shells and her quivering limbs were delivered to the flames (Gibbon, *Decline and Fall of the Roman Empire*, Chapter XLVII).

It hardly matters that there is no basis in the scanty historical record for the film's contention that Hypatia died for her defense of scientific rationalism in the face of religious fanaticism.[42] On the contrary, as with Alexander's deconstructive reading of the Galois legend, Amenabar's film is less interesting for its history of Hypatia than for what it tells us about our cultural moment: that we need a martyr to truth and beauty, or to the "science et []amour" of Leconte de Lisle's 1847 poem *Hypatie*. The following verses could serve as a point-by-point illustration of Alexander's characterization of the mathematician "absorbed by [her] own inner flame and a vision [she] perceives far beyond":

> *Sans effleurer jamais ta robe immaculée,*
> *les souillures du siècle ont respecté tes mains:*
> *tu marchais, l'oeil tourné vers la vie étoilée,*
> *ignorante des maux et des crimes humains.*
> *L'homme en son cours fougueux t'a frappée et maudite,*
> *mais tu tombas plus grande!* *

There is a kind of satisfying symmetry that helps us understand what is peculiar about mathematical madness as it appears in the wider culture. Those mathematicians who turn their eye, gleaming or otherwise, to the "starry life," neglecting the material world, are martyred in the flesh like the Hypatie of Leconte de Lisle's poem, or like Archimedes or Galois, victims of their devotion to their science. Those who assert the

---

* "Never brushing your immaculate robe / the century's stains respected your hands / you walked, your eye turned to the starry life/ignorant of human evils and crimes / Man in his impetuous course struck and cursed you / but you fell all the greater!" *Hypatie*, Leconte de Lisle.

Figure 6.8. Hypatia turning her eyes to the "starry life." The image appears in Kids Britannica online and elsewhere (Photos.com/Thinkstock).

primacy of the human mind, like Amenabar's Hypatia, or of the individual mind, like Frenkel's Mathematician, are again martyred in the flesh, victims this time of the "forces of Evil" that seek to extend their control over the material world. Madness, on the other hand, is martyrdom in the spirit, the fate of mathematicians who focus their minds too closely on what must remain unseen, like the protagonists of *Pi* or the Georg Cantor in Amir Aczel's *The Mystery of the Aleph*,[43] or who pursue their intuitions so far into abstraction that they cannot find their way back, like the Gweneth Paltrow character's father in *Proof* or Cantor (again) in *Logicomix*. With a bit of imagination, we can even squeeze these alternatives back into that handy structuralist double binary opposition scheme, as follows:

Lose touch with shared material world
(Galois, Archimedes)

Lose touch with shared mental world
(*Proof*, *Logicomix*)

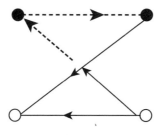

Oppose material powers
(Amenabar's Hypatia, Frenkel)

Oppose prohibition limiting mental powers
(*Pi*, *Mystery of the Aleph*)

But there is also a persistent asymmetry that brings us back to the scandal with which this essay began, insofar as it is possible to lose one's mind while one's body remains intact, while profane history records no instance of the reverse . . . except perhaps when the spectator identifies with the martyred characters on screen:

> [T]he spectator is absent from the screen. He cannot identify to himself as object. . . . In this sense, the screen is not a mirror. But this other whom he observes and hears in the film is a psychic prosthesis, an imitation with which it is possible to identify.[44]

In western metaphysics, the specialist in survival of the spirit despite martyrdom in the flesh is Jesus, who famously declared, "I am truth." In the Greek of John 14:6, the word for truth is *aletheia*, translated into Russian as *istina*. This is how the word becomes central to Russian Orthodox theology.[45] Readers familiar with *pravda* as the Russian word for "truth" may have been wondering why Frenkel instead chose the word *istina* for the icon hanging beside Mariko's futon. The answer is that the two terms have different roots, with *pravda* in the semantic complex, including the words for law, justice, and rules, while *istina* carries the sense of the adjective *istinnii*, genuine; thus in Lermontov's novel, Pechorin's lover could write "no one can be so truly [*istinno*] unhappy as you, because no one endeavors so earnestly to convince himself of the contrary." The *pravednik* is the righteous man one occasionally encounters in the Bible, but we have seen that *istina* is the religious truth Jesus claims to embody; *istina* is also the technical word for mathematical truth, as in Gödel's theorem. Here's Maria Kuruskina's advice, on the AllExperts Web site, for those of you wondering which word to choose for your next tattoo:

> 'Istina' is a great deal more pathetic and I'd say lofty or elevated than 'pravda'. In sentences like "Neo, you're the One because you know the Truth" . . . Russians would use 'istina'.
> 'Pravda' is neutral. It's used in phrases like "you must tell me the truth".
> For a tattoo i'd recommend 'istina'.[46]

What survives Frenkel's martyrdom is his formula, *istina*, tattooed above his lover's[47] womb. I will resist the temptation to read this as a promise of the mathematician's spiritual reincarnation and will instead turn to the practical, cinematic, and theological question of how mathematical truth can burst into our lives with the devastating consequences illustrated in *Rites*.

• •

> The true (*istinniy*) artist does not want his own truth, whatever the cost, but rather the beautiful, objectively beautiful, that is the artistically incarnated truth (*istina*) of things. . . . If the truth is there, then the work establishes its own value. (Pavel Florensky)[48]

I have never consciously attempted to find an equation for love, nor have I knowingly appeared in a movie. But in preparing this chapter, I learned that my namesake, the actor Michael Harris, starred as the evil half of a pair of identical twins in the film *Suture,* described as a meditation on mind-body duality and featuring a plastic surgeon named Renée Descartes. And on numerous occasions during the feverish year when I wrote my PhD thesis, I woke up vainly trying to recall the formula or the diagram that in my dream had solved one of my many pressing material problems, often with love as primary focus.

Presenting his film at the IHP, Frenkel confirmed Merel's intuition, saying that the film had been conceived as an "allegory" and a "fairy tale." That may be accurate as a description of the film's tone, but not of its narrative arc: a typical (Russian) fairy tale has a thirty-one-part structure, according to Vladimir Propp's classic study *Morphology of the Folktale*, and usually a happy ending to boot. Nor, the *Liebestod* notwithstanding, is *Rites of Love and Math* a tragic romance on the model of *Tristan and Isolde*. In Mishima's *Yûkoku*, Reiko announces to her officer "I know how you feel, and I will follow you wherever you go," and she keeps her word, dying over the beloved warrior's body, as Isolde did before her. No such symmetrical closure concludes *Rites*. Condemned as a sort of human parchment to outlive her fallen lover, Mariko must maintain her mind as a shrine to the lost Mathematician while her body preserves the fatal formula for eternity. Frenkel's second cinematic venture, a screenplay with Thomas Farber entitled "The Two-Body Problem," promises a parade of women in bikinis as a "visual trope,"[49] but the inconveniently tattooed Mariko will never be among them.

Attempts to reconcile the plot with what life has taught me about the material world confronted my mind with unwelcome questions. The Mathematician's mind is safely beyond the reach of the "forces of Evil," but how will Mariko, her skin still stinging from the pain of the tattoo needle, dispose of his body without attracting their attention? Because the Mathematician "wants his formula to live," Mariko has consented to keeping her own body healthy, her skin taut, forever; but how will she explain the formula to her doctor? And what will she tell them at the gym?

Focusing on such questions to fill in gaps in limited information, a precious logical skill in the hands of mathematical crime fighter David Krumholtz in the TV show *Numb3rs* or of Elijah Wood and John Hurt in the film *The Oxford Murders*,[50] is useless in the face of allegory and totally inappropriate on the Noh stage. But if love in *Rites* is to be more than a symbol, a four-letter word illustrated in red and gold and conventional iconography, the viewer cannot help wondering how the equation is actually supposed to work. Eugene Wigner's canonical version of the mind-body problem concerns "The Unreasonable Effectiveness of Mathematics in the Natural Sciences"; mathematics stands at the mental pole of the antinomy, the universe at that of the body. What troubles me is this: how am I supposed to understand the unreasonable erotic effectiveness of the Frenkel-Losev-Nekrasov love equation?[51]

Setting aside the odd interpretation on the Russian website *Woman Today*,

> [i]t is fully realistic, according to Frenkel, to predict life expectancy precisely, if you have at your disposal the statistics of a person's sexual acts.[52]

I ask: are the equation's "magic powers" like the magic formulas in Marlowe's *Doctor Faustus*?

> The iterating of these lines brings gold;
> The framing of this circle on the ground
> Brings whirlwinds, tempests, thunder and lightning;
> Pronounce this thrice devoutly to thyself,
> And men in armour shall appear to thee,
> Ready to execute what thou desir'st (Mephistopheles to Faust, Scene V).

Does the love formula bring new love into being or reveal existing love that is concealed? Is it actually a sex formula and can it, like a date-rape drug, force "love" on the unwilling? Or is that one of the "sinister goals" of the forces of Evil? Is the Frenkel-Losev-Nekrasov equation of figure 6.9 a record postfacto that explains how your seductions fit into a pre-ordained order, like Pasolini's *Teorema*? Or is it the profane translation of a "Let there be love!" from an alternative Scripture? And does it act on minds, or bodies, or both? Does it act on all bodies or minds at once or on a few at a time? (And in that case, how many bodies and how many minds?)

Writing having stalled while I puzzled over these questions, I sent a partial draft to Amir Alexander, who wrote back suggesting I read Frenkel's character not as one of the archetypes of Alexander's book but rather as a "classical Renaissance magus," like Dr. Faustus. "A magus is a manipulator of abstract signs, . . . not for their own sake (as is the case for the romantics) but rather for earthly power. He may use it for the good, as Frenkel's hero hopes to do, but his power is always suspect. . . ."

Before he became a filmmaker, Ed Frenkel was known as a brilliant product of the incomparable Moscow mathematical school[53]—one of the last mathematicians to be trained in that school before it broke up, along with the Soviet Union, when many of the best-known Russian mathematicians left for Europe, Israel, and (especially) North America. In their

$$\int_{\mathbb{CP}^1} \omega F(qz, \overline{q}\,\overline{z}) = \sum_{m,\overline{m}=0}^{\infty} \int_{|z|<\varepsilon^{-1}} \omega_{z\overline{z}} z^m \overline{z}^{\overline{m}} \mathrm{d}z\mathrm{d}\overline{z} \cdot \frac{q^m \overline{q}^{\overline{m}}}{m!\overline{m}!} \partial_z^m \partial_{\overline{z}}^{\overline{m}} F \bigg|_{z=0}$$

$$+ q\overline{q} \sum_{m,\overline{m}=0}^{\infty} \frac{q^m \overline{q}^{\overline{m}}}{m!\overline{m}!} \partial_w^m \partial_{\overline{w}}^{\overline{m}} \bigg|_{w=0} \cdot \int_{|w|<q^{-1}\varepsilon^{-1}} F w^m \overline{w}^{\overline{m}} \mathrm{d}w\mathrm{d}\overline{w}.$$

Figure 6.9. The Frenkel-Losev-Nekrasov Love Equation, conscious (from Frenkel et al. 2011). For the embodied version, see the trailer to *Rites of Love and Math*, http://vimeo.com/15492339, 1′44″.

important book *Naming Infinity*, Loren Graham and Jean-Michel Kantor trace the rise of Moscow mathematics to the influence of the Russian Orthodox doctrine of *name worshipping* [*imyaslavie*], whose theology and religious practice are based on the postulate that "The name of God is God Himself."[54] Treated as a heretical sect by the Orthodox hierarchy and persecuted by Soviet authorities, the name worshippers included Dmitri Egorov and Nikolai Luzin, considered by Graham and Kantor to be the founders of the Moscow school, as well as their colleague, the mathematician, engineer, philosopher, Orthodox theologian (and priest)—and political martyr[55]—Pavel Florensky.

Just as the name worshippers believed that the repetition of the name of God in the Jesus prayer would bring them into the presence of God, the Moscow mathematicians, according to Graham and Kantor, maintained that mathematical objects were brought into being in the course of giving them names. It would appear from this description that name worshipping goes further in breaking through the mind-body barrier than mathematics, whose naming process remains on the mental side. But Moscow mathematician A. N. Parshin, a contemporary interpreter of Florensky, views both word and meaning in purely mental terms:

[T]he meaning of a word represents itself as a wave of (intelligible) light, located in supersensible space. . . . the perception of a word is not merely connected to the perception of light, it *is* the perception of light, but a mental light.

And

vision arises from the sense of touch. Florensky liked this idea very much. The primary sense, the primary mode of knowledge, is touch of the surface, while vision is touch by means of the retina. . . . we may conclude that re-flection in the act of knowing takes place by means of reflection with respect to the surface of the body, that is all the surface of the body is a *mirror*.[56]

The name worshippers pronounced the name of God—many more times than thrice—not for earthly power but to bring about a spiritual experi-ence. The Moscow mathematicians went beyond manipulation of their abstract signs by giving them names. The theory behind the "love equa-tion" is analogous: invocation of the formula suffices to bring "eternal love" (as well as youth and happiness).

From this perspective, *Rites* is not a fairy tale but rather an allegorical expression of theological mysticism. But everything about the film's imagery suggests that its sphere of action is not limited to the mind but, like the Faust legend, encompasses the realm of the senses in its full materiality, not merely the realm of mental perception to which Parshin alludes. Is this the transgression for which Frenkel's Mathematician must pay with his life? Or are the filmmakers reaffirming the principle the Abbé Suger had inscribed on the doors of the Basilica of St. Denis?

Mens hebes ad verum per materialia surgit

[The dull mind rises to truth through that which is material.][57]

Florensky challenges our dull minds with a series of cryptic defini-tions of the body. Each definition illuminates a different aspect of the mind-body problem while reinforcing Parshin's conception of the body as mirror: "The body is the most spiritualized substance and the least active spirit, but only as a first approximation." "The body is the realiza-tion of the threshold of consciousness." "The body is a film that sepa-rates the region of phenomena from that of noumena."[58] Of particular relevance to *Rites* is the following:

That which is beyond the body, on the other side of the skin, is the same striving for self-revelation, but hidden to consciousness; that which is on this side of the skin is the immediate presence of the spirit, and thus not extended beyond the body. In being conscious, we clothe ourselves [literally "robe," as in the church], and ceasing to be conscious, we lay ourselves bare.[59]

The passage is obscure, but I will risk a possible interpretation. Parshin has written that "reflection with respect to" the mirror that is the surface of the body "exchanges the internal and the external."[60] Florensky seems to be saying that when we are (figuratively speaking) dressed, the surface of the body acts as in Parshin's model, exchanging the region of noumena—the intelligible [*umopostigaemii*]—with that of phenomena—the sensual [*chuvstvennii*], both of which are located within the mind; this is what it means to be conscious. And when we're undressed—well, when consciousness is absent—the body is doing its thing, and who knows what *that's* about?

I mean, who *really* knows?[61] Who can know well enough to express it all in a formula? Maybe what the existentialists couldn't understand in the monographs they caressed was what their formulas and arcane language had to do with human freedom, responsibility, and desire. Lacan's heckler was right: "*L'homme ne peut pas se résoudre en équation* [man cannot be expressed in an equation]." Nor could that couple in figure 6.6 really be reduced to canonical form. At least that's what Florensky thought. Writing about love, he explains what he sees as Spinoza's category mistake (not so different from the category mistake in philosophy of Mathematics):

> For love is directed toward a person, whereas desire is directed toward a thing. But the rationalistic understanding of life does not distinguish . . . between a person and a thing . . . it has only one category, the category of thingness, and therefore all things, including persons, are reified by this understanding, are taken as a thing, as *res*.[62]

You might say—and you would hardly be the first to say it (see chapter 8)—that love, as a subjective experience, is exactly what cannot be expressed in a formula. That's what makes it subjective rather than a variant of "thingness." The body can be tattooed from head to foot with an owner's manual, rules for operation (*pravila expluatatsii*) sent down from the mental heavens, but neither alone nor in combination will they produce the experience of true love (*istinnaya lyubov'*) that springs out with our bodies from the earth. Frenkel's Mathematician's Ultimate Formula of Love, subordinating the body to the calculating mind, can work only by means of a category mistake. In other words, it can't work. And that's why the Mathematician has to die. Otherwise no one would ever believe it possibly could have worked!

Florensky saw not only love but also the relation to truth [*istina*] as essentially personal. "Christ said: I am Truth," Parshin explained to me; "the truth is a personality, not a mere object." In Florensky's philosophy, "an act of knowledge is a communication or relation, even a kind of 'friendship' between the two persons, the one who studies and the one who is studied." For pure mathematicians in the romantic mold, Parshin's mirror model—physical reality as mediation between sensual experience and the intelligible world, our gleaming eyes turned, like Hypatie's, toward the stars, there to find the "land of truth and beauty" of which Alexander wrote—seems just about right. The mirror is mirrored by the words of Simon McBurney, director of London's *Théâtre de Complicité*, explaining how mathematics works as a metaphor for love when no formula is possible:"Infinity is a way to describe the incomprehensible to the human mind. . . . In a way it notates a mystery. That kind of mystery exists in relationships. A lifetime is not enough to know someone else. It provides a brief glimpse."[63]

Goethe reincarnated the magus Faust as a romantic hero, as Mephistopheles was well aware:

> Fate hath endow'd him with an ardent mind,
> Which unrestrain'd still presses on for ever,
> And whose precipitate endeavour
> Earth's joys o'erleaping, leaveth them behind.*

And though Faust eventually died, his Truth survived him, and the angels cheated Mephistopheles of his soul:

> Those that damn themselves,
> be healed by Truth;
> so that from Evil
> They gladly release themselves.†[64]

---

* *Ihm hat das Schicksal einen Geist gegeben,*
Der ungebändigt immer vorwärts dringt,
Und dessen übereiltes Streben
Der Erde Freuden überspringt.
† *Die sich verdammen,*
Heile die Wahrheit;

Sometimes I think the whole mad/martyr/mathematician angle is a ruse to trick the forces of Evil, to cheat the boardroom Mephistopheles who are skeptical that our efforts will bring them tangible returns. Our readiness to sacrifice our minds and bodies to our vocation is the ultimate proof that what we are doing is important, even if—as far as any observer can see—we never leave our side of the mirror.

Daß sie vom Bösen
Froh sich erlösen.

P. A.: Are you deliberately trying to be thick?

N. T.: How nice of you to stop by! But I'm afraid I have no idea what you are talking about.

P. A.: Frenkel is not claiming his formula actually induces or explains or calculates love in any way. It's a metaphor, silly. As Florensky might have said, his "act of knowledge" is a kind of "friendship" between the mathematician and his formula. The feeling of the mathematician about his formula is analogous to the feeling of love.

N. T.: *Aha*! So Frenkel is saying that mathematics feels like love. Are you convinced?[65]

P. A.: At least in both cases there is a feeling of transcendence of one's state as an individual. That's why love is described as an out-of-body experience. The proof is that I can represent love on the stage to people who are not bodily present.

N. T.: Now you're trying to compete with me in paradox. Frenkel did feel compelled to represent bodies on the screen.

P. A.: And you felt compelled to write more than was strictly necessary about those bodies. We should agree that you can have an out-of-body experience only if you have a body to begin with.

N. T.: Just as you need a mind to go out of if that's where you intend to go.[66]

P. A.: That's just the challenge mathematicians have lately been posing to performing artists, to represent a personality, perhaps with an exaggerated work ethic, but otherwise totally nondescript, on the verge of insanity.

N. T.: From what I've heard, it shouldn't be such a challenge for a performing artist to impersonate a borderline personality. But since you brought it up, you might be interested to know that transcendence is also amenable to mathematical analysis.

P. A.: I suspect you are trying to reassert your seniority in matters of paradox. For me, the transcendent is precisely whatever escapes rational analysis of any kind. As Levinas wrote, for example, "the idea of infinity is a thought which at every moment thinks more than it thinks."[67]

N. T.: Be patient; then you can decide whether or not your transcendence and mine have any hope of making contact.

P. A.: It would have been pointless to have gone out of my way at such an hour if there were no hope at all.

N. T.: If that's how you feel, we should get your disappointment out of the way immediately. As far as I can tell, there is exactly one *transcendence theory*, which as it happens is a branch of number theory. But let me switch off the lights and I'll go back to the big screen.

chapter β.5

How to Explain Number Theory at a Dinner Party

IMPROMPTU MINISESSION: TRANSCENDENTAL NUMBERS

We have been talking about roots of polynomial equations, and I think you are now willing to admit that the root of a polynomial $f(x)$ is an answer to a question about numbers and is, therefore, a number. For example, if

$$f(x) = x^3 - px + q,$$

then the question

For what α is $f(\alpha) = 0$?

has three answers. You remember we gave formulas for these answers and called them Olga, Masha, and Irina. You also remember that there is usually no formula when f is a polynomial of degree 5 or more; nevertheless, the roots are answers to a question similar to the one we just considered and, therefore, are numbers. According to the terminology that goes back to Leibniz, roots of polynomial equations are called *algebraic numbers*; all the other numbers are called *transcendental numbers*, and transcendence theory is the study of transcendental numbers.

The first obvious question is this:

Question 1: Are there any transcendental numbers?

The first transcendental number was exhibited by Joseph Liouville in 1844; it was a number λ he concocted for just that purpose, but it answers Question 1 and, therefore, certainly qualifies as a number. When I say Liouville "exhibited" the number, I mean that he wrote down a description and used this description to show (very ingeniously) that there is *no* polynomial f for which $f(\lambda) = 0$.

If all transcendental numbers were as artificial as Liouville's λ, transcendence would never have grown into a mathematical theory. Methods were soon developed, however, to show that some familiar numbers are transcendental. For example, $e = 2.718.\ldots$ (the basis of the natural logarithm) was proved to be transcendental in 1873; $\pi = 3.141.\ldots$ (the ratio of the circumference of a circle to its diameter) was proved to be transcendental in 1882, and the list continues to grow. The easiest ways to exhibit transcendental numbers is by studying *transcendental functions*—like the cosine and sine functions familiar from trigonometry and the exponential function. Leibniz introduced the word *transcendence* when he proved that such functions are not algebraic *functions*, and his proof made use of the fact that they are solutions to equations—differential equations, not polynomial equations.

There is a notion of algebraic differential equations that expands the class of numbers that admit finite descriptions far beyond the class of solutions of polynomial equations. Connected to this is a vast framework of conjectures predicated on the expectation that transcendental numbers have a structure and are not simply characterized negatively by their failure to be algebraic. One of the most sophisticated conjectures in transcendence theory belongs to Grothendieck's theory of *motives*, to which we return in the next chapter. There is practically no evidence for this conjecture and not the slightest hope that it will be resolved in the next few centuries, but my sense is that everyone working in the field would like it to be true.

In the meantime, Cantor's theory of orders of infinity had shown that practically all numbers are transcendental. Cantor had introduced the notion of the *cardinality*, or size, of an infinite set. His methods show that the sets of positive integers, of even integers, of all integers, and of rational numbers all have the same cardinality: they are all *countable*, or *denumerable*. The set of real numbers, on the other hand, is not denumerable; it is a much bigger set. Cantor established this by what we can call a *trick*. A countably infinite set is a set whose members can be listed in numerical order: the first one, the second one, and so on (forever . . .). Cantor reasoned by contradiction: assuming the real numbers could be so enumerated, he exhibited a member that was not on the list.

Just about every popular book about mathematics explains Cantor's diagonalization trick,[1] so I'll omit the details. The upshot is that Cantor's methods show easily that the set of algebraic numbers is countable, which implies that the transcendental numbers are uncountable. You can remove all the algebraic numbers from the number line and hardly notice the difference. But Cantor's ideas are much more disturbing. The transcendental numbers relevant to Grothendieck's conjectures are all related to algebraic differential equations and are, likewise, examples of what Maxim Kontsevich and Don Zagier call *period numbers*.[2] These are called *periods* for the same reason a number like

$$\int_2^3 \frac{dx}{\sqrt{x^3 - 25x}}$$

is called a period—in other words, for a historic reason, the mention of which would add nothing to the present discussion. The period numbers are in turn examples of what Alan Turing, in his paper on machine computation,[3] called *computable numbers*. The latter are the numbers that can be calculated to any degree of accuracy desired by the class of theoretical calculating devices Turing called "universal computing machines" and we call *Turing machines*.

Turing created the theoretical foundations of computer programming in his landmark paper in the process of defining his computable numbers. In so doing, he also proved that his class of computable numbers is a countable set. This is because you need a computer program in order to compute a computable number, but a computer program is of finite length, so you can enumerate all the computer programs by listing them in a kind of modified alphabetical order. Cantor's trick then shows that the set of all real numbers is much bigger.

In fact, no matter how we choose to describe numbers, the set of descriptions will be countable. (By a *description* I mean something more or less informal, like the parenthetical descriptions I provided earlier for *e* and π. For example, Gregory Chaitin described a number he calls Ω that is designed not to be computable, but since he writes down a description, it can be given its place in modified alphabetical order.) This means that we simply have no way of talking about practically all numbers—not individually, in any case.[4]

• •

P. A.: I was hoping you would say that transcendental numbers answer transcendent questions, because I have a few of those.

N. T.: Well, Vladimir Voevodsky gave a public lecture in which he said that "first-order arithmetic is totally a creation of human minds, there is absolutely no reason for transcendental forces to . . . ensure its consistency by transcendental means."[5] Since he, and not Frenkel, is performing in the next chapter, I would rather say that the indescribable numbers are the answers to the questions that can't be asked. But one has to be careful about what one means by that.[6] I was careful not to describe the indescribable numbers, though I was in no danger of describing them by accident.

P. A.: I was thinking of Levinas's "idea of infinity" . . . but also about the transcendent experience of simultaneously being a character in a play, following the rules set down by the author, and a performer endowed with human volition. Everyone who performs on stage is constantly tempted "to wonder at unlawful things . . . to practice more than heavenly power permits," or at least to practice more than what's written on the page. And we need to yield to that temptation. Otherwise we're just meat puppets.

N. T.: Doesn't that very rational account of the experience contradict your definition of transcendent as "whatever escapes rational analysis of any kind?"

P. A.: What I just said doesn't count as a rational account of the experience any more than calling a number indescribable counts as a description.

N. T.: In his lecture, Voevodsky admitted the possibility of transcendental knowledge of *natural* things. But he didn't give any details, so if that's the sort of thing you have in mind, I'm afraid we'll have to work it out on our own.

P. A.: Let's finish with numbers, then. On the one hand, you explain how a description can specify a number; on the other hand, you talk about numbers that can't be described at all. But how do you know two descriptions can't specify the same number?

N. T.: You don't. Any number has multiple descriptions, infinitely many descriptions. For example, "the positive number whose square equals 2" and "the diagonal of the square of length 1" are two ways to describe $\sqrt{2}$. And the word *the* in "the square of length 1" is infinitely ambiguous. Moreover, if you just start writing down descriptions at random, you're likely to find yourself describing things that are not numbers. This bottle, for instance, promises *un bouquet puissant empreint d'épices de framboises et de notes grillées*, which is an example of a description of something that is not a number in any obvious way and that I don't claim to understand.

P. A.: Tasting wine is a perfect example of a transcendent experience, impossible to capture in words. Whoever wrote that description shouldn't have bothered. And you number theorists should reserve the word *transcendental* for the numbers that are answers to questions that can't be asked.

N. T.: I would take it up with the relevant authorities, but they're all long dead. Besides, even understanding algebraic numbers in their natural habitat, which is the continuum, requires setting aside time for repeated acts of transcendence. For Hermann Weyl, "The continuum appears as something which is infinitely in the making inside," a "medium of free becoming."[7]

P. A.: Free becoming is a lovely description of the performing artist's state of being.

N. T.: You do have a gift for paradox. You'll have to tell me more about your transcendent experiences. As far as indescribable numbers are concerned, the only way to talk about one is to create one, for example, by writing down an infinite string of digits. But to be sure that we're creating one of these indescribable numbers, we must commit ourselves to writing down the digits in a completely pointless manner, what L.E.J. Brouwer called *free choice sequences*, for all eternity.

P. A.: We could take turns . . .

N. T.: At best, we would have just created one of unimaginably infinitely many numbers. And even then there's no guarantee, unless we have prepared a table of computable numbers in advance, which would fill infinitely many shelves, that we're not accidentally writing the digits of a computable transcendental number like $\sqrt{\pi} + 22$.

P. A.: I can see how that would get tiresome fairly quickly.

N. T.: And if we ever decided to quit, all we would have done is write a long formula for a probably boring rational number. So we may as well not get started. Anyway, I have a more transcendent idea. Let me switch off the lights again.

chapter 7

The Habit of Clinging to an Ultimate Ground[1]

> *Every scientific 'fulfillment' raises new 'questions'; it asks*
> *to be 'surpassed' and outdated. . . . In principle, this prog-*
> *ress goes on* ad infinitum. *And with this we come to inquire*
> *into the* meaning *of science. . . . Why does one engage in*
> *doing something that in reality never comes, and never can*
> *come, to an end?*
>
> —Weber, *Science as a Vocation*

The notions of real interest to mathematicians like myself are not on the printed page. They lurk behind the doors of conception. It is believed[2] that they will some day emerge and shed so much light on earlier concepts that the latter will disintegrate into marginalia. By their very nature, they elude precise definition, so that on the conventional account they are scarcely mathematical at all. Coming to grips with them is not to be compared with attempting to solve an intractable problem, an experience that drives most narratives of mathematical discovery. What I have in mind harbors a more fundamental obscurity. One cannot even formulate a problem, much less attempt to solve it; the items (notions, concepts) in terms of which the problem would be formulated have yet to be invented. How can we talk to one another, or to ourselves, about the mathematics we were born too soon to understand?

Maybe with the help of mind-altering drugs? That's what you might conclude from the snippet of dialogue from *Proof* quoted in the last chapter, but you would be wrong. It seems *Proof*'s author got that idea by reading Paul Hoffman's biography of Paul Erdős, the Hungarian mathematician famous for finding unexpectedly rich structures in apparently elementary mathematics; for spending nearly all his life wandering across the planet, crashing in the homes of mathematicians and inviting them to join him in solving problems, sometimes in exchange for small monetary rewards; and for saying "a mathematician is a machine for transforming coffee into theorems."[3] Heraclitus considered most people "oblivious of what they do when awake, just as they are forgetful of what they do asleep."[4] Hoffman depicts Erdős, in contrast, as the Dr. Gonzo of mathematics: "for the last twenty-five years of his life. . . [Erdős] put in nineteen-hour days . . . fortified with 10 to 20 milligrams of Benzedrine or Ritalin, strong espresso, and caffeine tablets," with the result that he spent the last quarter of the twentieth century fully focused on the kind of mathematics at everyone's disposal, seeing things there that perhaps he was the only one awake enough to see.[5]

Grothendieck, by his own account, spent much of his life dreaming his way across the same landscape, forcing a path into a conceptual future that ceaselessly receded, riding not a drug habit but rather a refined mathematical minimalism. "He seemed to have the knack," wrote number theorist John Tate, "time after time, of stripping away just enough. . . . It's streamlined; there is no baggage. It's just right."[6] Grothendieck's search for what his colleague Roger Godement once called "total purity" encompassed his physical existence as well: in São Paolo in 1953–1954, he subsisted on milk and bananas, and years after his withdrawal from mathematics, during a forty-five-day fast in 1990, he went so far as to refuse to drink liquids, bringing himself to the point of death in an apparent attempt to "compel God . . . to reveal himself."[7] And yet what he was really seeking, in life as in mathematics, continued to elude his grasp.

INCARNATION

Sometimes the curtain can be persuaded to part slightly and the elusive items can be approached with the help of *punctuation*—quotation marks, for example, which mathematicians use frequently, especially in expository writing, in at least four distinct (though overlapping) ways:

1. For direct quotations (the way everyone uses them).
2. For implicit quotations (of something that is often said in the field, as in the first two examples in item 3).

3. To substitute for the cumbersome procedure of formal definition, in other words as a synonym for "so-called." Thus Eric Zaslow writes:[8]

> We call such a term a "correlation function."

> Such a case is known as an "anomaly."

and

> The quantum Hilbert space is then a (tensor) of lots of different "occupation number Hilbert spaces."

The last example is more complex than the first two—visibly, Zaslow is making it clear to the reader that he has in mind a notion deserving a formal definition, but that it would be distracting to present it here and unnecessary as well, since the reader can probably figure out what's intended.

But the really interesting use of quotation marks is

4. When an analogy that is, strictly speaking, *incorrect* offers a better description of a notion—a better fit with intuition—than its formal definition. For example,

> you can think of actions as "molecules" and transitive actions as the "atoms" into which they can be decomposed.[9]

Since the sentence has exactly the same meaning without the quotation marks, one has to assume that they have been inserted to stress that the invitation to (chemical) intuition is also explicitly an invitation to relax one's critical sense. Some seminar speakers write such quotation marks on the board; others make the quotation-mark gesture, wiggling two fingers on each hand while uttering the problematic word; still others preface an informal explanation by something like *Let me explain this in words,*[10] which is literally meaningless unless understood as a warning that formal syntactical rules are temporarily suspended, to be replaced by some other kind of mathematics.

More piquant examples can be found in situations where normal semantics do not suffice for communication among specialists. Since I remember that it was advisable, before the 1991 two-week workshop on *Motives* in Seattle, to use the term *motive* only between scare quotes—I did so myself—I am gratified to see that Grothendieck made repeated use of type 4 quotation marks when explaining why he introduced the notion, in his unpublished 1986 manuscript *Récoltes et Sémailles* (*ReS*):

> One has the distinct impression (but in a sense that remains vague) that each of these [cohomological] theories "amount to the same thing," that they "give the same results." In order to express this intuition . . . I formulated the notion of "*motive*" associated to an algebraic variety. By this term, I want to suggest that it is the "common motive" (or "common *reason*") behind this multitude of cohomological invariants attached to an algebraic variety, or indeed, behind all cohomological invariants that are a priori possible.[11]

Cohomological, or its noun form *cohomology*, is the technical name for a method, properly belonging to topology, that introduces an *algebraic structure* in an attempt to get at the "essence" (a word some philosophers write in scare quotes) of a shape. So if the topological essence of the infinity sign (see figure 7.1) is that it consists of two attached holes, cohomology is a way of doing arithmetic (addition, subtraction, multiplication) with these holes. For example

$$4(\text{left holes}) - 5(\text{right holes})$$

is a legitimate formula in the cohomology of figure 7.1, which is also an (unorthodox) picture of the *cubic* equation $y^2 = x^2 (x - 1)$ (see chapters 2 and γ), whose holes are, therefore, meaningful in number theory.

Structure, already encountered in the second paragraph, is a loaded term. Historians stress the Bourbaki group's role in making structures central to their mathematical architecture, following the structural revolution in algebra undertaken in the 1920s and 1930s by Emmy Noether and her protégé Bartel Van der Waerden.[12] Grothendieck was an active Bourbakiste during the late 1950s, and the word *structure* occurs on 124 of the 929 pages of *ReS*. On page 48, for example: "among the thousand-and-one faces form chooses to reveal itself to us, the one that has fascinated me more than any other and continues to fascinate me is the **structure** hidden in mathematical things" (my translation; emphasis in the original). Weil was a founder of Bourbaki; his insight mentioned in chapter 2, the one that converted me to number theory, was that the geometric essence of a problem in number theory could be grasped by

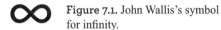

Figure 7.1. John Wallis's symbol for infinity.

means of the algebraic and topological structure of cohomology. And the *motive* Grothendieck hoped would provide the proper setting for Weil's conjectures is a structure that clenches this essence even more tightly. Grothendieck explains the choice of term by musical analogies:

> *Ces différentes théories cohomologiques seraient comme autant de développements thématiques différents, chacun dans le "tempo", dans la "clef" et dans le "mode" ("majeur" ou "mineur") qui lui est propre, d'un même "motif de base" (appelé "théorie cohomologique motivique"), lequel serait en même temps la plus fondamentale, ou la plus "fine", de toutes ces "incarnations" thématiques différentes (c'est-à-dire, de toutes ces théories cohomologiques possibles).**

Note the word *incarnation*. A philosopher might understand this on the model of an example or instance that points to an "essence" whose place in Grothendieck's text is occupied by the "common motive." On the same page, and several more times over the course of his rambling, lyrical— and often irascible—manuscript, Grothendieck uses the term *avatar* in scare quotes, possibly for the same reason:

> Inspired by certain ideas of Serre, and also by the wish to find a certain common "principle" or "motif" for the various purely algebraic "avatars" that were known, or expected, for the classical Betti cohomology of a complex algebraic variety, I had introduced towards the beginning of the 60s the notion of "motif". [*En m'inspirant de certaines idées de Serre, et du désir aussi de trouver un certain "principe" (ou "motif") commun pour les divers "avatars" purement algébriques connus (ou pressentis) pour la cohomologie de Betti classique d'une variété algébrique complexe, j'avais introduit vers les débuts des années soixante la notion de "motif".*]

---

* These different cohomological theories would be like so many different thematic developments, each in its own "tempo," "key," and ("major" or "minor") "mode," of the same "basic motif" (called a **"motivic** cohomology theory"), which would at the same time be the most fundamental, or the "finest," of all these different thematic "incarnations" . . . (that is, of all these possible cohomological theories) (Grothendieck, *ReS*, p. 60).

Motives still attract quotation marks of this type. In 2010 Drinfel'd could write that

> [m]orally, $K_{mot}(X, \mathbf{Q})$ should be the Grothendieck group of the "category of motivic $\mathbf{Q}$-sheaves" on X.[13]

(Note the philosophically fraught word *category*, which has a precise meaning in mathematics; we'll be seeing it again.) Drinfel'd adds parenthetically that "the words in quotation marks do not refer to any precise notion of motivic sheaf."

Placing quotation marks around motif or motivic sheaf—or for that matter using the word *morally*, as mathematicians often do, as a functional equivalent of scare quotes—permits invocation of the object one does not know how to define in much the same way that the name-worshipping Moscow set theorists brought sets into being by giving them names. But readers of the French mathematics of a certain generation, especially that associated with the Bourbaki group, can't help but notice that a taste for Indian (rather than, say, Russian Orthodox) metaphysics inflected their terminology. Weil referred to "those obscure analogies, those disturbing reflections of one theory on another"—an early avatar of Grothendieck's avatars—writing that "[a]s the *[Bhagavad]-Gita* teaches, one achieves knowledge and indifference at the same time." Pierre Deligne, whose work did much to embolden mathematicians to remove the scare quotes from "motive," explains the use of the word *yoga*, using the visual metaphor of *panorama*:

> In mathematics, there are not only theorems. There are, what we call, "philosophies" or "yogas," which remain vague. Sometimes we can guess the flavor of what should be true but cannot make a precise statement. . . . A philosophy creates a panorama where you can put things in place and understand that if you do something here, you can make progress somewhere else. That is how things begin to fit together.[14]

"The sad truth," Grothendieck wrote in a letter to Serre in 1964, was that he didn't yet know how to define the [abelian category of] motives he had introduced earlier in the same letter, but he was "beginning to have a rather precise yoga" about these nebulous objects of his imagination. One might argue that Grothendieck in fact made precise definitions and formulated equally precise conjectures that would retrospectively justify the "distinct . . . but . . . vague" impression of which he wrote, and that the use of the words *yoga* and *avatar*, like the quotation marks, is a theatrical effect that serves merely to make the impression more vivid. One might, with equal justification, ask why just these rhetorical techniques do, in fact, enhance vividness. Working with the various "cohomological theories" of which Grothendieck wrote—just as Deligne did, viewing them as avatars of a more fundamental theory of motives—may leave no logical trace but does represent a different *intentional* relation to the cohomological theories in question:

> A mathematical concept is always a pair of two mutually dependent things: a formal definition on the one hand and an intention on the other hand. He or she who knows the intention of a concept has a kind of "nose" guiding the "right" use of the formal concept.[15]

The author of this quotation, who seems to be using the word *intention* in the way I intended,[16] did well not to lift the veil of (type 4) quotation marks from that which is better left to the reader's imagination. Attempts at greater precision lead straight to the threshold of the abyss of speculative philosophy, where one seeks to explain what it means to take phenomena (impressions, mental images) as symptomatic of something that remains concealed. This applies to intentions as well as to avatars—which is not to say that the (intentional) relation I have in mind is purely subjective. If anything is peculiar about the use of *intentional* in connection with avatars, it is that the concealed "underlying theory" of which the phenomena are supposed to be symptomatic does not yet exist, as if the mathematician's role were to create the source of the shadows they have already seen on the wall of Plato's cave.[17] I limit my speculation to claiming that it matters to mathematicians what they think their work is **about**, *whether or not it matters to the work*, and that this ought to be a matter of concern for philosophers.

In chapter 3, I suggested that the goal of mathematics is to convert rigorous proofs to heuristics—not to solve a problem, in other words, but rather to reformulate it in a way that makes the solution obvious. As Grothendieck (*ReS*, p. 368) wrote:

> They have completely forgotten what is a **mathematical creation**: a vision that decants little by little over months and years, bringing to light the "ob-

vious" [*évident*] thing that no one had seen, taking form in an "obvious" assertion of which no one had dreamed ... and that the first one to come along can then prove in five minutes, using techniques ready to hand [*toutes cuites*].

"Obvious" is the property Wittgenstein called *übersichtlich*, synoptic or perspicuous.[18] This is where the avatars come in. In the situations I have in mind, one may well have a rigorous proof, but the obviousness is based on an understanding that fits only a pattern one cannot yet explain or even define rigorously. The available concepts are interpreted as the *avatars* of the inaccessible concepts we are striving to grasp. So, I agree with Grothendieck and disagree with Wittgenstein when the latter writes (in *On Certainty*):

> the end is not certain propositions' striking us immediately as true, i.e. it is not a kind of *seeing* on our part, it is our *acting*, which lies at the bottom of the language game.

In mathematics this separation of seeing and acting seems artificial; seeing and conveying what one has seen is as important as any other form of acting as a mathematician.

∞ ∞

The word *avatar* was used in English as early as 1784 and in French by 1800 as a form of the Sanskrit word *avatara*, denoting the successive incarnations of the Hindu god Vishnu.[19] By 1815 the word was so familiar that Sir Walter Scott could refer to Napoleon's possible return from Elba as the "third avatar of this singular emanation of the Evil Principle." The meaning relevant to mathematics of "transformation, manifestation, alternative version" first appeared in French in 1822 and in English in 1850.

Que d'avatars dans la vie politique de cet homme! Cette institution va connaître un nouvel avatar (one reads in the *Dictionnaire de l'Académie Française*). Educated French speakers were comfortable using the word in this way in the 1950s, if not earlier. Deligne may have been the first to give it its modern mathematical sense, in a widely read paper from 1971, where he wrote (in French) *"One should consider this as an avatar of the projective system (1.8.1)."* Physicists Sidney Coleman, J. David Gross, and Roman Jackiw used the word slightly earlier, and no doubt independently, to roughly the same end.[20] Today, Google Scholar has hundreds of quotations about avatars in geometry—*an algebraic-geometric avatar of higher Teichmüller theory*—or topology—*this operadic cotangent complex will serve as our avatar through much of this work*—or mathematical physics—*topological avatar of the black-hole entropy*—or any other branch of algebra or geometry. The incorporation of the word in the standard lexicon of mathematicians likely has little or no relation to the development of video games or virtual reality experiences like Second Life and can be traced rather to its use by Grothendieck in *Récoltes et Sémailles*, to identify cohomology groups as symptomatic incarnations of the objects of an as-yet inaccessible *category* of motives.

The next time I use the word *category*, I will begin to explain what it means, along with the word *structure*, with which it goes hand in hand, but which I have deliberately left undefined. For the moment, bearing in mind that a motive is a certain kind of algebraic structure, the expression "category of motives" should suggest that such a structure can be grasped only in relation to other structures of the same kind and that the category provides the formal unifying framework in which such relations are made manifest. And just as the introduction of Galois theory brought about a change of perspective, in which the search for solutions of polynomial equations was replaced by a focus on the new structure of the Galois group, you may anticipate that the design of a category of motives meeting Grothendieck's specifications will likely usher in a new mathematical era, in which the motives themselves—not to mention the equations whose "essence" they capture—will no longer be central. Attention will turn instead to the structure of the *category* to which the motives belong,[21] along with other structures to which it can be compared. This is nothing as brutal as a paradigm shift; each generation's new perspective is meant to be more encompassing, as if mathematicians were collectively climbing and simultaneously building a ladder that at each rung offers a broadening panorama and the growing conviction that the process will never end—"knowledge and indifference," as Weil wrote, alluding to the *Gita*; or *ataraxia*, the absence of worry at which Pyrrhonian skepticism aimed in conceding the fruitlessness of the quest for ultimate truth.

EVERYTHING

Faust: Ich fühl's, vergebens hab ich alle Schätze
Des Menschengeists auf mich herbeigerafft,
Und wenn ich mich am Ende niedersetze,
Quillt innerlich doch keine neue Kraft;
Ich bin nicht um ein Haar breit höher,
*Bin dem Unendlichen nicht näher.**

Physicists like Steven Weinberg can "dream of a final theory," but mathematicians can realistically dream only of an endlessly receding horizon. String theory, for example, the preferred approach at the IAS to unification of fundamental physics, is often described as a *theory of everything*. But what is "everything"? Those of us living in a different part of the universe or in an alternate, possibly virtual, reality have the leisure to reflect on this question, even though we may be ignorant of the physical laws governing your universe. Philosophy begins either by distinguishing things from one another or by affirming an underlying unity behind the diversity of appearance.[22] Thus Aquinas devotes questions 30 and 31 of *Summa Theologica, I,* to the triune unity of three persons in the divine essence; Vedantic monists identify *atman* with *brahman*, the personal self with the universal principle; and Nāgārjuna, whose *Mūlamadhyamika-kārikā (Fundamental Stanzas on the Middle Way,* henceforth abbreviated *MMK*) is a founding document of the Madhyamaka school of Buddhist philosophy, propounded the identity of *samsāra*, the cycle of deaths and rebirths in the world of suffering, with *nirvana*, the escape from that very cycle. Are there lots of different things or just One Big Thing (an elephant, for example, or the universe) viewed under different guises? Keep asking this question for a few thousand years and you are liable to invent *set theory*; and you might be tempted to call set theory the *theory of theories of everything*, but it's really only a first baby step toward such a theory.

---

* Faust: 'Tis true, I feel! In vain have I amass'd
Within me all the treasures of man's mind,
And when I pause, and sit me down at last,
No new power welling inwardly I find;
A hairbreadth is not added to my height,
I am no nearer to the Infinite (from the 1865 Theodore Martin translation).

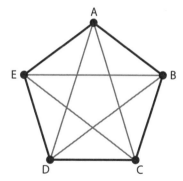

Figure 7.2. The complete graph on five vertices, from Wikimedia Commons (labels added): http://en.wikipedia.org/wiki/File:RamseyTheory_K5_no_mono_K3.svg.

The definition of a set as a collection of elements raises more ontological questions than it answers; it would be more forthright to define a set as a "collection" of "elements." When the sets are finite, though, the scare quotes can be safely discarded. Over 400 of Erdős's 1417 papers indexed by MathSciNet, the biggest single chunk of his papers, are in the field of *combinatorics*, the branch of mathematics largely concerned with counting elements of finite sets. Many of these papers study properties of *graphs*: a graph has a collection of points (*vertices*), the elements of the set, and lines (*edges*) connecting the points, like the one in figure 7.2. This graph has five vertices and 10 edges; it is a *complete graph*, which means that there is an edge connecting each pair of points, like a road running between any pair of cities on a map. Here is *Ramsey's theorem* about complete graphs:

Theorem. For every positive integer k, there is a positive integer $R(k)$ such that if the edges of the complete graph on $R(k)$ vertices are all colored either black or gray, then there must be k vertices such that all edges joining them have the same color.

Tim Gowers discusses Ramsey's theorem at length in his article *The Two Cultures of Mathematics*, on the differences between *problem solvers* and *theory builders*, which is convenient for our purposes because Erdős and Grothendieck exemplify the two cultures almost to the point of caricature. As it happens, "one of Erdős's most famous results," in Gowers's estimation, is that the Ramsey number $R(k)$ has to be at least $2^{k/2}$. If you know $R(k)$, you have solved the *party problem*: Erdős' theorem implies that if you want to hold a dinner party where either (at least) k people already form a clique or (at least) k people don't know each other yet—don't ask why—then you need to invite at least $2^{k/2}$ people. This is not very informative for small numbers: if $k = 3$ it says you need at least 2.828 guests, which you already knew; but for $k = 20$. Erdős tells you to invite at least $2^{10} = 1024$ guests to your party.[23]

Most graphs are not complete. If we erase the gray edges in figure 7.2, we're left with a graph in the form of a pentagon, where each partygoer knows his or her neighbors but not those sitting across the table. Given a big enough blackboard, we can draw a graph whose vertices are all the mathematicians who have ever lived; mathematicians A and B are connected by an edge if they have *collaborated* by publishing a paper together. Four hundred ninety-eight edges radiate from Erdős, making him the most connected mathematician in history. No edge connects Erdős to Grothendieck, who hardly collaborated at all; but you can get from Grothendieck to Erdős by a link of three edges, by way of Serre and Sarvadaman Chowla. This is the shortest such link; in other words, Grothendieck's *Erdős number* is 3—he has 3 degrees of separation from Erdős. My colleagues mostly have Erdős numbers and know what they are: your Erdős number is 1 if you have collaborated with Erdős, 2 if you have collaborated with someone with Erdős number 1 but not with Erdős himself, and so on; only Erdős has Erdős number 0. My own Erdős number is 3, as illustrated by the graph in figure 7.3.

Figure 7.3 is a graphic representation of a *communications network*: mathematicians who have collaborated are considered to be in communication. Results as apparently pointless as Ramsey's theorem have extensive applications in communications and computer science. Did the animators working on James Cameron's *Avatar* consult graph theorists in designing the organic network of branches and vines with ports to plug in the pigtails built into every living being on the moon Pandora?

The message of Gowers's lucid and thoughtful article is not merely that problem solvers, and Erdős in particular, are as deserving of respect as theory builders,[24] but also that problems whose statements are simple enough to be understood by readers with no specialized training in mathematics, even problems about finite sets, can nevertheless reveal a rich underlying *structure* (his word). Gowers makes this structure "obvious" in an eight-line proof of Erdős' theorem that $R(k) \geq 2^{k/2}$. It's a proof many mathematicians would be tempted to call beautiful. But can such a proof

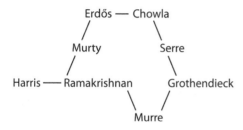

Figure 7.3. A small part of Erdős's collaboration graph, including Grothendieck and the author of this book. Chowla and Murty are among the 498 mathematicians with Erdős number 1.

be *sublime* in the sense of Kant and Edmund Burke? Can meditating about finite sets inspire awe and terror? Or are these sentiments reserved for the contemplation of infinity?

Cantor's diagonalization trick, mentioned in the last dialogue, shows how every infinite set S engenders a set $P(S)$ of an even bigger infinity ... bigger in the precise Cantorian sense that their elements can't be matched side by side like two decks of fifty-two cards. Think it over for a moment and you'll realize that this fundamental property of Cantor's *transfinite arithmetic* means no theory of everything is even remotely possible. . . because a gambler employing Cantor's trick will see your everything and raise you with an incommensurably bigger everything. Like many of his distinguished predecessors, Cantor resolved this hopeless paradox by giving names to the One Big Everything that you meet when you climb this endless ladder of everythings to its unattainable summit. He called it "Absolute"—"the 'Actus Purissimus' which by many is called 'God.'" Joseph Dauben's study of Cantor, from which this quotation is taken, also records how the inventor of set theory maintained an extensive correspondence with Catholic theologians in a largely successful effort to convince them that, precisely because any human attempt to comprehend the full sequence of infinities collapses in paradox, his transfinite arithmetic does not detract from the sublimity of the Absolute and therefore poses "no danger to religious truths."[25]

Voevodsky, worrying that the foundations of mathematics may be inconsistent, has occasionally advised mathematicians to welcome potentially contradictory foundations. Artificial intelligence finds this less problematic than one might think. Automated expert systems have been designed to take contradictory input—for example, conflicting diagnoses of a cancer patient—and to use them as guides to concrete action, rather than to submit to the paralysis of conventional logic, which deduces from any contradiction that every statement is simultaneously true and false.[26] Suspending attention to logical niceties when faced with a life or death decision needs no justification, but when is concrete action urgent in mathematics?

∞ ∞

The question of identity would seem to have been settled long ago in mathematics by the adoption of the = sign as a standard item in the lexicon used to construct meaningful mathematical propositions. But there is a rich philosophical literature on this question. Leibniz's principle of the *identity of indiscernables* states roughly that $A = B$ if everything that is true of (or can be predicated of) A is true of B, and vice versa—if A and B have the same *attributes* in any sense that can be given to this term. Cantor's Absolute, impervious to meaningful predication, resembles the God of the church fathers on the grounds of Leibniz's principle:

> According to the classical theism of Augustine, Anselm, Aquinas and their adherents, God is radically unlike creatures in that he is devoid of any complexity or composition, whether physical or metaphysical. . . . There is also no real distinction between God as subject of his attributes and his attributes. God is thus in a sense requiring clarification *identical* to each of his attributes, which implies that each attribute is identical to every other one.[27]

By insisting that "The essence of entities is not present in the conditions, etc." and that "*nirvana* is uncompounded. Both existents and nonexistents are compounded," the *MMK* excluded *nirvana* from Leibniz's world system. Nāgārjuna's commentator Candrakīrti considered "*nirvana* . . . cognitive nonsense . . . a 'scandal' to logic. In order for *nirvana* to be cognitively affirmed or denied, it must be reduced to *saṃsāra*, to existence in causes and conditions."[28] Nāgārjuna's equation of *saṃsāra* and *nirvana* has been translated as the identity of "the Phenomenal" and the "Absolute."[29]

Identity in Leibniz's sense is the property captured by the equal sign in set theory. But since A is called A and B is called B, this does not quite suffice to unravel $A = B$. In his 1892 article *Über Sinn und Bedeutung*, Frege introduced the distinction between *sense* and *reference*, giving the example of the morning star and the evening star, both of which turn out upon inspection to be the planet Venus: two senses for a single reference. Frege's attempt to solve the problem of identity remains influential among philosophers but is not nearly subtle enough to account for the gradations of identity that inevitably arise in mathematical practice.

Heidegger, meanwhile, was able to spin the tautological equation $A = A$—which we were just claiming was unproblematic for set theory—into an essay on fundamental ontology, starting with the claim that the proper formulation of "$A = A$" is that "every A is itself the same with itself."[30] Heidegger's doubts about the transparency of identity were anticipated (by nearly 2000 years) by Nāgārjuna, who wrote that "I do not think those who teach the identity or difference of self and things are wise in the meaning of the teaching."[31]

"Sameness," according to Heidegger, "implies the relation of 'with,' . . . the unification into a unity." For Voevodsky (following the traditions of *homotopy theory*; see the following), the equation $A = A$ is a *space*, the space of all ways A can be identical to itself. In this he is faithful to Grothendieck, the first mathematician to be guided by the principle that **knowing a mathematical object is tantamount to knowing its relations to all other objects of the same kind**, meaning objects that belong to a well-defined class of structures. The formal language for putting together objects of the same kind is that of *categories*: objects of a given kind—such as sets, or groups, or topological spaces—form a category, and the relations between them are called *morphisms*. The language of categories, with its deliberate echo of Aristotle and Kant, was invented by Saunders MacLane and second-generation Bourbakiste Samuel Eilenberg by abstracting common features of methods of reasoning that were simultaneously appearing all over mathematics, especially in algebra and topology. Chapter 4 of Bourbaki's treatise on set theory, written by Bourbaki's first generation, had attempted to make structures into a rigorous notion at the center of mathematics by proposing an axiomatic treatment of "structures" as such. But mathematicians of the postwar period paid no attention, preferring by far the categorical way of thinking, especially as expanded by Grothendieck, so that today when mathematicians refer to structures, they are usually thinking of objects in a specific category.[32]

Like any mathematical formalism, that of categories is based on a list of axioms; for example, every object, among its many self-relations,

enjoys the relation of *identity* with itself; if you have a relation (mor-
phism) from A to B and a relation from B to C, then you can string them
together to obtain a relation from A to C; and that's practically all you
need. So, according to the principle that Grothendieck used to great effect, "knowing" a set A means "knowing" where A fits in the category
of sets, which amounts to "knowing" the structured web of its relations
(the morphisms in this case are just *set-theoretic functions*) to all other
sets, including itself.

This principle, valid in any category, is known as *Yoneda's lemma*,
and it is so formulated as to be obvious to prove, but the experienced
will be aware that we are skating on the edge of paradox—of Russell's
paradox, more precisely—by talking about things like "all other sets."
Russell showed early on that careless talk about all sets—talking about
the "set of all sets that don't contain themselves," for example—leads
to instant fatal contradiction. That's why one refers to the *category of
sets* rather than the *set of all sets*; in the former, Russell's noxious deductions are not permitted. But this comes at a cost: unless one makes additional restrictions, we have lost the principle of identity; it is not appropriate to say that two sets A and B are *equal*. "The essential lesson
taught by the categorical viewpoint" according to Barry Mazur "is that
it is usually either quixotic, or irrelevant"[33] to insist on equality in a
category. Yoneda's lemma is in this sense quite different from Leibniz'
principle. If we think of the properties of relation with a given object in
the category as a *predicate*, then to say that two objects are indiscernable
with respect to Yoneda's lemma is not to say that they are identical but
that they are *isomorphic*—which means A and B are equal to all intents
and purposes but not really equal.

For one thing, as their distinct names make clear, A and B are not in
a tautological relation of identity, unlike A and itself. You can put A and
B into a relation of identity, like two decks of cards, placing them side
by side, one card at a time, in a long row of fifty-two cards. But you can
pick up the second deck, shuffle it, and then lay the cards down alongside
A again: the relation of identity has been shuffled, and the new one is
just as good as the old one, at least if you ignore what's on the faces of
the cards. With this version of identity, a deck of cards is identical to any
other fifty-two-element set (for example, the set of letters of the alphabet, upper- and lowercase). When identifying two decks of cards, you

can also insist on paying attention to *structure*; you might insist that cards in deck A be matched with cards in deck B in the same suit or that number cards be matched with number cards and face cards with face cards. This is the idea behind the Galois group, as we explained in chapter β: there is no uniform way to label the roots of a general polynomial, but a labeling is allowed only if it accounts for some underlying structure.

Different notions of equality are reviewed on the nlab Web site, at http://ncatlab.org/nlab/show/equality, with a warning that "it hasn't all been said here yet." So far so good; but this is where things begin to get complicated—or interesting, depending on your viewpoint. From the *equality* page you can link to the *identity type* page, where you learn that in homotopy type theory, the "incarnation of equality" is "an **identity type** . . . 'the type of proofs that $x = y$' or 'the type of reasons why $x = y$.' "[34]

"All things have one nature," wrote Nāgārjuna, "that is, no nature." This is interpreted to mean that "everything is empty of essence and of independent identity," that the nature of all things is to be "dependently co-arisen," in other words, to be in relation to other things.[35] We can read this as a paradox "arising at the limits of thought" or as a precursor of Yoneda's lemma. The *MMK* and its commentators seem not to have realized that they may have anticipated *higher-category theory*. If identity is subordinated in ordinary category theory to relatedness, higher-category theory takes relations between relations, and relations between relations between relations, and so on, to infinity if you choose to go there. Higher categories, such as *n-categories*, are central in one way or another to most of Grothendieck's work after he left the IHES and are the framework of Voevodsky's univalent foundations. The *n-Category Café* is a blog run jointly by mathematicians, physicists, and philosophers, all higher-category enthusiasts; and the *nLab* is a hyperlinked online "collaborative wiki" that records information relevant to the "n point of view" (nPOV), which is that "category theory and higher category theory provide . . . a valuable unifying point of view for the understanding of" concepts involved in mathematics, physics, and philosophy.[36] We will soon accompany them a short way up that ladder, but we're not quite ready yet.

Existence

> [I]t's a race to see who can venture out furthest into the borderlands of the nonexistent (Pynchon, *Against the Day*, p. 535).

Facing an audience of philosophers, Langlands confided that

> [i]t is . . . very difficult even to understand what, in its higher reaches, mathematics is, and even more difficult to communicate this understanding, in part because it often comes in the form of intimations, a word that suggests that mathematics, and not only its basic concepts, exists independently of us. This is a notion that is hard to credit, but hard for a professional mathematician to do without.[37]

Professional mathematicians can hardly avoid using the word *exist* in specific contexts, in a particular way, that does not readily lend itself to resolving differences between the standard options of platonism (or realism) and nominalism (or conventionalism). Philosopher Emily Grosholz reads Leibniz's thesis of the existence of intelligibles to mean, in particular, that mathematical objects "must exist if other things are to be possible for thought."[38] For the nominalist philosopher Jody Azzouni, the mere use of the word *exist*, in particular in mathematical practice, does not entail adherence to a specific philosophical position: "Ordinary speakers find the word 'exist' useful—even indispensable—in contexts where they either aren't committed to the subject matter, or where ontic commitment isn't at issue."[39] Foundations do provide some protection against ontological existence proofs, along the lines of Descartes' proof that God must exist because (1) Descartes has a (clear and distinct) idea of a supremely perfect being; (2) necessary existence is a perfection; and, therefore, (3) a supremely perfect being, that is, God, exists. Ontological proofs had already been proposed by St. Anselm in the eleventh century, and by the tenth century, Hindu logician Udayana, who reasoned that " 'God exists' makes sense because there is no means to establish 'God exists' makes no sense." Kant's rejection of Descartes' version, on the grounds that existence is not a predicate, is very similar to mathematicians' refusal to admit notions like the set of all sets, or Cantor's Absolute, in spite of any clear and distinct ideas one likes to think one has on the subject.[40]

Virtual reality offers an appealing metaphorical alternative to the futile opposition between platonism and nominalism, but it will not soon make the word *exist* obsolete in mathematics. Paradigmatic of its acceptable and indispensable use are the so-called existence and uniqueness theorems for various kinds of differential equations (see chapter 4). Conceptually speaking, such a theorem typically asserts that a (system of) differential equations of a certain type has (or fails to have) exactly one solution when something like *initial conditions* are specified, and this solution includes information about behavior in the long run. Thus when you toss a ball in the air or launch a rocket, the initial conditions, which in this case amount to the projectile's initial speed and angle, determine whether it will either fall back to earth at a predictable location (like the V-2 rockets in *Gravity's Rainbow*), or orbit around the earth at a predictable height (as Venus orbits the sun in *Mason & Dixon*), or hurtle off endlessly into boundless space (like the Chums of Chance at the end of *Against the Day*). In practice one has to specify more conditions, such as weather variables and properties of the materials constituting the rocket, in order to obtain an accurate answer, but the rocket's trajectory and the unique solution to the equations determining its motion are the same thing; so if you fail to bring the space shuttle home safely, it's the fault of the equipment, not the equations.[41]

When mathematicians establish the *existence* of objects, we usually want to understand their *uniqueness* at the same time. Most of us would agree the rocket's trajectory is unique, the path it really takes—that's how we understand the reality of reality, and this understanding must account in no small measure for why mathematicians are attached, as Langlands says, to the sense of the independent existence of mathematics. But branches of mathematics that cannot rely on the physical world for guidance are subject to the problems of identity. When we solve a problem—in algebraic geometry, for example—we want to be able to point to the answer unambiguously. In the categorical framework, the best we can do is say that the solution is "unique up to unique isomorphism." This means that when you and I set out to solve the same problem, we know we will be making choices at the outset, based on our individual perspectives; unique means there is a way to translate my solution into yours, and up to unique isomorphism means there is only one way to do it. But finding that translation means solving another problem!

Problems of translation pile up as one climbs the n-categorical ladder:

> [O]ne seems caught at first sight in an infinite chain of ever 'higher,' and presumably, messier structures, where one is going to get hopelessly lost, unless one discovers some simple guiding principle.[42]

Although we are free not to climb it, this ladder forces itself upon our attention when we think carefully about questions of identity and relations in a fixed mathematical context. The avatar ladder, in contrast, emerges when we observe mathematicians looking not down for reassurance as to the solidity of their foundations, but rather into the future in the search for ultimate meaning, which may well entail the scrapping of the foundations that are conventionally assumed to have allowed us to reach this point of reflection. But the two ladders run epistemologically parallel in the practice of *categorification*, to which we turn at the end of the chapter.

Erdős is sometimes called a platonist because he frequently referred to a register of optimal proofs, called THE BOOK, which could be consulted only by God, in whom he did not believe and whom he called SF, "the supreme fascist." But Erdős did not mean this dogmatically; "in a way," he said, "mathematics is the only infinite human activity."[43]

∞ ∞

But I'm now suddenly in a position to tell you what the Langlands program is about, and since this may not happen again, we had better take advantage of the opportunity. Very briefly, then: the Langlands program postulates a correspondence between *Galois representations*—structures derived from the symmetries of roots of polynomial equations, normally studied by number theorists (cf. chapter β)—and *automorphic forms* (or *automorphic representations*)—structures derived from solutions to differential and difference equations with an unusual degree of symmetry (cf. chapter 4), normally studied by geometers or mathematical physicists. Langlands conjectured (and in many cases proved) such tight relations between the two kinds of structures that one can only hope to be successful in either field today by learning to see each side of the Langlands correspondence as the avatar of the other—as exhibiting the other side's otherwise inaccessible properties.[44]

This is probably also my last chance to say something about the ideas Langlands calls "reckless," and I really ought to, because Grothendieck has been getting far more good lines than Langlands. The ideas themselves are too technical to present here—they have been attracting a lot of attention because they are the only proposal with the potential, however remote, of resolving all of Langlands's conjectures simultaneously—but I can say a word about just how reckless they are. How Langlands estimates their likelihood of success can be inferred from the anecdote that provided the epigraph for his first exposition of these ideas. Langlands has studied and written in many languages, among them Turkish, to which he has shown a special attachment. So it is not so surprising that Langlands would have chosen for his epigraph the expression *Ya tutarsa*, associated with one of the stories about the Turkish folk hero Nasreddin Hoca (pronounced "hodja"). Nasreddin is staring at a lake. When his neighbors ask him what he is doing, he explains that he has put some yogurt in the lake and that he is now waiting for it to ferment so that the whole lake will turn to yogurt. His neighbors laugh at him and tell him that the lake will never turn to yogurt. "*Ya tutarsa*," Nasreddin replies—"What if it did?"

Even when one is not in a position to prove Langlands's conjectures, a standard procedure is to reason on one side of the correspondence to deduce interesting and surprising consequences on the other side and then to prove these consequences directly. Specialists call this a "guide to intuition," as if to confirm that a "non-logical cognitive phenomenon"[45] is involved. The most spectacular example was undoubtedly the series of deductions that led to the discovery by Gerhard Frey and Ken Ribet, in the mid-1980s, that Fermat's Last Theorem could be proved by establishing one part of the Langlands correspondence. Since the Langlands program had already shown such exceptional explanatory power that it had grown "too big to fail," this dramatically intensified the belief among number theorists in the truth of Fermat's Last Theorem and gave Andrew Wiles the confidence he needed to spend seven years in an ultimately successful effort to prove as much of the Langlands correspondence as he needed.

The Langlands program, meanwhile, flourishes as never before, in large part thanks to the new ideas Wiles brought to bear on a problem that had motivated number theorists for centuries but that is now well advanced along Weil's cycle of knowledge and indifference. One of the most fruitful approaches to proving Langlands's conjectures was developed by the Soviet-born mathematician Vladimir G. Drinfel'd, using the full range of techniques developed by . . . Grothendieck . . . , which is fitting, since the Galois representation side of the Langlands correspondence is naturally interpreted as an avatar of Grothendieck's motives, and vice versa.[46] Having absorbed earlier work of Goro Shimura, Deligne, and Langlands, Drinfel'd defined several new (Grothendieck-style) geometries to bridge the two apparently unrelated "structures" that matched Langlands' predictions and, in so doing, launched the *geometric Langlands program* and incidentally began the process of aligning these structures into categories. Drinfel'd and Alexander Beilinson, and later Ed Frenkel and his collaborator Dennis Gaitsgory, were among the first to construct a Langlands correspondence as a relation between categories.

Drinfel'd's geometries looked extremely strange when he defined them; did they preexist his definition? Are they avatars of a platonist Langlands correspondence or do they bring the Langlands correspondence into being? In designing his geometries, Drinfel'd was clearly guided by the hope of applying Weil's topological insight; his definition's merit was that it provided just the *fixed points* he needed. In 1990 I was sitting in Manin's seminar in Moscow when Beilinson stood up and interrupted the speaker to explain that the objects studied in algebraic geometry were an illusion, *maya*, to hold fixed points together. Of course he said no such thing, but that's what I heard, perhaps because I had been reading secondhand Indian metaphysics—and whatever he actually said, this was the perspective behind a dream I had two years later. . . .

SECOND LIFE

Möge der Traum den wir das Leben nennen . . . * (Gauss, quoted by Mehrtens, p. 26)

Jaron Lanier, who invented the term *virtual reality*, described the "uncanny moment" when "your brain starts to believe in the virtual world

* May the dream we call life . . .

instead of the physical one." Jerry Garcia and Timothy Leary had compared early VR to a psychedelic experience. "The body and the rest of reality no longer have a prescribed boundary," writes Lanier. "So what are you at this point?"[47]

Many readers will have first experienced a single-sense version of this transition when they put on 3D glasses to watch a film called *Avatar*. The VR of mathematics is no less dependent on the experience of bodies in space. Gilles Châtelet uses the spatial metaphor of gestures to account for what Grothendieck called "obvious": " 'Obviousness' for Grothendieck is not linked to the proximity of two terms in a deductive chain but rather to the effect of 'naturalness' linked to the abolition of the space between the symbol that captures and the gesture that is captured."[48] One need add only that this "abolition" is a never-ending process. The socialization outlined in chapter 2 had less to do with assimilating norms, values, and respect for a transhistorical hierarchy than with learning to use the word *exist* appropriately and effortlessly, to display a pragmatic "belief in" the world of mathematical objects adequate for seamless interaction with the community. When you don the gloves and goggles of contemporary geometry, differential equations, number theory, or algebraic geometry, boundaries with the rest of reality no longer matter.

Maybe Châtelet's *geste* is a literal description in VR, not a metaphor:

[T]he participant in Second Life is not inventing a fiction about his avatar; he is performing fictional actions through his avatar. . . . No wonder, then, that animating an avatar in Second Life would come so naturally to you. . . . You have been animating an avatar all along: . . . the avatar consisting in your body, and you have been doing it whenever you act.[49]

Mehrtens thought the early-twentieth-century Foundations Crisis was not about foundations at all; rather, it marked "a shaking [*Erschütterung*] of the concepts of truth, meaning, object, and existence in mathematics. These concepts are not so much epistemological foundations of the science as the overall orientation markers by means of which the . . . identity of the science defines itself. . . . Mathematics is a language and the work thereon, and so its identity cannot be established [*herstellen*] on an 'object'. . . . "[50]

∞ ∞

Science today should be entrusted to men of great spirituality, to prophets and saints. But instead it's been left to business talents, chess--players, athletes. They don't know what they're doing.

—V. Grossman, *Life and Fate* (modified from
Robert Chandler's translation, p. 693)

Grothendieck, like Serre, Tate, and Godement, was (briefly) a leading member of Bourbaki's second generation and agreed with the claim in Bourbaki's "programmatic manifesto" that (thanks largely to Bourbaki) mathematics "possesses the powerful tools furnished by the theory of the great types of structures; in a single view, it sweeps over immense domains, now unified by the axiomatic method, but which were formerly in a completely chaotic state."[51] In *Récoltes et Sémailles*, Grothendieck takes Bourbaki's structuralism a step further, revealing himself as a kind of structural platonist—a structure is not something to be invented, "we can only bring it to light patiently, humbly make its acquaintance, '**discover**' it. . . . [structures] didn't wait for us to be, and to be exactly what they are!"[52]

But ruthless elimination of superfluous assumptions in the quest for maximum generality is no less important than structures in the Bourbaki legacy. (Ontological?) minimalism is the hallmark of the Bourbaki style. Godement reports that, as a young man, "Grothendieck wore only a shirt, pants and sandals winter and summer, and these were very worn-out (as he was totally penniless), but always incredibly clean, as was his room." All this struck Godement as "forming part of a search for total purity,"[53] the search that in mathematics led Grothendieck to "strip [. . .] away just enough," as Tate put it.

The hypothetical category of motives is supposed to be built out of *pure motives*. Grothendieck suggested a specific recipe to define pure motives in terms of algebraic geometry, and this is the recipe we have in mind when we think about motives, though after more than forty years, it has been shown to work as predicted only in the very simplest cases. Motives that are not pure are *mixed* (rather than impure—Deligne reports that Grothendieck chose this terminology). Current theories of mixed motives—Voevodsky's is the most influential[54]—are a stunning monument to their creators' persistence and are rich enough to have solved outstanding problems, but they fall far short of Grothendieck's original expectations. They are still avatars of the desired theory: one has access to various *derived* categories but not to the ulterior category.

A human being using categories inevitably has to make choices that would be redundant from an Absolute standpoint. We can still prove that the properties of interest don't depend on these choices: they are "unique up to unique isomorphism." But then, we've seen, new problems arise: *how* the properties don't depend on choices. As Plato reminds us in the *Phaedo*, "when the soul makes use of the body for any inquiry. . . then it is dragged by the body to things which never remain the same, and it wanders about and is confused and dizzy like a drunken man."[55]

Grothendieck contrasted Serre's elegance—on at least four occasions in *Récoltes et Sémailles*, Serre is depicted as the "incarnation of elegance"—in applying the "hammer-and-chisel method" to solve a problem, with his own approach: making the problem disappear by letting the sea rise to "submerge and dissolve" it. "Grothendieck creates truly massive books with numerous coauthors, offering set-theoretically vast yet conceptually simple mathematical systems adapted to express the heart of each matter and dissolve the problems."[56] Grothendieck lets the sea rise by eliminating dependence on a choice of perspective. As in Einstein's theory of special relativity, all perspectives are equally valid. Choices have been eliminated in the process, axiomatizing the possibility of choice as a new and higher level of mathematics up the category ladder. "When one follows Grothendieck's work throughout its development, one has exactly [the] impression of rising step by step towards perfection. The face of Buddha is at the top, a human, not a symbolic face, a true portrait and not a traditional representation."[57]

In an all too familiar trade-off, the result of striving for ultimate simplicity is intolerable complexity; to eliminate *too-long proofs* we find ourselves "hopelessly lost" among the *too-long definitions*. The Madhyamaka school thematized this very trade-off and rejected infinite regress without resorting to an Aristotelian Prime Mover. Nor did they have patience with those who hoped to take refuge in comfortable illusions—for example, with the help of mind-altering drugs. The *Prasannapada*, Candrakīrti's seventh-century commentary on *MMK*, "considers as sick that form of life which seeks expression through metaphysical thinking . . . the metaphysics of 'is' and 'not-is' [is] not a solution to the problem of suffering but a drug invented by those who love to 'get high.' "[58]

Grothendieck lived in communes off and on during the 1970s, but he reportedly wasn't interested in drugs either. In fact, I've found only a few documents suggesting that drugs can enhance mathematics, and the evidence is ambiguous. The Irish mathematician William Rowan Hamilton, the inventor of quaternions whose spirit haunts Pynchon's *Against the Day*, was an alcoholic for the last third of his life. Contemporary biographers claimed that he drank for the sake of his work:

> To continue to the end a task, in which good progress had been made, required, as he was convinced, support and stimulus for the brain, and this he administered to himself in the injurious form of porter taken in small sips as he felt fatigued. The need thus experienced, connected as it was, with his disinclination to be disturbed at his work by regular meals, was, according to his son's testimony, the principal cause of his recourse to alcoholic stimulant, for which he admits that his father had besides a constitutional proclivity, as well as a disposition, arising from his genial nature, to conform to the prevailing custom of the time when he first entered into social life.[59]

When his friend Ronald Graham bet Erdős $500 he couldn't quit amphetamines for a month, Erdős won the bet but complained, "I didn't get any work done . . . I'd have no ideas, like an ordinary person. You've set mathematics back a month." Erdős went back to taking pills and writing papers.

Hamilton and Erdős indulged their drug habits, it's said, to maintain their stamina. California chaos theorist Ralph Abraham is the only mathematician I know who has claimed drugs actually affected the contents of mathematical research, and for the better. In a 1991 interview with the style magazine *GQ*, Abraham claimed, "In the 1960s a lot of people on the frontiers of math experimented with psychedelic substances. There was a brief and extremely creative kiss between the community of hippies and top mathematicians. I know this because I was a purveyor of psychedelics to the mathematical community." The interview is maddeningly short on specifics. Molecular biologist Kary Mullis "seriously doubt[s]" that he would have invented the PCR technique for which he won the Nobel Prize if he hadn't taken LSD.[60] Timothy Leary wrote in 1977 that he expected "the new wave of turned on young mathematicians, physicists, and astronomers . . . to use their energized nervous systems . . . to provide new correlations between psychology and science."[61] Can any Fields Medals be traced to psychedelic inspiration? Will we ever know?

Pharmacological enhancement may be redundant if mathematics is itself the drug. Mathematicians stress the addiction even more than the buzz:

Marcus du Sautoy: "Doing mathematics is like taking a drug. Once you have experienced the buzz of cracking an unsolved problem or discovering a new mathematical concept, you spend your life trying to repeat that feeling."

Marie-France Vignéras: " . . . it happens suddenly: one direction becomes more dense, or more luminous. To experience this intense moment is the reason why I became a mathematician."

Joan Birman: "[T]he moment I learned about an unanswered question that involved braids, I was hooked!"

Grothendieck: *C'est en termes cohomologiques, . . . que Serre m'a expliqué les conjectures de Weil, vers les années 1955—et ce n'est qu'en ces termes qu'elles étaient susceptibles de m' "accrocher" en effet.* [It's in cohomological terms . . . that Serre explained the Weil conjectures to me, around 1955—and it was indeed only in these terms that they could have "hooked" me.]

Misha Gromov: "You become a mathematician, a slave of this insatiable hunger of your brain, of everybody's brain, for making structures of everything that goes into it."[62]

Grothendieck was briefly drawn to Buddhism, but his affinity was with Buddhist pacifism rather than its metaphysics. According to the mathematician Winfried Scharlau, who has spent years assembling documents and interviewing Grothendieck's acquaintances in an effort to reconstruct his life, he "was never especially interested in [Buddhist] scriptures . . . and never studied them in depth." He "understood meditation to be pretty much the exact opposite of what it means to Buddhism." Far from seeking to empty his mind to realize Nāgārjuna's "no nature," or the *ataraxia* of the skeptics, or "the rich treasure of inner peace" that the seventeenth-century Spanish mystic Miguel de Molinos promised to those contemplating *la nada*, Grothendieck used meditation as a form of "intellectual work, penetration of one's own ego, revelation, perception, knowledge, and understanding of psychological events"[63]—miring himself ever more inextricably in *saṃsāra*.

Erdős, meanwhile, comes across in Hoffman's book as "a weirdo," one of the most eccentric humans ever to walk the earth. "He lived out of a shabby suitcase and a drab orange plastic bag from . . . a large department store in Budapest. . . . He was twenty-one when he buttered his first piece of bread . . . 'I took him to the Johnson Space Center to see rockets,' one of his colleagues recalled, ' but he didn't even look up.' " But his theology is refreshingly down to earth. Had Jesus paid him a visit, the Jewish atheist Erdős would have asked him whether Cantor's continuum hypothesis is true—whether Cantor's procedure for creating an endless string of increasingly big infinities doesn't leave any out. Kurt Gödel and Paul Cohen proved that the question as stated is meaningless: it can't be decided within the standard axioms of set theory (ZFC: Zermelo-Fraenkel + axiom of choice). Yes and no are, therefore, both acceptable, but Erdős thought Jesus might have a nicer answer: "The Father, the Holy Ghost, and I have been thinking about that long before creation, but we haven't yet come to a conclusion."[64]

∞ ∞

FAUST: . . . *Da muß sich manches Rätsel lösen.*
MEPHISTOPHELES: *Doch manches Rätsel knüpft sich auch.**

Not only is it OK to write $A = A$ in a category, it's even one of the rules that any A has a special identity relation to itself, usually just called "identity." Writing $A = B$ is another matter, unless B happens to *be A*. In a category there can be no end to the Bs that are indistinguishable from— isomorphic to—A but are not the same as A. This is the price you pay to be able to talk about the category of all sets: it contains not just one set of real numbers, for example, but (at least) one for each time the set of real numbers is invoked—two new ones just as a result of this sen-

---

* FAUST: surely there / Will many a riddle be untied.
MEPHISTOPHELES: And many a riddle be knotted too (from the 1865 Theodore Martin translation).

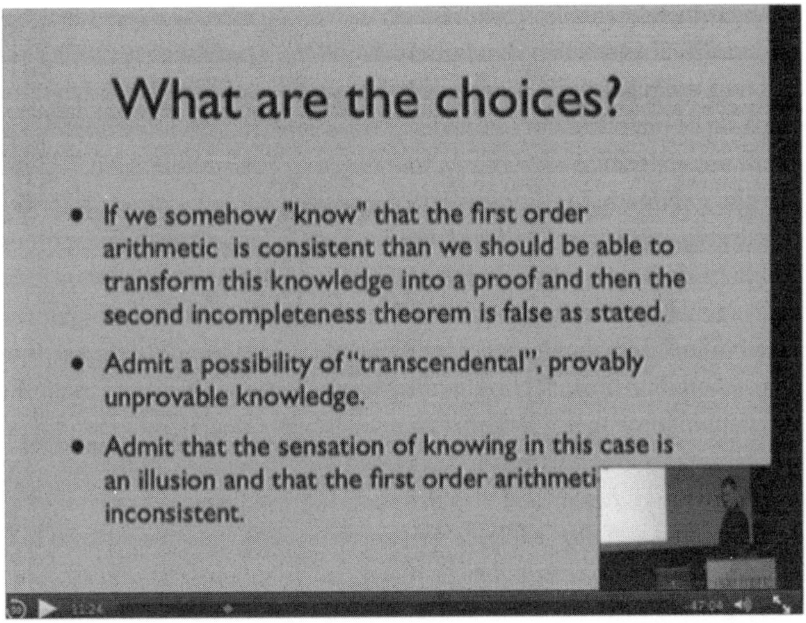

Figure 7.4. From Voevodsky's lecture on September 25, 2010. Voevodsky is in the lower-right corner (courtesy of the Institute for Advanced Study).

tence—all equivalent for all intents and purposes, all (uniquely) isomorphic, no two equal.

The next step is to replace $A = A$ by something more complicated, like a space. If you add the right axioms, the result is called an ∞-*category*.[65] Actually there are many ways to define ∞-categories, each useful for different purposes, none of them prior to the others, so that they can all be seen as avatars—the word *incarnation* is actually used in the literature—but when one asks avatars of what? the only sensible answer is that they are avatars of the theory of ∞-categories mathematicians need.

∞-categories serve as a model for Voevodsky's new foundations. Recall Manin's observation that two kinds of foundations are at issue in mathematics: the foundations that represent the starting point for building what one wants to build and the (logical and philosophical) Foundations, without which the entire structure will supposedly crumble. Grothendieck had strong feelings about the former, alluding in a 1983 letter to Daniel Quillen to "people such as yourself . . . who . . . may not think themselves too good for indulging [*sic*] in occasional reflection on foundational matters and in the process help others. . . ."[66] But it's only a matter of habit to call these foundations as if they were necessarily below our feet; since the sole stipulation is that we actually work at eye-level, there's no reason not to situate foundations of this sort up above our heads.[67]

What Grothendieck's letter to Quillen calls "foundations" is really a common language sufficiently rich and precise to address the problems that may arise in the development of the theory that one was going to develop in any case. Weil (and Bourbaki more generally) used the word in the same way. Such a language has more in common with the satellites and transmitters that carry the signals permitting electronic communication than with the 120 meters of concrete attached to the base of the 452-meter-high Petronas Towers in Kuala Lumpur to protect them from destabilization by the forces of nature.[68] The pragmatist philosopher C. S. Pierce understood this well:

> [T]he intellectual powers essential to the mathematician [are] "concentration, imagination, and generalization." Then, after a dramatic pause, he cried, "Did I hear someone say demonstration? Why, my friends," he continued, "demonstration is merely the pavement on which the chariot of the mathematician rolls."[69]

It can be argued whether, in the long war at the heart of the Indian epic *Mahabharata*, Arjuna the archer or his charioteer Krishna, avatar of the god Vishnu, is the true hero. The case could even be made[70] that the chariot and Arjuna's bow, not to mention the horses, should be granted equal standing. But only a certain philosophical cast of mind would think of giving top billing, Foundational as it were, to the pavement. . . .

The philological exegesis of *foundation* is not meant to reveal a buried secret history but rather to emphasize that foundation is a metaphor, one of several possible, even for the Euclidean axiomatic approach. Metaphors are useful as reminders of turning points in the development of a field, but most of the time mathematical practice speaks for itself. A starting point for the development of higher categories is the branch of topology called *homotopy theory*. Whereas two spaces are topologically equivalent (*homeomorphic*) if they can be transformed into one

$$\varnothing \rightarrow \oslash \rightarrow \textcircled{\mid} \rightarrow \infty$$

Figure 7.5. A homotopy from the empty
set sign to the infinity sign.

another by stretching or shrinking without tearing and without loss, they
are *homotopy equivalent* if one can be transformed to the other, still
without tearing, but certain kinds of loss are allowed. Thus in figure 7.5,
the two spikes on the empty-set sign can be retracted (in the first step),
and the vertical line can be pinched to a point (in the third step) without
changing the homotopy type; the *obwarzanek* in figure 7.6, which is
three dimensional (and edible), is homotopy equivalent to an imaginary
one-dimensional infinity sign making a circuit around the two holes.
Two homotopy equivalent spaces are not considered topologically the
same, but they have the same cohomology—the two holes in the cases
illustrated in figures 7.5 and 7.6—which is to say they have the same
algebraic structure. This is one reason homotopy theory provided the
source of Voevodsky's intuition in constructing his category of mixed
motives.

But working rigorously with homotopies presents new problems. To
make it a mathematical theory, we have to imagine each homotopy as a

Figure 7.6. An *obwarzanek* (Polish soft pretzel), emblem of the sixth European Congress of Mathematicians held in Krakow in 2012, homotopy equivalent to the infinity
sign. (© Polish Mathematical Society)

stretching/shrinking procedure that takes place in a fixed time interval, say one second. A problem is already apparent in figure 7.5: each of the three steps is supposed to last one second, but the whole procedure is also a homotopy and, therefore, wants to last one second. The traditional solution is to speed up the intermediate steps to make the total come out to one second. But there are (infinitely) many ways to do this, and the set of all such ways is itself a topological space: how do we choose the right one? The traditional answer is that there is no right answer: for the riddles topology was originally invented to solve, it doesn't matter which way you choose.

As Mephistopheles reminds us, however, solving one set of riddles leaves us knotted in a whole new tangle of riddles. Starting in the 1960s, topologists concerned with keeping track of all the intermediate choices sought (small-f) foundations for higher category theory. "Ontic commitment" was not at issue; it rarely is for mathematicians. "Usually, number-theorists (like me) neither understand, nor care about such foundational matters, and questions about them are normally met with a shrug." Thus spoke a prominent number theorist on MathOverflow, using the word *foundations* with an implicit capital F. Even as he took care to avoid triggering a Russell paradox in his massive opus on ∞-categories, Jacob Lurie referred to such questions as "a nuisance." "[T]he open secret," writes Barry Mazur, "is that, for the most part, mathematicians who are not focussed on the architecture of formal systems per se . . . somehow achieve a sense of utterly firm conviction in their mathematical doings, without actually going through the exercise of translating their particular argumentation into a brand-name formal system."[71]

To steer clear of set-theoretic paradoxes, Lurie chose to work in the setting of a *Grothendieck universe*. This gadget, one of Grothendieck's rare gifts to Foundations as such, is designed to outrun Cantor's trick for making a bigger set out of any given set, and it guarantees that one never meets an inherently contradictory concept like the "set of all sets." But it requires a special axiom of its own that (unfortunately?) cannot be proved consistent with the ZFC axioms: the existence of what set theorists call an *inaccessible cardinal*. Even if you don't know ZFC, it should be obvious that you can't use it to deduce the existence of an inaccessible cardinal; otherwise it would be accessible! Voevodsky's Univalent

Foundations require not just one inaccessible cardinal but an infinite string of cardinals, each inaccessible from its predecessor.[72]

In a move that is standard when Foundational matters briefly peep through the mathematical scenery, Lurie reassures his readers that "none of the results" of his book "will depend on this assumption in an essential way." This already qualifies inaccessible cardinals as "disputed objects," in the terminology of Jessica Carter. Before laying a foundation, wrote Kronecker, "a rational builder" will want to be carefully informed about the structure for which it is intended; he reminds us that "with the richer development of a science the need arises to alter its underlying concepts and principles" [*die ihr zu Grunde liegenden Begriffe und Prinzipien*], but there is no risk to rationality: "important results turn out to be completely independent" of "explanations of basic mathematical concepts." Jody Azzouni agrees:

> Once we've established the usefulness of a mathematical system, and ensured (to the best of our ability) its consistency both internally, and with respect to the mathematics and the empirical subject matter it's to be, respectively, applied with and applied to, *we're done*—epistemically speaking. There is nothing more to find out.[73]

Voevodsky shares Kronecker's and Azzouni's flexible attitude to Foundations but not their epistemic confidence. Even if you do not worry, like Voevodsky, that mathematics may be inconsistent, you have to wonder: is the Univalent Foundations program motivated by attachment to the notion of the independent existence of mathematics, and our consequent obligation to render it faithfully, or by the sense that its existence is contingent on our providing adequate Foundations?

Bourbaki's "manifesto" mentioned earlier, written by Jean Dieudonné (later Grothendieck's scribe), compares mathematics to a "big city, whose outlying districts and suburbs encroach incessantly . . . on the surrounding country, while the center is rebuilt . . . each time in accordance with a more clearly conceived plan and a more majestic order, tearing down the . . . labyrinths of alleys and projecting toward the periphery new avenues, more direct, broader and more commodious." The text's title is *The Architecture of Mathematics*, but the word *foundation* is conspicuously absent; the image is modernist and echoes Wittgen-

Figure 7.7. Historical Monument of the American Republic, by Erastus Salisbury Field. (Michele and Donald D'Amour Museum of Fine Arts, Springfield, Massachusetts. The Morgan Wesson Memorial Collection. Photography by David Stansbury)

stein's comparison in the *Philosophical Investigations* of language to an "ancient city," notably in its emphasis on horizontal expansion rather than vertical growth. A Bourbaki who grew up in today's parallel networked world, at ease with the blue Na'vi inhabitants of the distant moon Pandora of Cameron's film, where mountains float free of gravity, might trade in Dieudonné's image of a "universe" centered around a "nucleus" of "mother-structures" (groups, ordered sets, topological structure) for a three-dimensional network of intercommunicating roots, branches, and ladders.[74]

My private images of the intertwined avatar and categorical ladders remind me insistently of the Pandoran rainforest. Like the imperialist adventurers from a ruined earth, philosophers of Mathematics seek their "unobtainium" in the form of reliable Foundations, seemingly unaware that, just as Pandora's unobtainium neutralized gravity, upward movement along the ladder of avatars abolishes foundations—not as a common reference, which will indeed be retained as long as it facilitates communication—but as a restraint on our imagination, or as anything other than a "nuisance."

∞ ∞

The danger pinpointed by Ivor Grattan-Guinness as "royal road to me"–type history consists in projecting present notions abusively on historical texts, a danger to which he finds some mathematicians particularly insensitive.[75] The "royal road to me" is abusive retrospection; from the historian's perspective, on the other hand, the *avatar* is a (possibly abusive) projection of a speculative future mathematics on current practice. It would be odd to do history without any consideration for the past—but perhaps not for the mathematician, whose understanding is not directed to history but rather to other mathematicians.

Jamie Tappenden hints at the avatar perspective in his chapters for *The Philosophy of Mathematical Practice*—a book dedicated, as its title makes clear, to philosophy of small-*m* mathematics.[76] Recounting how a property of prime numbers migrated to algebra, he revives Aristotle's distinction between *nominal* and *real definitions*, so that the avatar would correspond to versions of the latter. "The core motivation is that in mathematics (and elsewhere) finding the proper principles of classification can be an advance in knowledge." "Putting together a useful, metaphysically uncontentious doctrine of real definition seems promising if we take the relevant improvement to partially involve finding 'a key that will explain a mass of facts.' " (Note Tappenden's type 4 quotation marks!)

Tappenden refers to the problem of "carving [reality?] at the joints." In 1900 Hilbert wrote:

> If we do not succeed in solving a mathematical problem, the reason frequently consists in our failure to recognize *the more general standpoint* from which the problem before us appears only as a single link in a chain of related problems (my emphasis; note the definite article).

But is this recognition or creation? Grothendieck's revolutionary rethinking of what mathematicians mean by "space" is often compared to Bernhard Riemann's nineteenth-century invention of a new framework for geometry, of which his non-Euclidean geometry is only the best-known illustration. Tappenden asserts that, in contrast to the competing Weierstrass school, for Riemann, "What is to count as fundamental in a given area of investigation has to be *discovered*."[77]

The avatar perspective blurs the border between ontology and epistemology; it would be of no interest were it not for the possibility of convincing the audience of its fruitfulness. Good examples are regularly provided by string theory, whose practitioners discover often astonishing properties of familiar mathematical objects on the basis of something called "physical intuition," only later confirmed rigorously by mathematical proof. Here I suppose the inaccessible mathematical theory (yet to be formalized within algebraic geometry, for instance) can be intuited through its avatar in physics. For example, Maxim Kontsevich's *homological mirror symmetry conjecture* postulates an equivalence between the category of *B-branes* on a complex manifold M and the category of *A-branes* on a symplectic manifold M'. The category of B-branes is a familiar object from Grothendieck's geometry, whereas the category of A-branes is usually said to be related to the *Fukaya category*, for which there is, unfortunately, no rigorous definition in general. This may or may not trouble physicists, but it makes communication between physicists and mathematicians problematic.[78]

ŚŪNYATĀ

When I was a Princeton undergraduate, the vast woods behind the IAS were a favorite place to explore alternate realities. These days it's done more officially in engineering labs on campuses all over the world, even on the IAS campus itself. All mathematicians spend our working lives exploring the *virtual* realities gradually furnished and decorated by generations of our predecessors. True creators like Grothendieck design not only our virtual surroundings, but also the high-level heuristics we use to perceive them; the rest of us play with their avatars.

Insofar as the A-branes of the previous section are well-defined objects of mathematics rather than of physical intuition, this is largely due to the work of Andreas Floer. Before his suicide at the age of 34, Floer had taken several steps up the ladder, inventing a version of homology in which holes like those in figures 7.5 and 7.6 are seen as avatars of infinite-dimensional holes in a family of infinite-dimensional spaces originally studied in the settings of the differential equations of mathematical physics.

When he was a professor at Berkeley, Floer reportedly lived at Barrington Hall, a student cooperative famous for its periodic "wine parties," where the punch was laced with LSD.[79] Most mathematical venues offer more conventional refreshments. By next morning's breakfast, mysterious hands will have whisked away the dozens of beer bottles that still littered each table when the Oberwolfach conference center's last guests retired to their rooms, between 1 and 2 a.m. The often-prodigious consumption of alcohol starts just after dinner and certainly contributes to social integration—and, thus, over the long term to fruitful scientific collaboration; but its direct effect on mathematical creativity is questionable. On my last visit I pegged 10 p.m. as the tipping point beyond which scientific coherence can no longer compete with the beer drinker's propensity to entertain increasingly controversial and less clear and distinct topics. It's much the same at the endless round of champagne receptions in France: mathematical notes are compared for the first glass or two, after which conversation reverts to university politics and gossip.

Nāgārjuna dissolves phenomenal reality with his doctrine of the emptiness of all phenomena, "including, most radically, emptiness itself."[80] The *MMK* uses the Sanskrit word *śūnyatā* for emptiness, and since *śūnya* is the term for zero in early Indian mathematics, it has been speculated that the Indian invention of zero can be traced to classical Indian metaphysics of emptiness, specifically that of Nāgārjuna.[81] From this standpoint, it is appropriate that (as recalled in chapter α) all of set-theoretic mathematics is founded on the act of counting to zero.

∞ ∞

One might attempt to dismiss the avatar perspective in mathematics as merely a fashionable name for the practice of assuming a familiar conjecture (including it as an axiom) and deriving consequences. For example, it is standard in analytic number theory to assume the Riemann hypothesis (see chapter α) and see where it leads. As with Grothendieck's hypothetical motives, reasoning on the basis of an unproved hypothesis can be naturalized as a reflex brought to bear in specific circumstances, and the practice can be derided as philosophically irrelevant on the grounds that reasoning with the unproved hypothesis follows the same rules as reasoning without it. That only begs a more interesting question: what circumstances lend themselves to the recourse to this reflex? From the standpoint of solving the problem, this is irrelevant; but the admission of a hypothesis or the adoption of the avatar perspective is seen in a different light if we allow that the goal is to make the solution obvious.

The practice I am describing here, working back from the avatar to the "underlying principle" (quotation marks of type 4) is precisely opposite to the practice of deduction.[82] In this it resembles, and overlaps, the increasingly familiar practice called *categorification*—rooted in Grothendieck's work on the Weil conjectures—in which one seeks structures one or more steps up the categorical ladder whose reflections are the familiar structures one seeks to understand.[83] This is a creative act, not a deductive procedure, since the categorified object is more complex, or richer, or somehow more *meaningful* than the avatar it is meant to explain; and, of course, the act can, in principle, be repeated indefinitely. If you were to ask for a single characteristic of contemporary mathematics that cries out for philosophical analysis, I would advise you to practice climbing the categorical and avatar ladders in search of meaning, rather than searching for solid Foundations. The philosophy of Mathematics that dominates university departments dispenses with meaning and sees actually existing mathematicians as avatars of formalized theorem provers. Thus, the ulterior entity of which a given mathematician is an incarnation is a sort of divine Turing machine, and the Turing test for mathematicians is beside the point.

"Without the capacity for symbolic transcendence," writes Robert Bellah, "for seeing the realm of daily life *in terms of* a realm beyond it . . . one would be trapped in a world of what has been called dreadful immanence."[84] But remembering Erdős, I need to qualify what I just wrote about the search for meaning. Erdős sought symbolic transcendence not at the apex of an endless categorical ladder, but rather in the inexhaustible activity of being a mathematician: for Erdős "the aim of life . . . is to prove and conjecture."

In other respects, Erdős had more than a few things in common with Grothendieck. Both men were extraordinarily devoted to their mothers. Both were Central European Jews displaced, irreversibly, by World War II: Erdős left Hungary and just kept traveling, while Grothendieck remained stateless for many years by choice.[85] While Grothendieck's premonition of the avatar ladder reaches ceaselessly skyward, Erdős built a no less tangled horizontal network of collaborations.

∞ ∞

Voevodsky's Univalent Foundations program was the focus of a highly successful yearlong program at Princeton's IAS in 2012–2013. Some of the participants will be reconvening in Paris in the spring of 2014 for a three-month program on "certified mathematics"; Voevodsky is slated to make an appearance. It's impossible to overestimate the consequences for philosophy, especially the philosophy of Mathematics, if Voevodsky's proposed new Foundations were adopted. By replacing the principle of identity by a more flexible account modeled on space, the new approach poses a clear challenge, on which I cannot elaborate here, to the philosophy underlying "identity politics"; it also undermines the case for analytic philosophy to seek guidance in the metaphysics of set theory, as in W.V.O. Quine's slogan "to be is to be the value of a variable." By a ricochet effect, the new Foundations might also destabilize the currently fashionable computational theory of mind. When a cognitivist like Steven Pinker confidently asserts that "the brain processes information, and thinking is a kind of computation,"[86] he is not claiming that the brain works like an electronic computer but rather drawing on an established repertoire in philosophy of mind, especially in English-speaking countries. *Information*, as cognitivists understand it, is native to set theory and propositional logic; *type theory*, on the other hand, like the logic underlying Voevodsky's univalent foundations, opens the way to a potentially endless expansion of logic along the lines of geometric intuition and is often described as a plausible foundation for intuitionistic logic. To say, for example, that "the brain is a topos[87] and thinking is a way of creating new Types" may in the end have the same meaning as Pinker's sentence quoted earlier, but it would signal a different *intention*.

Within mathematics itself, Voevodsky's proposal, if adopted, will create a new paradigm. In his "fairy tale" and some of his other papers, Langlands made deft use of categories and even 2-categories, but number theory is only superficially categorical, and so is the Langlands program. In the event that Univalent Foundations could shed light on a guiding problem in number theory—the Riemann hypothesis or the Birch Swinnerton-Dyer conjecture, which is not so far removed from Voevodsky's motives—then we could easily see Grothendieck's program absorbing the Langlands program within Voevodsky's new paradigm. As a number theorist, I find such a scenario unlikely—in number theory, prime numbers are the main players, and they are so different from one another that their aggregate behavior is more statistical than categorical—but there's no denying that curiosity about homotopy type theory is growing in other branches of mathematics. The topic has appeared in several guises on MathOverflow; proof theorists as well as algebraic topologists are paying close attention.[88]

Sets, categories, and higher categories are all types in Voevodsky's system; but propositions, and even "true" and "false," the bottom rungs of logic, are types as well, and one reason some find the program appealing is that it encompasses the language and the matter of mathematics—syntax and semantics—in a common framework. I'm intrigued by this, and even by the suggestion (see chapter 3) that it will facilitate "computerization of mathematics," but I can't say it appeals to me. I would like to say that the ideal-type of a mathematical proposition is not a string of symbols in a formal language but a *declarative sentence*, in natural language, in which every word means exactly what it says, so that quotation marks are disallowed; and this may even be a characterization of mathematical propositions. So in the statement of a theorem, you are not allowed to say *If X is a "motive"* but only *If X is a motive*, and this *motive* has to mean or to be able to mean the same thing to any potential reader. But I immediately have to qualify this attempt at a characterization, because I want the sentences about "motives" to be included in mathematics as well. Mathematics unrolls on two tiers at least, the one we know and the one where what we know is revealed as an avatar of what we expect to know—but beyond that there is always another tier.

We have learned to see the roots of a polynomial as an avatar of the Galois group, which is, in turn, an avatar of the (category of) Galois representations, which is an avatar of the category of motives, which is an avatar of the automorphic side of the Langlands correspondence. Some suggest that Langlands will find his correspondence is itself an avatar of a still more abstract correspondence with other avatars in algebraic geometry, in physics (where it overlaps with mirror symmetry, whose *A*-models and *B*-models can be treated as avatars of one another), and even in homotopy theory. My research on this chapter provided unexpected preparation for the latest developments in the *geometric Langlands correspondence*, which cannot be formulated without reference to ∞-categories. So it's not absurd to speculate that something like Voevodsky's Univalent Foundations might reveal all the programs now associated with Langlands' name to be avatars of a common theory, itself soon to be caught up in the cycle of "knowledge and indifference." Bourbaki's insistence on the unity of mathematics, taken to its grimly logical conclusion, suggests that if the development is sufficiently tight, the whole infinite cascade of Langlands dualities, and every other mathematical relation yet to be conceived, could be derived step by step from One Big Theorem at the infinite level—something on the order of *samsāra = nirvana*. "Buddha is said to have remarked that *śūnyatā* is to be treated like a ladder for mounting up to the roof of *prajña* [wisdom, understanding]. Once the roof is reached, the ladder should be discarded." Did Wittgenstein have Buddha in mind when he wrote in the *Tractatus* that "anyone who understands me eventually recognizes [my propositions] as nonsensical, when he has used them—as steps—to climb up beyond them. (He must, so to speak, throw away the ladder after he has climbed up it.)"?[89] The metaphor of foundations does not suit mathematical *concepts*; they are rather suspended from more remote concepts, and when after a long and arduous climb "suddenly . . . one direction becomes . . . more luminous" and we finally find our way to the concepts we had been seeking, we see that they are in turn the avatars of concepts we glimpse only dimly, and so on without end.

chapter 8

The Science of Tricks

Banish the tunes of Cheng and keep clever talkers at a dis-
tance. The tunes of Cheng are wanton and clever talkers are
dangerous.

—Confucius, *Analects* 15:11[1]

Individuals who never sense the contradictions of their cul-
tural inheritance run the risk of becoming little more than
host bodies for stale gestures, metaphors, and received
ideas.

—Lewis Hyde, *Trickster Makes This World*, p. 307

MY *KUNSTGRIFF*

In January 2010 it was revealed that I am a trickster. Toby Gee, a young
English mathematician, broke the story. Before a packed Paris auditorium, Gee explained how he and two even-younger colleagues had
found a new way to exploit what he called "Harris's tensor product
trick"—implicitly allowing that it might not be the only item in my bag
of tricks, that I'm not necessarily a one-trick pony—to improve on my
most recent work. I had first employed the trick in a solo paper, but it
had been recycled to much greater effect in my joint paper with Richard
Taylor and two of his students, who in turn recycled themselves as Toby
Gee's collaborators. During his talk, he hinted that there was more to the
story. At the break Gee let me know that by March, he and all three of
my erstwhile collaborators would have combined a version of my trick
with a few new "key ideas" to prove a big new theorem that incorporates
most previous results in the field—notably all the results to which my
name was attached.

The trickster, in one of many disguises—the Yoruba Esu-Elegbara or
his African American equivalents, the Signifying Monkey and Br'er

Rabbit; the Winnebago Wakdjunkaga, and the Coyote of the American Southwest; or Hermes, Prometheus, and Loki from European mythology—attended nearly every skirmish of the 1990s culture wars but had already tired of running laps around the postcolonialist circuit by the time the archetype's presence as a mathematical figure was abruptly brought to my attention. And here I am barely into the second paragraph and already playing tricks on you—at least two, if you're keeping count, with more on the way. Most blatantly, there's no such thing as a trickster in mathematics. There are plenty of mathematical tricks, enough to fill a "Tricks Wiki" or "Tricki," a "repository for mathematical tricks and techniques."[2] But even the rare mathematicians skilled at inventing trick after trick—the "mathematical wit" Paul Erdős comes to mind—are not known as "tricksters." What my colleagues call a trick is a kind of mathematical gesture or speech act; and what I'm calling a trickster is a role, a persona, one of several, an attribute not of the author who invents or employs it but rather of the text in which this kind of argument or idea or style of thought appears.[3]

As for the trick Toby Gee mentioned in his lecture, it wasn't really even mine, although I was the first to notice its relevance in the context of mutual interest. I learned about this particular class of tensor product tricks from a few of my contemporaries (when I was a not-yet-old dog— just around Toby Gee's age as I write this) and I've applied them—we don't "play" tricks, much less "turn" tricks, in mathematics—to solve problems in at least three completely different settings. So it would be ironic[4] if "Harris's tensor product trick" were to be the last of my marks to fade from number theory, like the Cheshire cat's smile. And as "tricky Dick" Nixon's career reminds us, it is a mixed blessing to be remembered primarily for one's tricks.

You don't really want to know the details of my trick, but you may be wondering what makes it a trick rather than some other kind of mathematical gesture. I ought to be able to answer that question if anyone can, because I'm the one who called it a tensor product trick! Whatever was I thinking? Frankly, I wish I knew. To determine when I began to view this kind of argument as a trick would require an experiment in personal intellectual archaeology or at least a time-consuming scouring of my old hard drives; but it's too late to deny my responsibility for the terminology.[5] There's my usual false modesty, of course, labeling my new invention a (mere) trick in the hope of eliciting warm praise and protestations that "it's more than a trick, it's a game-changer."[6] But the hypothetical mathematical ethnographer will need to understand that not every mathematical speech act deserves to be called a trick and will want to know why.

A straightforward calculation, for example, is certainly not a trick. Nor is a syllogism, a standard estimate of magnitude, or a reference to the literature. Can I be more precise? Probably not. While capital-M Mathematics is neatly divided among axioms, definitions, theorems, and proofs, the mathematics of mathematicians blurs taxonomical boundaries. A mathematical trick, like a trickster, is a notorious crosser of conventional borders; a "lord of in-between" like the Yoruba trickster Eshu, "who dwells at the crossroads," a mathematical trick simultaneously disturbs the settled order and "makes this world," to quote the title of Lewis Hyde's classic study. I would suggest that a trick involves drawing attention to an intrinsic element of a mathematical situation that *appears to be, but is not, in fact, irrelevant* to the problem under consideration. Alternatively, since a trick need not be subordinated to a preexisting problem, it provides an *unexpected point of contact*, like a play on words, between two domains not previously known to be related.[7] Thus *Lieberman's trick*, the first trick I saw identified as such, roughly consists in the use of multiplication in a situation when only addition seems relevant; the *unitarian trick* of Adolf Hurwitz, Hermann Weyl, and Issai Schur introduces a measure to solve a purely algebraic problem; my trick, like other tensor-product tricks, uses the possibility of a kind of (matrix) multiplication to reveal a structure not otherwise visible. Chapter β.5 mentioned a trick connected with Cantor's theory of infinity. The reader (or listener) in each case undergoes a prototypical *Aha!* experience, the apprehension of a gestalt: the connection is obvious, but first you have to experience it.

The ambivalence of the word *trick*, whose associations include magic, prostitution, and deceit, neatly reflects the tension between satisfaction with a synoptic proof and disapproval of the shortcut that avoids the hard work, or the grappling with essential matters, without which recognition seems undeserved. As idealized by logical empiricist philosophers, Mathematics with a capital M is insensitive to the complex interplay of delight (a "neat trick") and disdain (a "cheap trick") that accompanies the revelation of a new mathematical trick and constitutes a privileged moment of pleasure, precisely like that afforded by magic tricks or like García Márquez's reaction, nearly falling out of bed when he read the first sentence of Kafka's *Metamorphosis*: "I didn't know you were *allowed* to write like that."[8] The trickster is a mathematical magic realist who exclaims, "You didn't know you were allowed to fly from peak to peak; you thought you had to trek, or at least calculate, your way across the wilderness. But look: a flying carpet!" This situates the trickster at the pole opposite to the *lumberjack*, who makes his one and only appearance as an incipient mathematical archetype in one of Langlands's most oft-quoted exhortations:

> We are in a forest whose trees will not fall with a few timid hatchet blows. We have to take up the double-bitted axe and the cross-cut saw, and hope that our muscles are equal to them.[9]

The ambivalence of tricks, the sense of getting something for nothing, persists in other languages. The Dutch word *truuk* (also spelled *truc*) is "[m]ostly used in connection with magicians and card tricks . . . a 'truuk' cannot be something very serious."[10] Russian mathematicians use the word *tryuk* (трюк), which in other settings can mean deceit or craftiness.[11] In French a mathematical trick is called an *astuce*, whose primary association with "cleverness" or "astuteness" seems to connote approval, quite unlike the word *tour* used for magic tricks; *jouer un tour* means to "play a trick," usually not a nice one.[12]

German mathematicians nowadays often use the English word *trick* for what traditionally was, and sometimes still is, called a *Kunstgriff*; this is how one properly refers in German to the unitarian trick of Hurwitz, Weyl, and Schur. Stüttgart professor Wolfgang Rump assigned the *Kunstgriff* a legitimate role in mathematics:

> [T]ricks precede a theory, they reach into the as yet unknown, connect what is apparently separate, so that after further reflection the latter finds its natural place in the general theory and thereby becomes known.[13]

Like the African American trickster High John de Conquer, a mathematical *Kunstgriff* "mak[es] a way out of no-way."[14] But the German word has its own unsavory connotations. Arthur Schopenhauer's unpublished 1830 manuscript, entitled *Kunstgriffe*, outlines thirty-eight rhetorical

Kunstgriffe—"stratagems" in English—included in the posthumous compilation entitled *The art of being right.* It could be described as a list of dirty tricks for winning arguments or, alternatively, a training manual for the "clever talkers" of the Confucian epigraph. *Kunstgriffe* of this sort—for example, number 24, which offers advice on "stating a false syllogism"—are uniformly unwelcome in mathematics, but number four on the list may provide insight into the construction of this book and of the present chapter in particular:

Kunstgriff 4

If you want to draw a conclusion, you must not let it be foreseen, but you must get the premisses admitted one by one, unobserved, mingling them here and there in your talk. . . . Or, if it is doubtful whether your opponent will admit them, you must advance the premisses of these premisses. . . . In this way you conceal your game until you have obtained all the admissions that are necessary, and so reach your goal by making a circuit.[15]

THREE FUNCTIONS

Mathematicians used to air their ambivalent feelings about tricks in public. One finds educators opposing tricks to knowledge in 1909:

Much time was spent in trying to find a simpler way [to solve an examination problem] until the "trick" required was found. . . . The question remains: should such questions, based upon the use of special artifices, be set in examinations . . . ? They are no test of the knowledge of candidates, and merely lead them into traps from which they emerge disheartened.[16]

As recently as 1940, the Mathematical Association of America (MAA) could publish an article by a college teacher who complained "that most of us proceed to teach certain sections of elementary mathematics in a way that discourages students by giving them the impression that excellence in mathematical science is a matter of trick methods and even legerdemain."[17] In contrast, in its advice to prospective authors of mathematical articles, the American Mathematical Society (AMS) gives tricks a positive valuation: "Omit any computation which is routine (i.e., does not depend on unexpected tricks). Merely indicate the starting point, describe the procedure, and state the outcome."[18] Readers are jaded by the *routine* but are always on the lookout for new tricks!

Voevodsky's vision of mathematical publication cleared by automated proof checkers, if realized, will require the AMS to countermand current policy. All routine computations will have to be included in order to satisfy the expert system gatekeeper who guards the border with no orders to distinguish tricks from the routine. Ambivalence to tricks, even the sense that tricks form a separate genre, will then be seen as yet another transitory feature of the mathematics of the human period.[19]

By implication, the routine is what keeps mathematicians occupied when we are being lumberjacks rather than tricksters. This is not quite right; the routine is actually only one aspect of normal mathematical practice. Ignoring neutral words like *method* and *procedure*—grounded etymologically, like *routine*, in movement, specifically movement along the road, a meaning too remote to have retained any enchantment value as metaphor—one finds the protagonists of mathematical articles—both their actual authors and the first-person pronouns that narrate the action, so to speak—applying *tools* and *techniques* to transform *ingredients* into *constructions*. Sal Restivo, materialist sociologist of mathematics, put it this way:

> Mathematical workers use tools, machines, techniques, and skills to transform raw materials into finished products. They work in mathematical "knowledge factories" as small as individuals and as large as research centers and world-wide networks. But whether the factory is an individual or a center, it is always part of a larger network of human, material, and symbolic resources and interactions . . . a social structure. Mathematical workers produce mathematical objects, such as theorems, points, numerals, functions and the integers.

Restivo was writing in the 1990s, but already in 1908 you could read in the MAA's *American Mathematical Monthly* that

> [m]athematics consists of ten thousand tools, but . . . Mathematics does not only develop a large number of simple tools . . . it especially emphasizes the putting together of these tools into powerful thought machines.[20]

Mathematicians actually use the word *machine* to refer to a procedure or, more often, a collection of procedures that yield a solution to a class of problems complete enough to be implemented mechanically. Pascal's *machine arithmétique*, or *Pascaline*, was a literal machine that first solved this problem for basic arithmetic in 1642.[21] The formulas illustrated in chapter β, which in constrast to the Pascaline are procedures rather than physical objects, can be described as "machines" for finding the roots to equations of degree 2, 3, and 4. We have already seen that Abel and Galois are famous for proving that there is no such machine for equations of degree 5 and greater. Contemporary algebraic topology has its *canonical delooping machine*, whereas number theory has an *eigenvariety machine*—blueprints courtesy of Toby Gee's thesis adviser Kevin Buzzard—that provides a mechanical solution to a small part of the Langlands program.[22]

It should come as no surprise that mathematicians—as opposed to Mathematicians—lose interest in problems that can be implemented by machines. Plutarch recorded Plato's irritation with Eudoxus, Archytas, and Menaechmus for introducing "constructions that use instruments and that are mechanical" [*organikas kai mekhanikas*] in the doubling of the cube.[23] Gauss wrote in 1850 that in the symbolic language and terminology of his time, "we possess a level, by which the most complex arguments are reduced to a certain mechanism. Thereby the science has won infinitely in richness, but in beauty and solidity . . . it has lost just as much."[24] At the other end of the production process, one finds not a sneering and exploitative factory owner but rather a consensual leader, a *genius*, chosen according to the principles of workers' self-management on the basis of sheer charisma, in the sense of chapter 2. At least that's what we would like to believe. Grothendieck and Langlands provided guidance not only through the example of their work but also, as we've seen, through the active and energetic recruitment of students and collaborators. Although Langlands once posed as a metaphorical *lumberjack* and has long had a deserved reputation for not shirking hard work, his influence has been exercised not only through the *Langlands program*, whose ambitions and methods can be identified with precision, but more diffusely through what continues to be called the *Langlands philosophy*. "The word 'philosophy' was fashionable in 1967, no longer so by 1979. There were lots of philosophies in 1967. . . . It was just the way people talked." Most participants active in the Langlands program dropped the word *philosophy* long ago, but beginners and nonspecialists continue to use it.[25]

Grothendieck is also frequently depicted as a philosopher: you can read about his "philosophy of the six operations," or his "philosophy of anabelian geometry," and especially his "philosophy of motives."[26] But the word more often associated with Grothendieck is *yoga*, as explained in Deligne's remarks quoted in the previous chapter (the "yoga of weights," the "yoga of six operations," the "yoga of de Rham coefficients"). Grothendieck may not have introduced the word into the mathematical lexicon, but he seems to have given it its current meaning:

> *Par "yoga" il [Grothendieck] entendait un point de vue unifiant, une piste dans la recherche des concepts et des démonstrations, une méthode qu'on pouvait réutiliser.*[27]

In between the routine and the exalted, one finds the level of normal mathematical problem solving, descriptions of which are dominated by the vocabulary of combat, as they have been since the beginning of my professional life. The metaphoric fields of mathematics become equally metaphoric battlefields, which is a little surprising given the generally pacific nature of the typical mathematician. Inspiration follows a strategy, a plan of attack, as in Piet Hein's rhyme:

> Problems worthy of attack
> Prove their worth by hitting back

An attacking mathematician identifies strategic objectives and the obstacles to their attainment and mounts the attack, usually armed with *concepts* rather than weapons. Richard Dedekind defended his conceptualism as follows:

> [I]t seems to me that . . . a theory based on calculation . . . does not offer the highest degree of perfection. . . . it is preferable to seek to extract proofs, not from calculation, but immediately from characteristic fundamental concepts, and to construct the theory in such a manner that . . . it shall be in a position to predict the results of calculation.

Though Dedekind did not express himself in martial terms, this is where E. T. Bell, writing in 1944, saw Dedekind formulating "the strategy of abstract algebra."[28] The *strategy of a proof* is then the step-by-step progression from hypotheses to conclusions, outlined in conceptual terms; the experienced speaker constructs a seminar talk on the basis of this outline, omitting routine details and departing from the script only when the strategy incorporates an unexpected feature—a trick, for example.

An attack in the absence of strategy is called *brute force*, which still follows a plan, but one based on the systematic use of routine methods. So the familiar contrast between *theory builders* and *problem solvers* is replaced by the metaphorical distinction between the *strategist* (who in the upper reaches shares some of the charisma of the philosopher) and the *technician* (who in the lower registers is scarcely more than a machine). Prestige is naturally associated with broad vision, so the latter term is typically used in a derogatory manner; in this context "technically very strong" is a double-edged compliment that often suffices to eliminate contenders for prestigious prizes. *Powerful*, on the other hand, applied to tools, methods, insights, or strategies, is a term of high praise, a word chosen expressly to focus the attention of hiring committees and granting agencies, as well as Powerful Beings of both public and private sectors.

It is tempting to map the descending scale of normal mathematical prestige onto the four traditional castes (*varna*) of Hinduism: the *brahmins* are the philosophers and yogis, and the *kshatriyas* are the problem solvers, with the techniques, constructions, and tools in the hands of *vaishyas* and *sudras*. In this way mathematical ethnography collides with Georges Dumézil's *trifunctional* theory of Indo-European mythology. The three functions of the *sovereign* (king, priest; brahmin), the *warrior* (knight, soldier; kshatriya), and the *producers of wealth* (merchant, peasant, artisan; vaishya, sudra) correspond in Dumézil's analysis to the "intellectual structure and mold of thought [*moule de pensée*]" characteristic of the social organization and mythological self-representation of all branches of Indo-European civilization.[29] Not coincidentally, their responsibilities—for justice and relations with the spirit world, war, and production of goods—provide a catalogue of the matters any society must take seriously in order to assure its survival.

The three functions should not be read as a description of a caste system within the mathematical profession. Few can exercise the charisma of a Langlands or a Grothendieck, but the roles of warrior (problem solver) and worker (technician, toolmaker) do not correspond to a functional division of labor within the profession; they rather identify distinct aspects of the work of most individual mathematicians, each associated with a specific lexicon in the ritualized metalanguage. We have already seen Langlands inviting his colleagues to use the tools and the brute force of the lumberjack; but he has promoted and continues to promote specific strategies to achieve intermediate goals and, more recently, to promote his "reckless" strategy that may potentially lead to a realization of his entire program.[30]

Mythological wars are not particularly memorable for their strategy. The action of the *Mahabharata*, the *Iliad*, and Ferdowsi's *Shahnama* is dominated by brute force and skill with weapons. When we look for hints of strategy, we instead find *dirty tricks*. The dirtiest trick in the western canon is undoubtedly the Trojan horse, devised by Odysseus *polumekhanos* [of many devices].[31] At least two of the *Mahabharata*'s key battles were won thanks to dirty tricks suggested by Krishna, the epic's chief trickster, over the protests of the Pandava brothers, with whom he was allied.[32]

There are obvious objections to the imposition of a trifunctional model on mathematicians' apparently informal but, in fact, deeply ritualized discourse. Leaving aside the global nature of contemporary mathematics, influenced by cultural strands having little or no connection to Indo-European roots, perhaps the most cogent objection is that mathematics is hardly the only activity to which a trifunctional analysis can be applied. In business circles we hear about *management philosophy*, *commercial strategy*, and *marketing tools*, while politicians have a *philosophy of government*, a *political strategy*, and *techniques of communication*. The distinctiveness of mathematics may lie, after all, in the nature of the *tricks* characteristic of each activity. Unlike business, politics, or war, mathematics has no place for dirty tricks. As my colleague Marie-France Vignéras put it, "as mathematicians, we play and dream but we don't cheat."[33] That is not because pure mathematicians are purer than everyone else but rather because cheating defeats the purpose of mathematics, however that may be construed. The pointlessness of cheating is, I submit, one of the very best clues to the peculiar appeal of mathematics as a human activity—to its similarity to *play* in Huizinga's sense, for example—and poses a special challenge to science studies as well as to philosophers of Mathematics, who may consider the matter settled once mathematical machines have been programmed not to cheat.

Archeology

No dirty tricks, then, but tricks nonetheless. Contemporary writers frequently read tricks into early mathematics; thus several familiar arguments from Euclid's book on arithmetic, including his proof that there are infinitely many primes, have independently received the name "Euclid's trick." But in my own attempts at archaeology, I've managed to dig down only as far as 1815 before I run out of examples of conscious use of the word, or its equivalent in another language, to designate the phenomenon described before.[34] The cuneiform tablets that represent the earliest-known mathematical texts are punctuated by "you see" (*ta-mar*) that seem designed to force an *Aha!* experience. Euclid's prose is pretty monotonous, but Reviel Netz pictures Archimedes and his "ludic" Hellenistic contemporaries setting "traps," employing "cunning surprise," and indulging in a "carnival of calculation," all with the intention to "dazzle and overwhelm" the reader and to provoke "gasp[s] of delight." The *Līlāvatī* written by Bhāskara II in twelfth-century India is notable for its playful tone. But none of these contains tricks in the contemporary sense.[35]

Like flying carpets, the habit of seeing tricks in mathematics may have come from the East. The evidence is tenuous but appealing. In the mathematical chapter of his *Catalogue of the Sciences* (*iḥsa' al-'ulum*), the tenth-century Baghdad philosopher al-Fârâbî listed algebra not as a branch of mathematics like arithmetic and geometry, but rather alongside mechanical devices, in a chapter on *'ilm al-ḥiyal*—"the science of al-ḥiyal" (singular *ḥila'*)[36]—an equivalent of the Greek *mekhane*, variously translated as "ingenious devices," "mechanics," or "tricks." A century earlier the Banū Mūsā brothers had published *Kitab al-ḥiyal*, a celebrated catalogue of mechanical devices, including automata.

The Banū Mūsā were among the founders of the Arabic mathematical school, and it is perhaps not an accident that in their text *Kitāb marifat masakhat al-ashkal* [The Book of the Measurement of Plane and Spherical Figures], "the terminology of arithmetic is perhaps for the first time applied to the operations of geometry," so that the ratio of the circumference to the diameter of the circle, what we now call π, was treated as a number.[37] The *Book of Science* [*Dāneš-nāma*] of Ibn Sīnā (Avicenna) listed algebra among the "secondary parts of arithmetic." Omar al-Khayyām thought otherwise: *"Those who think algebra is a trick (ḥila')* *to determine unknown numbers think the unthinkable; therefore you* *must not pay attention to those who judge by appearances and are of a* *different opinion."*[38]

If we dig a little deeper, we find Plutarch's account, already mentioned, of Plato's rejection of mechanical methods in mathematics, "a sort of foundation myth for the science of mechanics," which must have been familiar to al-Fârâbî and Ibn Sīnā, and "which explained the separation of mechanics from philosophy as the result of a quarrel between two philosophers."[39] The Aristotelian context for the controversy, interesting in its own right, is discussed in chapter 10. For the moment, we simply note that Gherard of Cremona's Latin version of al-Fârâbî 's catalogue translated *'ilm al-ḥiyal* by *ingeniorum scientia* (the science of *ingenium*, the Latin equivalent of *mekhane*); that *ingenium* also figured in Latin texts on mathematics as well as in the title of Descartes' early *Rules for the Direction of the Mind* [*ingenium*]—in more than one relevant way an exact inversion of Aristotle's value system; and that it admits a great variety of German translations, one of which is *Kunstgriff*.[40] The continuing associations of *ingenium* with *machines* (engineering) as well as *genius*, at the two ends of the trifunctional scale, neatly mirror mathematicians' ambivalence to tricks and incidentally suggest that anything a mechanical theorem prover could invent would be assigned ipso facto the status of trick.

Back in the present, my Taiwanese colleague Kai-Wen Lan tells me "there is no convenient [Chinese] translation for the word 'trick.' Depending on the context, one can find more than a dozen different translations, among which many do have unsavory connotations. However, it is not natural in our culture(s?) to mix them up." According to Teruyoshi Yoshida, the word *torikku* has been adopted in Japan and carries roughly the same meaning as its English homonym; but Japanese has an alternative that is especially intriguing in view of the possible medieval derivation from *mekhane*:

[T]he word "karakuri" (mechanism, device, strategem, trick, system), which goes back to the early Edo period (17c) and used . . . especially for the sophisticated puppet shows (Bunraku/Ningyo Joruri). In mathematics I hear people using the term: "I finally understood the karakuri behind this scary-looking proof" (i.e., how the proof works; idea + mechanism)." It sounds like "it was deceptively mysterious/curious but the actual mechanism is simple," but it doesn't have the negative implication of the English word—"trick" sounds as if the solution to (or avoiding) the profound-looking question was disappointingly easy.[41]

The earliest "trick" I've found in a mathematical journal article is a *Kunstgriff* in the second installment of a long article by Moritz (or Moriz) Abraham Stern, dated 1833, entitled *Theorie der Kettenbrüche und ihre Anwendung* [Theory of continued fractions and their application]. Since this is the first appearance of the word or one of its cognates in the oldest mathematical journal still in existence, the sentence in which it appears deserves to be quoted in full:

In all the transformations of continued fractions indicated above, one applied the trick [*Kunstgriff*] of treating a part of the continued fraction as if it were summed, designating this sum by a letter, and through the relation of this letter with the other members of the continued fraction finding new expressions.[42]

This *Kunstgriff* is recognizably a trick, the first of ten to be given that name during the nineteenth century in *Crelle's Journal*, the oldest mathematical journal still being published.

It took centuries to dissipate the negative connotation the French word *astuce* acquired in 1370, when the mathematician and philosopher Nicolas Oresme wrote, "*Et doncques se l'entention est malvese, tele puissance est appellée astuce ou malicieuseté*"* in the manuscript of his translation of Aristotle's *Ethics*. In his *Mer des Croniques* of 1532, Pierre Desrey described an *astucieux* as a person "who possessed an inventive mind for finding means of deception."[43] The *Dictionnaire historique de la langue française* of A. Rey traces the contemporary positive usage in the sense of "ingenious invention" or "amusing joke" to the middle of

---

* Loose translation: "And thus if the intention is bad, such a power is called *astuce*, or *malicieuseté*."

the nineteenth century and specifically to the "argot" of the grandes écoles, in particular the École Polytechnique. In this case, mathematics apparently led the way to the word's rehabilitation.[44]

I found these examples using online search engines. Many old journals and some old books have been converted into digitized databases that offer the possibility of searching for a word like *trick* or *astuce* or *трюк* or *Kunstgriff*, and that's what I did. For a few hours' work, the results are pretty satisfying. Specimens of trifunctional vocabulary were more difficult to come by, although one might have expected the three functions to be present since the beginnings of Indo-European mathematics.[45] Thus I find Sylvester writing "words are the tools of thought" in *American Journal of Mathematics*, 1886; and in 1870 one finds *Werkzeug*, the German word for "tool," in *Crelle's Journal*, used as a mathematician would today:[46]

$$(12.)\ \delta f(v) = \delta f_0(u) = \sum_a \frac{\partial f_0(v)}{\partial v_a} \delta x_a - \sum_a \frac{\partial f_0(u)}{\partial u_a} \delta x_a(0).$$

This equation, in which the quantities u_t and v_t are viewed as dependent on the elements x_a and $x_a(0)$, is the tool [*Werkzeug*] that will be used to study the normal form $f(du)$.

"Technique," or *Technik*, had acquired its contemporary meaning in mathematics by the 1830s,[47] and the use was standard in the review journal *Jahrbuch über die Fortschritte der Mathematik* practically from its foundation in the second half of the nineteenth century. Given the intense interaction between mathematicians and philosophers over the centuries, it is probably hopeless to use databases to distinguish trifunctional from literal uses of the word *philosophy* in mathematical texts. On the other hand, the earliest example I found of a mathematical "strategy" is in the 1944 E. T. Bell article quoted a few pages back; the expression "strategy of the proof" makes its first appearance in *Mathematical Reviews* in Jean Dieudonné's review of the *Séminaire Cartan* volume on the Atiyah-Singer index theorem[48]—and "brute force" appears at around the same time. Problems were being "attacked" at least as early as the mid-eighteenth century; the word starts to be used in mathematical journals a bit more than a hundred years later.[49]

A trained historian will find it easy to improve on my amateur findings, but even a superficial examination of the printed record makes it clear that the terms all became progressively much more common after World War II—*really* common: Google now returns nearly 5,000,000 sites for "strategy of the proof"—and reached their current configuration by 1970. Around that time, I began my own mathematical initiation, which included training in the use of the metaphoric vernacular as well as the specialized ("technical") vocabulary in which theorems and their proofs are written. It is as if when European and North American mathematicians acquired sufficient professional autonomy to constitute themselves as a tribe, their savage minds reclaimed the three functions of their Indo-European ancestors. The following strikingly trifunctional paragraph in Bourbaki's *The Architecture of Mathematics* (dated 1948, 1950 in English) is characteristic of Bourbaki's structuralist framework (emphasis is mine):

> The "structures" are **tools** for the mathematician; as soon as he [*sic*] has recognized . . . relations which satisfy the axioms of a known type, he has at his disposal immediately the entire **arsenal** of general theorems. . . . Previously . . . he was obliged to **forge** for himself the **means of attack** . . . their **power** depended on his personal talents and they were often loaded down with restrictive hypotheses. . . . One could say that the axiomatic method is nothing but the "**Taylor system**" for mathematics.

Having exhibited the mathematician as blacksmith (later to be joined by Langlands' lumberjack) and assembly-line worker, as well as military strategist, Bourbaki reminds us in the next paragraph of the (charismatic) first function:

> This is however, a very poor analogy; the mathematician does not work like a machine, nor as the workingman on a moving belt; we can not overemphasize the fundamental role played in his research by a special intuition . . . not the popular sense-intuition, but rather a kind of direct divination . . . which orients at one stroke in an unexpected direction the intuitive course of his thought, and which illumines with a new light the mathematical landscape.[50]

So where does the trickster fit in? Dumézil wrote a book about Loki but did not attempt to assign the trickster a consistent role in his trifunctional theory. The sector of French mathematics represented by Bour-baki had (and perhaps still has) little use for tricks. I was unable to find any trace of the word in Bourbaki's *Eléments des Mathématiques*, and Bourbaki's only allusion to tricks in *The Architecture of Mathematics* is dismissive:

> [T]he axiomatic method has its cornerstone in the conviction that, not only is mathematics not a randomly developing concatenation of syllogisms, but neither is it a collection of more or less "astute" tricks, arrived at by lucky combinations, in which purely technical cleverness wins the day.[51]

My provisional hypothesis is that the trickster serves as a *bridge* between *high* and *low* genres. This is the role often reserved for Indo-European tricksters: think of Krishna as simultaneously the god Vishnu and his human avatar; or of Prometheus, who descended to earth with the gift of heavenly fire. Hermes was not only the messenger of the gods but also the guide who presided over the passage of the soul to the underworld; Mephistopheles plays a similar role in the Faust legend. Esu, the Yoruba divine trickster, limped because his legs were of different lengths: "one anchored in the realm of the gods, . . . the other . . . in . . . our human world."[52] In mathematical settings the trickster predates logicist or formalist idealizations and continues to offer a shortcut bypassing the (Indo-European?) idealized route from human practice to inscription of theorems in the register of the eternals.

AFFINITIES

And this brings me to the question at the origin of this chapter. How can I explain the persistent tendency to class mathematics as a high genre, along with the fine arts and specifically with classical music? Why do you find a string trio and a Steinway grand piano in the music room at the Mathematisches Forschungsinstitut Oberwolfach but no electric guitar, no drum set, no scratch mixers, no samplers?

Popular books and articles have explored the supposed affinity between mathematics and music. The authors typically start with Pythagoras,[53] continue by way of the medieval quadrivium, where music was taught as a branch of mathematics (as also in Ibn Sīnā's *Book of Scien-*

tific Knowledge), through Kepler's *Harmonices Mundi* and Leibniz's claim that music was an "unconscious exercise in arithmetic," and arrive in the present day with anecdotes about musical mathematicians.[54] What is not mentioned so often is that at most times, the affinity between mathematics and "serious" music was markedly one-sided. Jean-Philippe Rameau, it is true, based his influential principle of harmony on a mathematical theory of harmonic overtones, the *corps sonore*; he wrote in 1722 that "only with the aid of mathematics did my ideas become clear." Rameau corresponded with the leading mathematicians of the day, including D. Bernouilli and Euler; he sought and obtained the approval of the French Académie des Sciences for his *Démonstration du principe de l'harmonie*; and his good relations with d'Alembert persisted until Rameau overreached, claiming priority for the *corps sonore* over "the other arts and sciences," geometry included.[55] It is also true that Ernst Krenek argued in 1939 that music, like geometry, needed to be based on axioms.[56] Iannis Xenakis wrote in the 1950s that

> [m]usic can be defined as an organization of these operations and elementary relations [i.e., Boolean algebra and predicate calculus] between sound events [. . .] mathematical set theory [is useful] not only for the construction of new works but also for . . . better understanding of works of the past. Thus even a stochastic construction or an investigation of history with the help of stochastics [is impossible] without the help of Logic or Algebra, its mathematical form, the queen of the sciences and even of the arts.[57]

And Milton Babbitt, who wrote a thesis in 1946 entitled "The Function of Set Structure in the 12-Tone System," taught mathematics at Princeton before switching to the music department, the first of many professors at leading American universities who have looked to such varied branches of mathematics as group theory, orbifold geometry, and topos theory to explain musical structure.[58] Even now IRCAM in Paris sponsors regular meetings on mathematical theories of music (see figure 8.1).

Outside the academy, however, such moments of convergence have been relatively rare. Long before Douglas Hofstadter wrote *Gödel, Escher, Bach*, C.P.E. Bach denied that his father had any mathematical leanings: "the departed [J. S. Bach] was, like myself or any true musician, no lover of dry mathematical stuff."[59] Hector Berlioz, in his 1836 eulogy of his teacher Anton Reicha, wondered whether the latter's at-

Musical covering and the associated simplicial complex

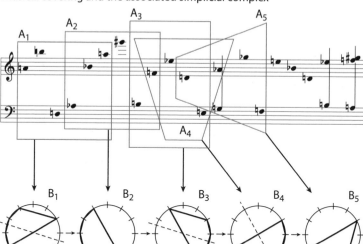

Simplicial complex (nerve)

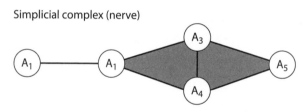

Figure 8.1. A reconstruction of one of the less elaborate slides from *Modern Algebra and the Object/Operation duality in music*, a presentation by Moreno Andreatta, Music Representations Team, IRCAM/CNRS.

tachment to mathematics didn't make his compositions "lose something in melodic or harmonic expression, in purely musical effect, what they gained (if to be sure, it was gaining) in arduous combinations, in conquered difficulties, in curious works made rather for the eye than for the ear?" Camille Saint-Saëns, referring to Berlioz, added that "One doesn't learn art as one learns mathematics." At the end of the century, a London music critic elaborated:

> The Brahms Sextet . . . is one of those odd compositions which at times slipped from the pen of Brahms, apparently in order to prove how excellent a mathematician he might have become, but how prosaic, how hopeless,

how unfeeling, how unemotional, how arid a musician he really was. You feel an undercurrent of *surds*, *of quadratic equations*, of hyperbolic curves, of the dynamics of a particle. But it must not be forgotten that music is not only a science; it is also an art. The Sextet was played with precision, and that is the only way in which you can work out a problem in musical trigonometry.[60]

Jazz age Paris saw arithmetic participate in the torture of the naughty child in Ravel's *L'enfant et les sortilèges*, with libretto by Colette:

LE PETIT VIEILLARD :
Deux robinets coulent dans un réservoir; / Deux trains omnibus quittent une gare,
A vingt minutes d'intervalle, / Valle, valle, valle!

. . .

L'ENFANT :
Mon Dieu c'est l'arithmétique!

LE PETIT VIEILLARD :
Tique, tique, tique! / Quatre et quatre dix huit,
*Onze et six vingt cinq, /Sept fois neuf trente trois.**

In 1930 music theorist Heinrich Schenker could just as well have written "mathematical" as "mechanical"—"the mechanical spreads throughout the entirety of average men like a poison gas"—when he contrasted Rameau's "paralysis" with the living tradition represented by German musicians. "How," he asked, "did the *paralysis of the theoretical tendencies* . . . manifest itself?" His answer was that "something mechanical . . . lay in the basic idea of Rameau from the start."[61]

Well into the twenty-first century, mathematics is still widely seen as music's diametrical opposite: "The pieces [Schoenberg] composed at this time were mystical, richly symbolic and emotionally charged, hardly pure mathematics," wrote one British critic, while another adds that "once you're inside music, all thoughts about mathematics become ir-

---

* LITTLE OLD MAN: Two faucets flow into a reservoir/two local trains leave a station/at 20 minutes interval / Val, val, val!

. . .

CHILD: My God, it's arithmetic!
LITTLE OLD MAN: Tic, tic, tic / 4 and 4 make 18 / 11 and 6 make 25 / 7 times 9 make 33.

relevant." Even Pierre Boulez, the founder of IRCAM, who elsewhere wrote that "music is a science as much as an art," complained at one point that "what is called the 'mathematical' . . . mania . . . gives the illusion of [music as] an exact, irrefutable science" and referred to "number-fanatics" who seek a "form of rational reassurance."[62]

Don't you think there's something pathetic about how certain mathematicians keep insisting that they are artists—we'll see more of this in chapter 10—when the artists want only to keep their distance? It's probably no accident that mathematics and classical music were not perceived as antithetical only during those periods when the notion of a "love formula" was not seen as oxymoronic. Rameau and Clairaut were contemporaries; Xenakis, Boulez, and IRCAM belong to the same world as Lacan and the structuralists, the world of an "objectivist" aesthetic of music in which "musical style . . . is eventually conceived in terms of statistics and music tends to lose its human significance."[63]

WHY SO SERIOUS?

In his *Emblems of Mind*, Edward Rothstein, *New York Times* critic—and former mathematics graduate student at Brandeis University—explores parallels between mathematics and classical music, ignoring the historically less well worn but, in principle, no less cogent affinity of mathematics with rock and roll or rap. Catherine Nolan's article in the *Princeton Companion to Mathematics* is written from the same perspective, as was a recent lecture in Paris by philosopher Alain Badiou, entitled *Mathématiques/Esthétiques/Arts* and sponsored by IRCAM. In the (otherwise very different) treatises of Herbert Mehrtens and Jeremy Gray on modernism in mathematics, the cultural references are Picasso, Stravinsky, Klee, or Apollinaire, not advertising, the development of a mass market, mass production, or the trade-union movement, much less the mechanically reproducible pop culture to which Walter Benjamin's celebrated essay (indirectly) drew attention.[64]

"Pop music," writes Rothstein, "fulfills a different function from art music and often has different ambitions." Just what are these functions, and why are mathematicians reputed to find the prestige of art music irresistible, in spite of the persistent indifference of its practitioners? The relegation of popular culture to an inferior status—analogous to that of tricks in mathematics—is anthropological in nature, in that it is not generally the object of an explicit proscription, nor is it ascribed to a rational judgment (except a posteriori), but neither is it merely a matter of individual taste. Names for distinct varieties of facetiousness come readily to mind, but how can we talk about seriousness? Why do routine mathematics and classical music exemplify this virtue more convincingly than mathematical tricks or popular music? As Heath Ledger's Joker asked in *The Dark Knight*, "Why so serious?"

Overlapping accounts of the origins of professorial seriousness in general stress its sources in the respect due to power,[65] the respect due to respectability,[66] the respect due to standards of perfection on which society depended,[67] or the respect due to norms of virtuous behavior.[68] These points are all valid and convincing and must be taken into consideration when trying to understand how social pressure is brought to bear upon the academic community, and upon scientists in particular, when enforcing standards of behavior. My goal in this chapter, however, is ethnographic, not historical or sociological, a description from the inside of a living culture, and therefore I want to understand how these social forces are internalized as norms within the specific culture of pure mathematics—and, no less importantly, to look for traces or tokens of these standards in the symbolic system used by members of the tribe. The ambivalent status of the trick and the supposed affinity of mathematics with classical music are my main exhibits. I am not defending the position that these characteristic attitudes originate in external social pressure, nor, on the contrary, that they arise naturally from the practice of mathematics. Though I have my opinions—musical taste looks to me more exogenous, the suspicion of tricks more endogenous—I'm not professionally qualified to make that determination. Here, then, are a few hypotheses that may deserve further study.

1. Recall the word *legerdemain* in the 1940 MAA article on tricks. Is the word *trick* perhaps an unwelcome reminder of the historic association between mathematics and magic, even witchcraft? The word *mathematicus*—probably the correct name for the figure in a painting by Luca Giordano (figure 8.2), pointing to what appears to be an astrological chart[69]—was used primarily for astrologers. The medieval scholastic

Figure 8.2. Luca Giordano, *El Matemático* [the *Mathematicus*?]. (From the collection of the Museo Nacional de Bellas Artes, Buenos Aires)

Roger Bacon, who condemned the mathematics that was "a subdivision of magic," nevertheless considered "astrology rather than astronomy . . . by far the most important and practical part of mathematics."[70] This state of affairs persisted until well into the seventeenth century at least: casting horoscopes was one of the duties of the chair of mathematics at the University of Padova, and Kepler and Galileo, both prominently fea-

244 • Chapter 8

tured on IBM's "Men of Modern Mathematics" chart, were occasionally employed as astrologers and had at least mixed feelings about the practice.[71] The *matemático* in figure 8.2 is thus in the same line of work as Johannes Faustus and his trickster sidekick, Mephistopheles:

> *Das gefiel Doctor Fausto wol speculiert vnnd studiert Tag vnnd Nacht darjnnen / Wolt sich hernach kein Theologum mehr nennen lassen / ward ein Weltmensch / Nennt sich ein Doctor Medicinæ, ward ein Astrologus vnnd* **Mathematicus.**[72]

In his work on magic, Giordano Bruno saw mathematics as a bridge between the higher and lower orders of being, "a link . . . between the celestial and the terrestrial":[73] just the role we assigned the trickster. John Dee, mathematician and court magus to Elizabeth I, described the bridge as follows:

> A marvellous newtrality have these things Mathematicall, and also a strange participation between things supernaturall, immortall, intellectuall, simple and indivisible: and things naturall, mortall, sensible, compounded and divisible.

Dee, a possible model for Marlowe's *Doctor Faustus* (and for Shakespeare's Prospero, in *The Tempest*) had been arrested in 1555 for his horoscope predicting Elizabeth's accession to the throne and was one of the most prominent mystics of his time; he had reportedly "on occasion transmuted base metals into gold."[74] He was also a leading promoter and popularizer of mathematics, a central transitional figure in the period before the elaboration of the modern scientific worldview, "one English author for whom both philosophy and mathematics figured together as vital parts of an intellectual programme." As mathematician, Dee's most important work was probably his *Mathematicall Praeface* to the 1570 English edition of Euclid; as *mathematicus*, a typical publication was his *Monas Hieroglyphica* (Antwerp, 1564; see figure 8.3), an early attempt at cosmological grand unification in the form of a list of numbered theorems, written in a arcane language apparently designed to prevent the acquisition of "an unlawful knowledge of mysteries" by "unworthy persons who"—very much like the "forces of Evil" in *Rites of Love and Math*—"would correctly interpret, but abuse, his all too candid exposition":[75]

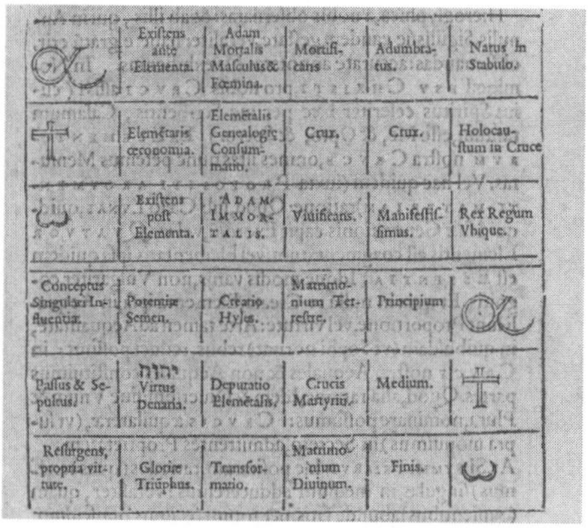

Figure 8.3. John Dee. (*Monas Hieroglyphica*, p. 23)

2. The "very serious man" Hermann Weyl did as much as anyone to legitimize the use of the word *trick* in the informal language of working mathematicians. He was the first to use the English word *trick* in each of the two leading U.S. mathematical journals of his day, and the word *Kunstgriff* appears in a number of his papers in German. But there are no allusions to tricks in his highly regarded book *Philosophy of Mathematics and Natural Science* and only two in a separate collection of his philosophical writings; in one, he finds that a "strange scholastic trick . . . shows up peculiarly in the disposition of [Newton's] *Principia* which is otherwise so rigorous." So Weyl also locates the *Kunstgriff* on the wrong side of the border between "low" and "high." Rump's theoretical analysis of mathematical tricks—the only one I have been able to find—justified the *Kunstgriff* explicitly as a step on the way to the "general theory" with its "natural places"; this is consistent with G. H. Hardy's attribution of what he called the " 'seriousness' of a mathematical theorem" to "the *significance* of the mathematical ideas which it connects."[76]

Hyde writes that "accidents happen in time, essences reside in eternity . . . [t]he [Norse] gods grow old and gray" when Loki's intervention gives the giants a taste of the Apples of Immortality.[77] The precarious status of tricks in view of Rump's criterion is well illustrated by Grothendieck's reported reaction to Deligne's 1973 proof of the last of the Weil conjectures (see chapter 2), now known as *Deligne's purity theorem* and considered one of the outstanding theorems of twentieth-century number theory, indeed of all time. Grothendieck made a special trip back to the IHES from his provincial exile to learn about Deligne's shortcut solution to the problem Grothendieck had posited as the culmination and ultimate justification of his remodeling of geometry. "If I had done it using motives," recalled Deligne, "he would have been very interested, because it would have meant the theory of motives had been developed. Since the proof used a trick, he did not care."[78] One assumes that Grothendieck felt the method had not revealed the motives at "the heart of the matter"[79]—just as al-Fârâbî had included algebra among the tricks. Years later, Langlands speculated that "perhaps [Grothendieck] could have drawn a different conclusion. . . . the last of the Weil conjectures was proven, by Deligne, in essence on the basis of a profound understanding of the étale cohomology theory accompanied by an observation arising in the theory of automorphic forms."[80]

Bourbaki had no use for tricks, as we've seen, and among the Bourbakistes, Grothendieck found them particularly distasteful. But in this case, what Grothendieck disdained as a mere trick is now seen by the majority of mathematicians as a brilliant insight Deligne derived from his study of the Langlands program. Deligne's proof indeed preceded a "general theory," namely, Langlands's theory of functoriality—but for Grothendieck it was the wrong general theory. Which goes to show that trickiness is not an intrinsic, much less quantifiable, property of a mathematical text.

3. In his *Moderne Sprache, Mathematik*, Mehrtens stresses that aspirations to seriousness were integral to the modernization process. "As a mathematician [German folklore figure Till] Eulenspiegel transformed himself from an anarchist fool to a theologian." Nor did mathematics merely leave its trickster outfits behind: the German modernizers were conscious of parallels between their goals and those of their artistic contemporaries. Mehrtens explains that, when the mathematician Heinrich Liebmann described mathematics as a *"freie, schöpferische Kunst"* ("free creative art") in his Leipzig inaugural address, comparable to that of the Berliner Secession painter Max Liebermann, the "freedom" Liebmann had in mind was to establish one's own *Qualitätskriterien* and not to be subject to the "quality criteria" of the Kaiser and similar Powerful Beings (in the case of artists) or of engineers, teachers, and philosophers (in the case of mathematicians).[81] So mathematical modernists were not only being elitist when they demanded autonomy, professional recogni-

tion, and authority comparable to that of the state; their elitism was precisely modeled on that of their fine arts counterparts. The Mehrtens book reproduces Liebermann's 1912 portrait of Felix Klein, a central protagonist of mathematical modernization.

A parallel can be drawn with the "fashioning" of English literature as a "serious discipline"[82] in the period following World War I. In 1877 it could be argued that the study of English literature "might be considered a suitable subject for 'women . . . and the second- and third-rate men who . . . become schoolmasters.' " Terry Eagleton writes that even in the 1920s, "it was desperately unclear why English was worth studying at all; by the early 1930s . . . English was . . . *the* supremely civilizing pursuit, . . . immeasurably superior to law, science, politics, philosophy, or history." This was largely the work of F. R. and Q. D. Leavis and their associates at Cambridge, whose passionate attachment to England's language and high literature were matched only by their contempt, displayed in Leavis' epigraph to this chapter, for "the philistine devaluing of language and traditional culture blatantly apparent in 'mass culture.' "

Speaking to the British Association in 1897, Andrew Russell Forsyth fretted about a new "attitude of respect . . . almost . . . of reverence . . . we mathematicians are supposed to be of a different mould . . . breathing a rarer intellectual atmosphere, serene in impenetrable calm. It is difficult for us to maintain the gravity of demeanour proper to such superior persons; and perhaps it is best to confess at once that we are of the earth." The adoption or reaffirmation of seriousness of demeanor as a characteristic virtue of the university professor plays itself out as a series of events in the history of the construction of the image of the professor as an authority figure. In his analysis of Liebmann's inaugural address, Mehrtens is calling attention both to the role of seriousness in establishing status in a hierarchical society during the period of breakdown of traditional authority and to an alliance, at least de facto, between mathematicians and practitioners of the fine arts. Mehrtens may have considered it superfluous to remind his German readers that if Till Eulenspiegel dressed like a clown, it was in part because the trappings of the higher strata were forbidden to him by German sumptuary laws. (Compare the "rude mechanicals" in *A Midsummer Night's Dream* or the father of Quevedo's picaresque hero *El Buscón*, who reassures his son that "thievery is a liberal art, not a mechanical art."[83])

ARTISTS IN RESIDENCE

Since serious authority nowadays is vested in *finance* rather than royalty, it's prudent to remember that, though Freud did claim that "the opposite of play is not seriousness, but—reality," Huizinga—who saw seriousness as a derivative concept, defined by the absence of play—reminded us that the opposite of play can also be *work*. Our exploration of virtual realism suggests that Freud's reading may be faithful to the inner life of mathematics, but work is what Powerful Beings, public or private, expect in exchange for sponsoring our research.[84]

Nevertheless, acceptance of recognized canons of serious demeanor is no longer necessary to establish authority. A new relation to seriousness may even now be under negotiation. Back in February 2011 at the IAS, the setting of so many of this book's scenes, I approached the term's first "After Hours Conversation"—artist-in-residence Derek Bermel had promised to talk about *Stravinsky, Hip-hop, and the Scientific Method*—with misgivings. The cultural offerings I sampled during previous Institute visits had been relentlessly lofty and Apollonian, and I could feel another Dionysiac episode coming on. Bermel, a classically trained composer and clarinet virtuoso, had brought his instrument, and he began his allotted ten minutes of remarks by illustrating metric displacement as a source of variation and development in Stravinsky's *L'histoire du soldat*. He then noted that the rapper Rakim uses a similar process, moving his texts across the poetic line within a fixed meter. To illustrate the process Bermel performed this text by Rakim (note the enjambment in the first and last three lines):

> Write a rhyme in graffitti in every show you see me in
> Deep concentration 'cause I'm no comedian
> Jokers are wild if you wanna be tame
> I treat you like a child then you're gonna be named
> Another enemy, not even a friend of me
> 'cause you'll get fried in the end if you pretend to be.

I nearly fell out of my chair. I didn't know you were *allowed* to do things like that at the IAS![85]

Bermel is at home in the classical repertoire of previous IAS artists-in-residence, but I can't imagine what the latter would make of Rakim.

"Performers in the classical tradition often have an extremely limited range," Bermel told me, "and in particular have trouble with unfamiliar rhythms." So are rockers and rappers[86] the new Pythagoreans? Consider Math Rock, which, according to Toby Gee, is "characterised by a somewhat 'angular' sound, and maybe some tricky time signatures." For example, *26 Is Dancier Than 4*, by the Oxford Math Rock group This Town Needs Guns, is written in 26/8 time; Don Caballero's *Slice Where You Live Like Pie* features guitars playing in 5/4 over a drumbeat of 7/8; *Calculating Infinity* by The Dillinger Escape Plan (founders of the variant of Math Rock called Mathcore) opens with a sequence of 3/8, 3/8, 3/8, 4/8, 3/8, 3/8, 3/8, 4/8, 3/8.

It's no accident that Toby Gee has resurfaced at this point of the narrative. If I chose him as my guide to math rock it is because we are members in good standing of the same mathematical clan. The word *tribe* used a few pages back is not a metaphor, nor is it a sly reminder of this chapter's anthropological theme.[87] Mathematics has not one, but *two*, independent formally recognized kinship systems. The first is based on *networks of collaboration*. Mathscinet, the online version of the AMS's *Mathematical Reviews*, provides a list of every collaborator with whom a given mathematician has published and calculates collaboration distance, on the model of the Erdős number mentioned in the previous chapter. Thanks to my trick, for example, my collaboration distance from Toby Gee has dropped to 2, one less than my Erdős number. The second kinship system is recorded in the *Mathematics Genealogy Project* (MGP).[88] *Genealogy* refers to the ascending series of dissertation advisers. It's common for mathematicians to refer to their advisers as "parents" and fellow students of a given adviser as "siblings." So I am doubly linked to Toby Gee through our mutual collaborator Richard Taylor, who can claim Gee as one of his grandchildren. Gee's lineage is particularly distinguished—Taylor was "begotten" by Andrew Wiles, whom (Cambridge professor) John Coates "begat"; before his retirement, Coates held the Sadleirian chair previously occupied by G. H. Hardy and was himself a student of Fields Medalist Alan Baker. The MGP counts me, but not Gee, among the 109,100 known descendents (as of June 2013) of Nilos Kabasilas, fourteenth-century bishop of Thessaloniki; my distant cousins include Michelangelo and Leonardo da Vinci, as well as both Grothendieck and Langlands.[89]

If authors of popular books about mathematics actually looked at mathematicians, they would find enthusiasm for and performance of popular music as common as among the general public. The interesting question is why they find it convenient to ignore this in favor of an exclusive insistence on the shaky alliance with classical music. Among my mathematical kin, Gee is noted for his serious interest in popular music, as is his "father," Kevin Buzzard, not to mention his "sibling," Dan Snaith (Manitoba, Caribou, Daphni), the most prominent indie rocker with a PhD in number theory. Skeptical, like Buzzard, about the term "math rock," Gee advised me to listen to *Atlas* by Battles, "by far the best thing that gets called math rock" that Gee had heard in recent years. And so I learned that, Pynchon's Weed Atman notwithstanding, rock's relations to mathematics are not straightforward. In a 2011 interview, Battles' members had this to say about the genre:

What about the term "math rock"?

[*drummer John*] *Stanier*: As a band we say we dislike it immensely, but there's really not much you can do about it. . . . it's lazy, like you can't think of anything better to say than math rock. I hated math.

[*guitarist Ian*] *Williams:* But I loved rock.[90]

In *Victims of Mathematics*, the punk rock group Grade argued that "applying fractions to modern day living is as useful as . . . handing gasoline to an arsonist." Cherry Ghost's *Mathematics* is about alienation: "hear the unforgiving sounds of cold mathematics making its move on me now." Rock's main gripe with mathematics seems to be that—once again—love doesn't conform to its equations. Little Boots' *Mathematics* put it this way:

By the time I calculated the correct solution
The question had escaped me and so did the conclusion . . .
I'll believe you 'cause your X is equal to my Y
But equations pass me by

Patti Smith and her late husband Fred weighed in as well:

No equation to explain
Destiny's hand

Moved, by love
Drawn by the whispering shadows
Into the mathematics
Of our desire

Russian rocker Boris Grebenshchikov, who studied mathematics at Leningrad State University, placed the relation of mathematics to music in a broader context:

As one scientist to another, I'll say to you frankly—no one, except the perfectly enlightened, can even begin to imagine how the world is really constructed. . . . One day, when I was studying . . . in the Faculty of Applied Mathematics and Control Processes, I started thinking and naively asked our professor of mathematical statistics, the marvelous Nikolai Mikhailovich Matveev: ". . . Statistics answers the question of the probability that some event will take place; but it seems to me much more important to know—why did this specific event happen and not some other one?" And he gently replied: "Young man, to answer that question, you will have to go to the Theological Academy."

With that, he conclusively solved the problem of my relation to science. From that moment mathematics lost all interest for me, though I didn't go to the Theological Academy, inasmuch as I was playing music—and probably in the depth of my soul I was certain that music would lead me to the answers to all my questions. Which is what happened.

In the same vein, filmmaker David Lynch, who recently recorded an album entitled *Crazy Clown Time*, once expressed "his belief that mathematics might be the only way of rationally describing what the real world is like, but he doubted that we would ever be capable of developing so complex a mathematics."[91]

We turn to African American popular music with relief: there's no stigma attached to mathematics! Thus Robert "Noise" Hood, cofounder of the minimal techno collective *Underground Resistance*, recorded a track entitled *Calculator* under the pseudonym Mathematic Assassins.[92] Contrast C.P. E. Bach's dismissive rejection of "dry mathematical stuff" with these lyrics from George Clinton's *Mathematics of Love*:

Cause any percentage of you / is as good as the whole pie
and any fraction thereof/brings dividends of love

Hip-hop culture, for its part, positively embraces mathematics. Kanye West's interlocutor is meant to be flattered, not insulted, when he informs her (in *Never Let Me Down*) that "[y]ou don't need a curriculum to know that you are part of the math." Mathematical physicist and jazz saxophonist Stephon Alexander grew up in the Bronx:

> I used to take the bus home from school and these guys would be on the back of the bus around, rappin' and battling each other—droppin' science, so to speak, and one day, you know, I heard someone say, "Yeah, man, 'cause I got enough mathematics."[93]

In hip-hop slang, "math" or "mathematics" can also mean a telephone number, as in *Brooklyn Masala* by Masta Ace: "She winked at me, and kinda laughed / Ripped the piece of the grocery bag and wrote her math." But mathematics is more than just an incredibly supple idiom for urban romance. In a song entitled *Mathematics*—the top hit for "mathematics" on YouTube—Yasiin Bey (better known as Mos Def) advises his "working-class poor" listeners to learn the numbers to survive:

> This new math is whippin motherfuckers ass
> You wanna know how to rhyme you better learn how to add
> It's mathematics[94]

Reading this entry from the online biography of Wu-Tang Clan associate, DJ/producer Allah Mathematics (also known simply as Mathematics)—

> Mathematics is the universal language. . . . It can cross barriers of language, color, religion, ethnicity, race, and more.[95]

—I am tempted to speculate that the universalism that comforts the hip-hop artist speaking in the name of a marginalized population, living in a town that certainly doesn't need more guns, is precisely what rockers like Grebenshchikov or Smith, anxious to affirm their uniqueness, find disaffecting.[96]

Now even those musicians—Rameau, Xenakis, or Yasiin Bey—who express admiration for mathematicians do not welcome us as fellow artists; their vision of mathematics is as exemplar of the true (or scientific) rather than of the beautiful. Nevertheless, bearing in mind that "signifying is the grandparent of rap" and that "It's Tricky to rock a rhyme, to rock a rhyme that's right on time," hip-hop's deepest affinities with mathematics may still be cultural rather than ethical or epistemological.[97]

UNDERGROUND RESISTANCE

Raymond Williams points to "a ready-made historical thesis" for what certain critics find disappointing in popular culture: "After the Education Act of 1870, a new mass- public came into being, literate but untrained in reading, low in taste and habit. The mass-culture followed as a matter of course."[98] The obvious corollary to this thesis has never lacked defenders. F. R. Leavis thought that being capable of a "discerning appreciation of art and literature . . . constitute[s] the consciousness of the race (or of a branch of it) at a given time. . . . Upon this minority [those capable of such discernment] depend the implicit standards that order the finer living of an age. . . ." "[T]here's this idea," Boulez once said, ". . . that *everybody* should make music, *every* discipline is equally worthy— whether it's rock, jazz, folk—all *"les musiques,"* plural. No, they're *not* equal, I don't agree. I believe in fine art, I believe in aristocracy, and I believe in elite [culture]."[99]

If advocates of elite culture disdain rock and rap, it's because, like the "wanton" music of Cheng, they are said to appeal to the body's lower sectors, unlike classical music.[100] Friedrich Schiller's poem *Archimedes and the apprentice*, which features the line "He who woos the goddess, seek in her not the maid," was a favorite among nineteenth-century German mathematicians, starting with Gauss. Bertrand Russell recommended mathematics as "a means of creating and sustaining a *lofty habit of mind*"—"like poetry or music," added G. H. Hardy.[101] Plato found mathematics lofty to the point of bypassing or transcending the body altogether: mathematical knowledge is stored in the soul, according to the *Meno*; involving the body leads, as we've seen, to confusion and dizziness.[102] This is demonstrably false. Mathematics is typically created not at the desk or blackboard, but during the strenuous Wednesday afternoon hike at the Oberwolfach institute or a leisurely stroll in the IAS woods. Well into his eighties, Jean-Pierre Serre (whom the Paris daily *Libération* compared to Mozart) still goes rock climbing regularly in the forest near Fontainebleau; Cédric Villani ("the Lady Gaga of mathematics," according to the Paris weekly *Télérama*) lists "walking" along with music as hobbies on his CV.[103]

But elitist standards continue to evolve, and Bermel—who has collaborated with Mos Def/Yasiin Bey—was exercising his faculty of distinction, rather than being provocatively lowbrow, when he insisted to me that "strong and mediocre music can be found in any genre." What matters is what Bermel (sounding very much like Bourbaki) called "coherence of structure" and what Laurence Dreyfus called the "respect for the inherent meaningfulness of the world," a recognition of how freedom and necessity are "inextricably linked."[104] Bracketing a novel invention as a trick is a way to maintain mathematical standards while admitting that, depending on the trick's "coherence" and "inherent meaningfulness," its status may change in time.

Schiller, author of the *Ode to Joy*, thought, romantic that he was, that "Man plays only when he is in the full sense of the word a man and he is only wholly Man when he is playing." "Plays are play," writes Alison Gopnik, "and so are novels, paintings, and songs." Cognitive scientist Steven Pinker has a more functional point of view. What pure mathematics and music have most clearly in common is that they both fit Pinker's definition of "cheesecake," "unlike anything in the natural world because it is a brew of megadoses of agreeable stimuli which we concocted for the express purpose of pressing our pleasure buttons. Pornography is another pleasure technology. . . . [T]he arts are a third."[105] Pure mathematics would appear to be a fourth, and what it shares with the arts, and with music in particular, is that, unlike cheesecake or even pornography, its variety is literally limitless.

Schiller's *playing*, whether it is opposed to seriousness, to work, or (anticipating Freud as well as Pinker) to reality, is Burghardt's "relaxed field," freedom from constraint. Reflecting on his use of a *truuk* to prove a conjecture Grothendieck had formulated thirty years earlier, Huizinga's compatriot Frans Oort admits that the "[tricky] aspect of mathematics" seen by some to lack elegance "has an appealing beauty to me . . . I find this exciting."[106] In the end the alliance between mathematics and music—and the arts more generally—is in freedom and creativity rather than beauty or formal analogy.

Some of us find Pinker's cognitive behaviorism too . . . mechanical . . . and are drawn rather to Lévi-Strauss's construal of myth and music as "instruments for the obliteration of time." Mathematics, whose proofs admit the impersonal imperative and the conditional but whose conclusions are stated in the eternal present tense, is another such instrument. What need has mathematics of the formal trappings of seriousness when, to quote Norbert Elias, it is "one of the symbolic structures in whose name one may claim, as [G. H.] Hardy does, to offer eternal verities outlasting death?"[107]

Goethe wrote in *Die Natur* that "Life is [nature's] loveliest invention, and death is her *Kunstgriff* for having much life." But outwitting death, though it occasionally backfires, is a trickster specialty; outlasting death must be the finest trick of all. And the conundrum of mathematical seriousness, such as it is, may resolve itself when viewed in a broader perspective. Reading popular culture's mathematician as a variant of the computer hacker or "wizard" who is by now a staple of thrillers of every genre, the recent fascination with mathematics is easily explained as a by-product of our deepening immersion in cyberspace as a second nature. Ed Frenkel's fictional persona, as mathematician, occupies the same moral culture of resistance as the hackers in *The Matrix*, or the plugged-in primitives of *Avatar*, or *The Girl with the Dragon Tattoo*. Nearly all the apparently mathematical protagonists of popular films work in areas closely associated with computing: prime numbers for *Proof*, graph theory for *Good Will Hunting*, numerical computation for *Pi*—it wouldn't be a stretch to add the rock-star mathematician's chaos theory in *Jurassic Park*. In the background we hear popular rather than classical music in these films—R&B in *Good Will Hunting*, a blend of (black and white) electronic dance music genres in *Pi*, while the romantic interest in *Proof* plays in a rock band—and these productions are best seen as borderline variants of a much more extensive genre of which the (ideal)-typical representative is *cyberpunk fiction*, whose protagonist is a film noir reincarnation of the nineteenth-century mathematical romantic and whose very name is an homage to the dominant rock ethos of the time.[108]

Like the (often inadvertently) anarchist heroes of Pynchon's novels, who struggle to maintain their values, their personal identities, even their lives in the face of the implacably deterministic projects of more or less precisely identified Powerful Beings, cyberpunk stages a confrontation between an artificial intelligence programmed to impose its version of certifiable seriousness and a (Neu-)romantic trickster hero whose quest is to preserve what remains of our common humanity. "I'mm goinng to do everrythinng I can to sstopp you fromm turrnning that poorr olld mman innto a piece of ssofttware in the bigg bopperrs' memorry bannks."[109] Circumstances force the typical cyberpunk protagonist into a life of underground resistance to a political and economic order projected onto a future dominated by AI but not so different, in the end, from the instrumental market rationality with which each of us must come to terms in this life. But this faceoff mirrors the confrontation within mathematics between two facets of the same split personality: mathematics is both the condition of alienation and the skill without which community cannot be restored.[110] The former is *ingenium* as *mekhane*, the relentless mechanical unfolding of an "unforgiving cold mathematics," the Matrix, "the man," the deterministic and Powerful Beings of *Gravity's Rainbow* known only as They[111]—and is represented in Hollywood by mathematicians with pronounced social pathologies. The latter is *ingenium* as *trick*, the way of Br'er Rabbit and Coyote, of the cyberpunk hackers of film and fiction—and of the eternal sun-drenched present of *Vineland*'s group-theorist and groupie-magnet Weed Atman, avatar of Galois if not Grothendieck, and his utopian People's Republic of Rock and Roll.

PART III

How to Explain Number Theory at a Dinner Party

THIRD SESSION: CONGRUENCES

The proof of Gauss' two square theorem, as well as its statement, is based on the distinction between even and odd numbers. But the statement also distinguishes between odd numbers of the form $4k + 1$ and those of the form $4k + 3$. One should imagine a cyclic classification of numbers into four classes, called *congruence classes modulo 4*:

$$0, 4, 8, 12, \ldots$$

$$1, 5, 9, 13, \ldots$$

$$2, 6, 10, 12, \ldots$$

$$3, 7, 11, 15, \ldots$$

These classes can be labeled by their first members: the class of 0 modulo 4, of 1 modulo 4, and so on. One can define congruence classes modulo any number; for example,

$$0, 3, 6, 9, \ldots$$

$$1, 4, 7, 10, \ldots$$

$$2, 5, 8, 11, \ldots$$

are the congruence classes modulo 3. Congruence classes modulo 12 are used to tell time: the clock reads the same after 3 hours as after 15 hours as after 27 hours, and so on. Formally, we say that the integers a and b belong to the same congruence class modulo the integer N if the difference $(a - b)$ is a multiple of N—in other words, if N divides $(a - b)$ without remainder.

The story of congruences is that a problem where the variables can take infinitely many values can be replaced by one in which the variables can take only finitely many values, and it is sometimes enough to solve the latter problem in order to solve the former. This is roughly how the proof of Gauss' theorem goes, though a great deal of algebra has to be established ahead of time to make this work.

Gauss' methods show that when p is an odd prime number, then

$$x^2 + 2y^2 = p$$

has infinitely many rational solutions when p is congruent to 1 or 3 modulo 8, none when p is congruent to 5 or 7 modulo 8. For example, if $p = 17$, take $(x, y) = (3, 2)$; if $p = 19$, take $(x, y) = (1, 3)$. Similarly,

$$x^2 + xy + y^2 = p$$

has infinitely many rational solutions when p is congruent to 1 modulo 3 but none when p is congruent to 2 modulo 3.

In the next session we're going to move up to equations of degree 3 in two variables, which is as high a degree as I would ever want to calculate at a dinner party. Motivation will have to wait until the next session; for now, I just want to explain how to *count* solutions to congruences. Something interesting already happens with equations in a single variable. Suppose p is a prime number and consider the equation of degree p:

$$x^p - x = 0 \tag{F}$$

It's obvious that $x = 0$ and $x = 1$ are solutions and if p is an *odd* prime number, then $x = -1$ is also a solution. It is less obvious, but also true, that these are all the solutions in rational numbers; the proof is based on the same principles as the proof of irrationality of $\sqrt{2}$. However, there is a notion of *approximate* solutions for each prime number p: we say x is an approximate solution to equation (F) for the prime p if the difference between the right-hand and left-hand sides of the equation is a multiple of p—in other words, if the two sides *belong to the same congruence class* modulo p. This is determined by dividing the difference between the two sides by p; if there is no remainder, then we have an approximate solution. For example, if $p = 7$, we can check that $x = 2$ is an approximate solution: the left-hand side is

$$2^7 - 2 = 128 - 2 = 126 = 7 \times 18$$

which is divisible by 7. But, in fact, we didn't have to calculate; we could simply have quoted the following.

Fermat's Little Theorem: Let p be a prime number. Then every integer a is an approximate solution to equation (F): $a^p - a$ is *always* divisible by p.

This curiosity, which is not difficult to prove, was discovered by Fermat in the early seventeenth century and is now presented at the very beginning of any course in modern number theory. We could say that there are infinitely many approximate solutions, since every integer is a solution. For the reasons to be explained shortly, it is more reasonable to count each congruence class modulo p only once; thus equation (F) has p approximate solutions modulo p for any prime number p.

Before we move on to equations in two variables, let's return briefly to equation (I) of chapter α. It turns out that the equation $x^2 = 2$ has two approximate solutions modulo p if $p^2 - 1$ is divisible by 16, and no approximate solutions modulo p otherwise. So if $p = 7, p^2 - 1 = 49 - 1 = 48$ is divisible by 16; and $x = 3$ and $x = 4$ are both approximate solutions to $x^2 = 2$ modulo 7. We check: if $x = 3$, the left-hand side is $3^2 = 9$, whereas the right-hand side is 2, and $9 - 2 = 7$ is a multiple of 7. Again, if $x = 4$, the left-hand side is $4^2 = 16$, the right-hand side is 2, and $16 - 2 = 14$ is again a multiple of 7. The diligent reader can check that there are no other approximate solutions with x between 0 and 6; and an even more diligent reader will check that $x^2 = 2$ has no approximate solutions modulo p when $p = 3, p = 5$, or $p = 11$, which is consistent with what we said previously, because $9 - 1 = 8$, $25 - 1 = 24$, and $121 - 1 = 120$ are not divisible by 16.

As for the equations $x^2 = 3$, $x^2 = 5$, $x^2 = 7$, and so on, the *quadratic reciprocity theorem*, proved by Gauss (in eight different ways), completely determines the number of approximate solutions modulo p for any prime p:

- If p and q are two different odd primes, at least one of which is of the form $4k + 1$ (so can be written as the sum of two squares), then $x^2 = p$ has two approximate solutions modulo q if and only if $x^2 = q$ has two approximate solutions modulo p.

- If both p and q are of the form $4k + 3$, then $x^2 = p$ has two approximate solutions modulo q if and only if $x^2 = q$ has *no* approximate solutions modulo p.

It looks like the riddle has been answered by another riddle, but the two preceding statements [together with the approximate solutions to equation (I)] contain enough information to solve the problem in all cases.

There is no (obvious) way to count the rational solutions to an equation in two variables, but as with equation (F), there is a notion of approximate solutions for each prime number p. We say x and y are an approximate solution to the equation

$$y^2 = x^3 - x \qquad \text{(E1)}$$

for the prime p if the difference between the two sides of the equation $x^3 - x$ and y^2 is a multiple of p. We can then count approximate solutions to equation (E1) and similar equations, for example,

$$y^2 = x^3 - 25x. \qquad \text{(E5)}$$

The number of approximate solutions for each p is written $S(p)$, as follows. I illustrate the procedure for $p = 2$, $p = 3$, and $p = 5$ and expect the reader to get the hang of it in general. Some of the calculations are indicated in the table in figure γ.1: Y means Yes (the pair is a solution), N means No (not a solution). The number of solutions is the number of Ys.

The point is that we have to try only one x in each congruence class and, likewise, one y; replacing x by another number in its congruence class—say, replacing $x = 0$ by $x = 2$ for $p = 2$—gives us no new information. For $p = 2$, we have to try only four pairs. When $x = 0$ and $y = 0$, the equation is

$$0^2 = 0^2 - 0,$$

which is true, so we have one solution—a genuine (not approximate) solution. When $x = 1$ and $y = 0$, we have another genuine solution. When $x = 0$ and $y = 1$, we have $y^2 = 1$ but $x^3 - x = 0$, and the difference is 1, which is *not* a multiple of 2. Finally, when $x = 1$ and $y = 1$, we have $y^2 = 1$ but $x^3 - x = 0$, which is again not a solution. We thus have found two solutions, and we say $S(2) = 2$.

E1, p = 3	x = 0	x = 1	x = 2
y = 0	$y^2 - (x^3 - x) = 0$, Y	$y^2 - (x^3 - x) = 0$, Y	$y^2 - (x^3 - x) = -6$, Y
y = 1	$y^2 - (x^3 - x) = 1$, N	$y^2 - (x^3 - x) = 1$, N	$y^2 - (x^3 - x) = -5$, Y
y = 2	$y^2 - (x^3 - x) = 4$, N	$y^2 - (x^3 - x) = 4$, N	$y^2 - (x^3 - x) = -2$, Y

E5, p = 3	x = 0	x = 1	x = 2
y = 0	$y^2 - (x^3 - 25x) = 0$, Y	$y^2 - (x^3 - 25x) = 24$, Y	$y^2 - (x^3 - 25x) = -42$, Y
y = 1	$y^2 - (x^3 - 25x) = 1$, N	N	N
y = 2	$y^2 - (x^3 - 25x) = 4$, N	N	N

E1, p = 5	x = 0	x = 1	x = 2	x = 3	x = 4
y = 0	Y	Y	N	N	Y
y = 1	N	N	Y	N	N
y = 2	N	N	N	Y	N
y = 3	N	N	N	N	N
y = 4	N	N	Y	N	N

E1, p = 7	x = 0	x = 1	x = 2	x = 3	x = 4	x = 5	x = 6
y = 0	Y	Y	N	N	N	N	Y
y = 1	N	N	N	N	N	Y	N
y = 2	N	N	N	N	Y	N	N
y = 3	N	N	N	N	N	N	N
y = 4	N	N	N	N	N	N	N
y = 5	N	N	N	N	Y	N	N
y = 6	N	N	N	N	N	Y	N

Figure γ.1. Counting solutions to some cubic equations.

Next, try $p = 3$. We still have the two genuine solutions, $x = 0, y = 0$ and $x = 1, y = 0$. But now we can try $x = 2, y = 0$. The left-hand side is $x^3 - x = 0$, whereas the right-hand side is $x^3 - x = 8 - 2 = 6$, and the difference between the sides is 6—which is a multiple of 3! In other words, $x = 2, y = 0$, while not a genuine solution, is an approximate solution, so we count it as well. We can continue to test all possible combinations where $x = 0$, 1, or 2 and $y = 0$, 1, or 2. The only solutions are the first three we found; therefore, $S(3) = 3$.

Now consider the case $p = 5$. We have to test x from 0 to 4 (after which the congruence classes repeat), and likewise for y. The first novelty is the case $x = 2$, $y = 1$. The left-hand side is $y^2 = 1$; the right-hand side is $x^3 - x = 6$. The difference between these two sides is 5, a multiple of 5. We find that $x = 3$, $y = 2$; $x = 4$, $y = 0$; and $x = 2$, $y = 4$ are all solutions, and there are no others. Thus $S(5) = 6$.

The last table in figure γ.1 shows the answers for $p = 7$, and one counts $S(7) = 7$.

The reader is invited to check the correctness of the table and to determine $S(p)$ for the next few primes: $S(11)$, $S(13)$, $S(17)$,. . . . The expressions y^2 and $x^3 - x$ grow quickly as p gets bigger, but not too big for a calculator to handle. But before you get started, you might want to know whether or not there is a pattern to the $S(p)$. The answer is literally: yes and no—explanation next time.[1]

chapter 9

A Mathematical Dream and
Its Interpretation

On sabbatical from my position as professor at Brandeis, I spent the 1992–1993 academic year in France, visiting colleagues and teaching courses at two universities—Université Paris 7, in the center of Paris, and Université Paris-Sud, in Orsay, a half-hour's train ride to the south—in preparation for a possible move to Paris. Boston was then and still is one of the world's great mathematical centers, and by attending Harvard's number theory seminar and the MIT representation theory seminar, I kept in touch with all the most important developments relevant to my own work in automorphic forms, on the border between these two subjects. Paris, however, was not only actively and consciously exercising its role as the natural headquarters of mathematical research in Europe, with the most extensive seminar schedule anywhere; it was also home to the world's largest concentration of specialists in automorphic forms, most of them roughly my own age. This meant that for nearly twenty years we had followed developments in the field and neighboring fields in the same sequence, had witnessed the same breakthroughs and had met one another repeatedly at the same international meetings, and were deeply familiar with one another's complementary contributions to a highly active, influential, and competitive branch of mathematics.

A year after I wrote a thesis in pure number theory in 1977, I switched to the field of *Shimura varieties*, a geometric structure invented by Goro Shimura (one of my professors at Princeton) in his work relating automorphic forms to number theory, an early inspiration for the Langlands program. In this way I gradually became a specialist in automorphic forms. My interests, reflecting my start in number theory, were somewhat peripheral from the standpoint of most of my Paris colleagues, who were mainly guided by the priorities of the Langlands program. Of all the possible techniques for proving his conjectures, Langlands preferred those connected with the Selberg trace formula and its vast generalization by (his student and fellow Canadian) James Arthur. In this, Langlands has been consistent through most of his career, frequently to the point of criticizing proofs of his conjectures that avoid use of the trace formula. The general idea of the trace formula is not hard to explain and is quite similar to the classical Lefschetz fixed point formula in algebraic topology, the source of Weil's topological intuition, which, vastly generalized by Grothendieck and his collaborators, was central in their approach to the Weil conjectures.

Both the Lefschetz and the Arthur-Selberg trace formulas can be seen as examples of *index formulas*, which arise in one form or another in most branches of pure mathematics as a means of deriving often unexpected consequences for large and complicated *global* objects—such as differential equations, solutions of polynomial equations, or Galois theory—from purely *local* data that are, in principle, elementary and amenable to calculation. It was one's attitude to this "in principle" that determined where one stood with respect to Langlands' program. To begin calculating the local data relevant to his program, one needed to assimilate a mass of complex, specialized notation and terminology that had mostly been developed by Langlands and his closest collaborators, some of them Parisians, much of it in articles by Langlands himself that were notoriously difficult to read. Arthur's version of the Selberg trace formula not only involved local data, but also presumed the solution of global problems in simpler situations and hence introduced a recursive structure into the problem with its own complications. Finally, the heart of Langlands' plan involved use not of Arthur's trace formula, but a hypothetical *stable* version thereof, whose construction of the stable formula relied on yet another series of difficult articles by Langlands, as well as the *fundamental lemma*, which, as mentioned in chapter 2, was not resolved by Ngô until fifteen years after my Paris visit, following the latter's work with (his former thesis adviser) Gérard Laumon, based on yet another new version of Grothendieck's trace formula.

For all these reasons—and also because my original interest in Shimura varieties derived from number theory and geometry, rather than from group theory—those parts of the Langlands program based on the trace formula were not to my taste, and I had avoided learning the relevant techniques, although inevitably I was exposed to the methods in international conferences and to the phenomena they were designed to explain in my own work. Langlands had been a frequent visitor to Paris, however. This was where he had first presented his vision of the stable trace formula in a series of lectures I had attended in 1980. At the time I had understood nothing whatsoever, but the Parisian specialists had studied these ideas over the years in their weekly automorphic forms seminar. A few of these specialists were recognized internationally as experts in the Langlands program; nearly all had made direct use of Langlands' techniques in one way or another in their own work.

Even had I wished to sign on to the Langlands program, I had no hope of catching up with specialists who enjoyed a fifteen-year head start. I preferred to remain on the margins, where my work would not be judged by comparison with a preestablished set of goals and milestones. By 1992 I had spent fourteen years using Shimura varieties to study *special values of L-functions*, which is, roughly speaking, a way of relating two different classes of transcendental numbers using algebra. I entered this field by accident, because I knew how to put together two kinds of methods, one from geometry and one from group theory, whose conjunction I had recognized under another form in an article by Shimura himself. During the intervening years I had learned to vary the specific ingredients, initially as a close reader of Shimura's work, and in this way I learned a good deal more geometry and group theory, but the kind of combination was invariably the same. I did not so much set myself specific goals as discover new problems similar to those I had already considered in the articles I read while pursuing my education or in conversations with my collaborators. Already in 1989, I was getting tired of this, and when I spent a year in Moscow I hoped to switch fields—or at least to look at a class of special values of L-functions I hadn't previously considered. Five years earlier this had been a major priority for the Russians with whom I was expecting to work, especially for Alexander Beilinson, but in the interim most of them had radically shifted their priorities, and although I returned from Moscow with a host of new ideas, they were very much along the lines of what had been on my mind when I had arrived and owed little or nothing to my interactions with my Russian colleagues.

The projects I had begun in Moscow were nearly exhausted by the time I arrived in Paris three years later. I was looking for something new, not only because I was tired of the old subject but because it was tired of me and had no new problems to offer. My most promising new departures of the previous years had turned out to lead to problems far beyond my powers to solve. While waiting in Paris for inspiration to strike, I had found the key to completing the last of my Moscow projects. Once again it was a matter of aligning geometric and group-theoretic ideas, and although I knew clearly enough what was involved, I could tell it would be long and tedious to write out the details, and I was not eager to begin. I must have been complaining to my friends, because one of them wrote me at that time:

Dear Michael,

It is time for a good book on special values of L-functions but I don't know if you really want to write one. On the other hand, it does seem perfectly rational to be dismayed at the prospect of another 30 years of results about period relations in the setting of Shimura varieties. So I understand your desire for change. . . . I don't think it can be that satisfying to be thought of as Mr. Coherent Cohomology and I can see the desire to get beyond that pigeonhole.

My only other project was a long shot, an attempt to understand yet another kind of special value of L-functions by comparing a version of the *relative trace formula*—in some respects a refinement of the standard Arthur-Selberg trace formula—with a sort of refinement of the Grothendieck-Lefschetz trace formula called *arithmetic intersection theory* (or *Arakelov theory*). The goal was to find an abstract framework to explain, and ultimately to generalize, the constructions in the landmark work of Benedict Gross and Don Zagier on the Birch-Swinnerton-Dyer conjecture, in which two infinite collections of terms (one geometric, one group-theoretic) are shown at the end of nearly one hundred pages of computations to match miraculously, term by term, with striking consequences.

This is the sort of fuzzy idea that inevitably occurs to a number of people when they have nothing better to do, and though I had nothing very definite in mind, two talks I had heard the previous summer had revived my interest in the question. At a conference in Jerusalem, I had heard Steve Rallis talk about his new and abstract approach to the relative trace formula; at a conference at the MFO in Oberwolfach, I had heard Michael Rapoport explain his new work with Thomas Zink on the p-adic properties of Shimura varieties, which I hoped would explain the geometric side of the Gross-Zagier formula.[1]

Rapoport, now at Bonn, at the time a professor at Wuppertal, was going to be in Paris for a few weeks in December in connection with his participation in a jury overseeing the Orsay thesis defense of one of Laumon's graduate students, named Alain Genestier. I had seen Rapoport at the Tuesday seminar at Orsay and had proposed that we meet for lunch, to talk about the work he had described in Germany. The lunch took place on Thursday, December 10, the day before Genestier's thesis defense.

Rapoport and I were both satisfied with the meal, which was unusual in itself. I raised my question; Rapoport expressed interest and had even brought some relevant documents to the table, but after a brief discussion, we agreed that neither of us was sufficiently prepared to go more deeply into the question and decided to postpone further consideration of the problem until we had had time to read the relevant background material. We made tentative plans for me to visit Wuppertal in January. The conversation then turned to matters connected with Genestier's thesis. Genestier was studying an analogue of Shimura varieties called *Drinfel'd modular varieties*, and he was doing so from the point of view of their *p-adic uniformization*. This was a property they shared with certain Shimura varieties, a fact that had only recently been established by Rapoport and Zink as an application of the ideas Rapoport had explained in Germany. This is not quite right. Actually, Genestier was not thinking about p-adic uniformization but about the *Drinfel'd upper half-space*, also known as Ω, an object with both geometric and group theoretic properties that needed to be understood before moving on to the more elaborate questions of p-adic uniformization. A primary motivation for studying Ω, Rapoport explained, was that it was widely expected to give a geometric model (cohomology) for a group-theoretic object— the local Langlands correspondence for $GL(n)$ of a nonarchimedean local field. For the kind of Ω treated in Genestier's thesis, this local Langlands correspondence had been established by Laumon and Rapoport in collaboration with Rapoport's Wuppertal colleague Ulrich Stuhler. But there was no geometric model for this correspondence, and the correspondence was still a major open problem for p-adic fields, where p-adic uniformization was known.

If these ideas were familiar to me at all, it was only in the vaguest way. Four years earlier I had given a presentation at a conference in Ann Arbor (as "Mr. Coherent Cohomology"), in which Henri Carayol, then as now at Strasbourg, had given several lectures on Ω, setting out the conjectures Rapoport sketched to me over lunch. I remembered Carayol's lectures and the notes he had distributed at the conference, but I could not relate what I remembered to what Rapoport was telling me. Rapoport advised me to reread the article based on Carayol's lectures in order to prepare for Genestier's thesis defense. I did so that night, motivated by the potential relevance of this work to my understanding of the Gross-Zagier formula. Carayol's article was written very clearly, and although the notions he treated were unfamiliar, I could see a parallel with other geometric constructions of group representations, specifically Wilfried Schmid's proof in the mid-1970s of a conjecture of Langlands to the effect that the discrete series of real Lie groups occur in L_2 cohomology of period domains—an object for which Rapoport and Zink claimed to have found an analogue for p-adic groups—and the Deligne-Lusztig construction of representations of finite groups of Lie type, originally inspired by work of Drinfel'd. A curious feature of the conjectures outlined by Carayol was that they involved the actions of *three* groups: the group GL(n, F), where F is a p-adic field; the multiplicative group J of a certain division algebra over F, and the Weil (or Galois) group W of F. Drinfel'd had proved in a very difficult paper that his Ω was the first of an infinite family of *unramified coverings* $\Omega_1, \Omega_2, \Omega_3, \ldots$ and Carayol's conjecture, an elaboration of earlier conjectures of Deligne and Drinfel'd, was that the natural action of $W \times J \times$ GL(n, F) on the cohomology (cf. chapter 7) of this family simultaneously realized the (conjectural) local Langlands correspondence between representations of W and GL(n, F) and the (known) Jacquet-Langlands correspondence between representations of J and (certain) representations of GL(n, F).

Genestier's thesis was a difficult piece of work, his defense was professional but intended primarily for experts, and I was not an expert. Nevertheless, I understood enough to be struck by the resemblance of his *irreducibility theorem* to a theorem I had studied as a graduate student, due to Ken Ribet. I wondered whether Genestier might not be able to derive his irreducibility result as Ribet had, by studying the action on the unramified coverings of the stabilizers of certain natural points on the base space. The reader will not be surprised to learn that Genestier's thesis defense was followed by the customary champagne reception, and I remember insisting on this idea in conversations over champagne with Laumon, with Rapoport, and with Genestier himself. It is more than likely that at the reception I drank more than four glasses of champagne, which I have learned in the course of many thesis receptions marks the border beyond which my remaining capacity for coherent thought is no longer equal to the demands of the profession.

The following morning was Saturday. My wife had an early appointment and we had set the alarm early. I drifted into consciousness with the certainty that I had just dreamt about the cohomology of unramified coverings of Drinfel'd upper half-spaces and that the dream had brought me an insight I could not quite recover but that I was certain I should not let slip away. Warding off my wife's attempts to rouse me completely, I remained at the edge of wakefulness for several minutes, until the insight solidified to the point of being expressible in words—or, more accurately, a combination of words and images to which I could associate mathematical content. Over the next few weeks my ideas grew clearer as I reread Carayol's article and discussed the problem with colleagues in Paris and Orsay, so that by December 29 the insight that came to me in my dream had taken the form of a research program that I described in detail in a letter to Rapoport.

Dear Michael—

It is probably a good time to think about organizing my visit to Wuppertal, if this is going to happen. Since I saw you I started thinking about a related but very different problem, namely, the one you mentioned of trying to construct the discrete series of GL(n) by imitating Atiyah-Schmid for the Drinfel'd modular varieties. I came up with a crazy idea that is impossible (for the moment) to put into practice but that is probably right nonetheless. It is inspired by Schmid, rather than Atiyah-Schmid, and actually more by Zuckerman's algebraic version of Schmid, and even more by recent work of Schmid and Vilonen on realizing the characters of discrete series by a Lefschetz fixed point formalism. The basic idea is the following. . . .[2]

. . . I think I can come to Wuppertal during the week of January 18.

Rereading this letter, it is clear to me that I had learned a phenomenal amount of mathematics from my colleagues in Paris during the last two weeks of December. At the time of my lunch with Rapoport, I had no precise idea of many of the notions described here; certainly I would not have written about them with such confidence. The allusions to Laumon's suggestions refer to conversations that took place the week after Genestier's thesis defense, mostly with Genestier in attendance. Laumon made clear his preference for a cohomology theory that "really exists," as opposed to the one, implicit in my letter, recently developed by Vladimir Berkovich, which did indeed "exist" but which did not obviously have all the properties needed to prove the a Lefschetz formula. The paragraph (see note 2) beginning *Hypothesis (d) would follow* is an almost verbatim account of what, after the dream itself, was the most uncanny incident of the whole experience: when Genestier asked me how I hoped to carry out the comparison in (d), I proceeded without the slightest hesitation to explain the argument involving Lubin-Tate groups, of whose possible relevance to the problem Genestier, Laumon, and I learned simultaneously from the unconscious source to which I had tuned in during the dream and which had not bothered to provide this detail until it was specifically requested.

By chance, that same week Laumon's Orsay colleague Guy Henniart was hosting two visitors, Phil Kutzko of the University of Iowa and Colin Bushnell of London's King's College. Bushnell, Henniart, and Kutzko were the three leading experts in the representation theory of $GL(n)$;[3] they had been interested for years in the local Langlands correspondence for p-adic fields and had begun writing a long series of articles on the subject, following Henniart's proof of the "numerical Langlands correspondence"[4] and the monumental book of Bushnell-Kutzko.

I had known Kutzko for some time and met Bushnell at a conference in the soon-to-be-former East Germany in December 1989, when the two of them presented the results that soon appeared in their book. Due to a misunderstanding, I had not realized this was the main point of the conference—which Rapoport also attended—and my own presentation was interrupted before the end by the main East German organizer, who loudly protested that it was not only incomprehensible to everyone else in the room but that it was irrelevant to the proceedings. In 1989 I had nothing to tell Bushnell and Kutzko, nor was I in any way able to appreciate their work. Three years later, though, I eagerly followed the two of them, and Henniart, to a *brasserie* in Montparnasse, where I spent much of the meal asking their opinions of what my dream had taught me.

The story of the dream is only halfway done, and although I will spare you most of it, I have not yet told you whether or not it has a happy ending, nor whether or not it is the one the text thus far seems to have prepared. But I already want to stress the point of this story, which is that it does *not* follow the standard account of the role of the unconscious in scientific thinking, as exemplified by Kekulé's (possibly apocryphal) dream about the benzene ring, or Poincaré's celebrated discovery of the relation between Kleinian groups and non-Euclidean geometry as he stepped onto the omnibus, or the dream of Robert Thomason to which I devoted a speculative article.[5] Max Weber wrote famously that

> [i]deas occur to us when they please, not when it pleases us. The best ideas do indeed occur to one's mind in the way in which Ihering describes it: when smoking a cigar on the sofa; or as Helmholtz states of himself with scientific exactitude: when taking a walk on a slowly ascending street; or in a similar way. In any case, ideas come when we do not expect them, and not when we are brooding and searching at our desks. Yet ideas would certainly not come to mind had we not brooded at our desks and searched for answers with passionate devotion.[6]

Kekulé, Poincaré, Thomason, and dozens of others have recounted the dreams and unconscious interludes that helped them solve problems that had long troubled them, perfect instances of Weber's ideal-type. The contrast with my situation could not be more striking: the dream I have described provided a strategy for solving a problem about which I had never brooded and to which I had devoted no passion, a problem I had considered altogether irrelevant to my interests one week earlier. And though I was unable to bring the dream argument to a successful conclusion, the dream and the interest it inspired in this question did change my mathematical priorities radically and was instrumental in my acquiring the degree of charisma to which I allude in chapter 2.

For three months I thought intensely about how to transform the research program proposed in my dream into rigorous mathematics. Concretely, this meant I read widely and spoke to all the colleagues I could reach in an attempt to solve problems (a)–(d) described in note 2. I did visit Wuppertal in January and explained my ideas at length to Stuhler as well as Rapoport.[7] I accepted an invitation to Strasbourg to visit Carayol a few weeks later, and although my lecture was on another topic, most of my conversations were again about the ideas of my dream. Carayol was in Strasbourg, as was Jean-François Boutot, and their papers on the subject were my main source of inspiration.

Carayol's Ann Arbor lectures clearly made the connection between his conjecture and Shimura varieties, and his article with Boutot was the main reference for Drinfel'd's Ω, other than Drinfel'd's original and ferociously difficult short note. I spent half my time looking back and forth between Genestier's thesis, which had developed a new way to calculate with the coverings of Ω, and the Boutot-Carayol article, unable to apply Genestier's methods to problems (a) and (b), especially the latter, but hoping that inspiration would strike. The other half of my time I spent catching up on fifteen years of work on the trace formula in connection with the Langlands program.

In the spring, I was teaching an undergraduate course at Orsay and sharing an office with Luc Illusie, who had been Laumon's thesis adviser and, by that token, Genestier's mathematical grandfather. Early in May, Luc arrived at the office one morning and announced, in English, "You're cooked!" He showed me a message he had received from a California colleague named Arthur Ogus: the German algebraic geometer Gerd Faltings, one of the most overpowering mathematicians of his generation, had just given a lecture in Berkeley on the cohomology of Drinfel'd's coverings of Ω based on an approach apparently very similar to mine, but he had claimed much more than I could dream of proving. "You and Genestier are both cooked!" I entertained hopes that there had been a misunderstanding until notes taken at Faltings's lecture arrived in Paris, including a calculation roughly equivalent to (c) and (d) of my message to Rapoport (note 2), and—much more importantly, to my mind—the announcement of a proof of a version of (b).[8] Question (a) was left as a conjecture except in the two-dimensional case. The details appeared a few weeks later when Ogus mailed photocopies of his notes, most notably: Faltings's effective use of the Berkovich cohomology theory, in which he was able to make sense of the Lefschetz trace formula; his ingenious (partial) solution of problem (b); and his very difficult solution to problem (a) in the first nontrivial case, that of dimension 2.

In the meantime I had written to Carayol. In my files, my message is dated May 12, 1993. The original was in French, but it loses nothing in an English translation:

Henri—

It seems that Ogus is at the origin of a rumor according to which Faltings has proved "Drinfel'd's conjecture" on the cohomology of coverings of the non-archimedean half-plane. You must be aware of this. No one has seen Ogus's notes, so we don't know what this is about. I can hardly imagine he has proved the local Langlands conjecture.

If by "Drinfeld's conjecture" one means the conjecture that all discrete series representations can be realized in the cohomology of the coverings, without specifying the multiplicity, nor possible non-discrete components, then the claim seems strange to me, because I had the impression that you had given a more or less complete argument in your Ann Arbor talk. In any case, I think I can complete your argument for supercuspidal representations, using the results of Kottwitz and Clozel on twisted unitary groups[9] . . . But I suppose you already knew how to do this. Nevertheless, if Faltings is really in the process of proving the conjecture (by local methods, perhaps), it would be useful to make the global proof public (for example, the people at Orsay don't know it).

There follows a final technical paragraph in which I sketch my approach to the "global proof." Rediscovering this message, I find myself a little surprised by the timing. I had thought the conversation with Illusie had taken place in March and that the ideas described in my message to Carayol were developed in the two intervening months, as a way of channeling the disappointment at learning the news about Faltings. My message to Rapoport described what I had hoped would be a completely new research project to occupy me for five years or more—a chance to give Mr. Coherent Cohomology a rest. Apparently the reality was quite different. Between March and May, I had convinced myself of the possibility of a global argument. But my dream's appeal lay precisely in the *absence* of global techniques. Devotion to an ideal of methodological purity led me to prefer a purely local approach to a problem that was itself purely local; and the problem offered what looked like the prospect of a five-year vacation from Shimura varieties.[10]

The news about Faltings put an end to these daydreams, and I reluctantly resolved to save what could be salvaged from the six-month apprenticeship. A few weeks later, Carayol came to Paris for a day or so (to see his dentist). In the interim I had described my global approach to Henniart, who quickly showed that my results implied that the cohomology of the coverings of Drinfel'd's Ω gave a "numerical correspondence"[11] that was both constructive and natural—what mathematicians call *canonical*, meaning insensitive to the ambiguities of identity that were explored in chapter 7. I met Carayol at a not particularly memorable café on the Place d'Italie and explained how the global argument to which I had referred in my message led to a natural candidate for the local Langlands' correspondence. He had predicted as much in his Ann Arbor talk, but at the café he denied he had thought through the consequences of the global argument.

The story lasted another eight years, and in a significant sense it has not yet ended. The ideas I worked out with considerable help from my French colleagues that spring (and many others later, not only French) were finally published four years later, as "an elaboration of Carayol's program." Several articles, several ideas, and several years later, Richard Taylor and I wrote a book containing, among other results, the first proof of the local Langlands conjecture for p-adic fields.[12] But the one we solved was only one of the many local conjectures, those formulated by Langlands himself and those proposed by analogy, and there are currently several active branches of number theory that derive in part from the ideas I first encountered in my dream in 1992.

Just over midway through my career to date, that dream set in motion developments that changed my life in more ways than I care to name. But I have recorded this story because I want to understand its uncanny side, and this particular incident is more uncanny, I believe, than the typical intervention of the unconscious in science. The literature on the role of the unconscious in creativity contains many striking examples of dreams providing solutions that had long resisted the persistent efforts of scientists' conscious minds. I know of no other example of a dream providing a strategy to solve a problem that had never previously laid serious claim to the dreamer's attention. The dream did help me solve what I may well have felt to be my most pressing scientific problem, escaping the role of Mr. Coherent Cohomology. But that hardly suffices to explain the dream's manifest content.

Not being inclined to seek supernatural explanations for life-altering events, I have been wondering for years how that marvelous idea found its way into my dreams and stayed there long enough for me to remember it. Just recently I have begun to piece together a tentative explanation. My theory does not show me in a particularly flattering light, but it is highly plausible. I suspect the unconscious drive behind my dream was, in a word, jealousy—long-forgotten jealousy, more precisely, directed at one person with whom I overlapped only briefly and a second who was a total stranger during the period in which the jealousy was experienced.

As a graduate student, I might not even have become aware of the local Langlands conjecture, were it not for the fact that Jerrold Tunnell's Harvard thesis containing the first proof of the conjecture in a nontrivial setting was written the same year as my Harvard thesis on a totally unrelated subject. Tunnell's thesis overshadowed all others that year; being naturally competitive, I suppose that must have made me uncomfortable. When I occasionally made use of Tunnell's other work in subsequent years, I don't remember any conscious residue of the jealousy of my last year as a Harvard graduate student. But, it's certain that the incident persisted as an unconscious memory, and it's plausible that it was triggered by the allusions to the local Langlands conjecture in my lunchtime conversation with Rapoport.

Even earlier, as an ambitious undergraduate math major in Princeton, I had been exposed to the local folklore of Princeton undergraduates who had realized their ambitions. The star shining on the distant horizon was John Milnor, the 1962 Fields Medalist, who at the time was a professor at the nearby Institute for Advanced Study but whose Princeton senior thesis was still being quoted by knot theorists. A more recent landmark was the undergraduate career of Wilfried Schmid. I no longer remember the stories told about him, but I was aware that, only a few years past his PhD, he was already recognized as a leader in two fields. While writing my own senior thesis, I looked up Schmid's in the Princeton archives, although the subjects had no relation whatsoever. Even earlier, I had attempted to read his article entitled *On a conjecture of Langlands*[13]—same Langlands, different conjecture, already briefly mentioned as an analogue of Carayol's Ann Arbor conjecture.

As an undergraduate, I was not able to make much sense of Schmid's article, but ten years later I studied it very carefully when I started working on coherent cohomology. I have already mentioned that Schmid's article made explicit use of local as well as global trace formulas, and if the strategy outlined in my dream struck me immediately as believable, it was precisely because of the analogy with Schmid's work, specifically with the article I just mentioned.[14]

Psychoanalytic dream interpretation is based as much on the dreamer's subsequent associations as on the content of the dream itself. It seems reasonable to conclude that my dream was motivated in part by an entirely unconscious but deeply buried wish to write an article like the one Schmid published in 1970 about a conjecture of Langlands. It's hard to deny that, from a conscious point of view, publication of one particular article in a "great journal" and the acquisition of an indelible aura of charisma reflected from the Langlands program represented a satisfying epilogue to the story that began with my dream in December 1992. But it is disorienting to speculate that my unconscious mind might have begun preparing this outcome more than twenty years earlier.

chapter 10

No Apologies

When as teenagers we began our initiation into the values and aspirations of research in pure mathematics, we never tired of quoting to each other—and to the uninitiated—from G. H. Hardy's *A Mathematician's Apology*, especially the parts where he insisted that "I have never done anything 'useful.' " It was our cliché that he had chosen number theory exactly "because of its supreme uselessness" and indeed that "[n]o one has yet discovered any warlike purpose to be served by the theory of numbers. . . ."[1] Nowadays the cliché opens with the same Hardy quotations but immediately veers off toward an "ironic" surprise ending:

> The irony of [Hardy's] life is that his "useless" work in obscure number theory and random numbers has found application in cryptography and encryption.

More precisely, "When you enter your credit details on the Internet, they are encrypted using pure mathematics so that only the dealer can decrypt your message and complete the transaction. The entire boom in e-commerce"—now worth $1 trillion worldwide—"would not have been possible without pure mathematics."[2] Number theory, in other words, is not merely useful: it is the bedrock of modern shopping.

A word like "useful" is useful not so much because it promises an unambiguous scale against which to measure our ambitions and priorities, but because it provides a marker to divide its users according to priorities that differ radically, depending on the positions they have chosen to occupy, or that have been chosen for them, in the social panorama. After I moved to France, I was grateful for the public-key cryptography, based on the congruences explained in chapter γ, that allowed me to order books in English online with only minimal risk of identity theft; but on each trip back to the United States I found that another bookstore had disappeared, and with it a neighborhood's cultural focus and the skilled work that kept the store alive, replaced by low-paid and infinitely replaceable jobs in filing, packing, lifting, and data entry. This is practical for many and profitable for a few, but is it "useful"?

> Book store owners and record store owners used to be oracles . . . ; you'd go in this dusty old place and they might point you toward something that would change your life. All that's gone.[3]

When I reread Hardy's *Apology* to prepare for writing this chapter, I was appalled to see what a hearty dose of unapologetic elitism I had imbibed along with Hardy's mathematical idealism when I was fourteen years old, too young to know any better. "[M]ost people can do nothing at all well," he wrote, but even a good mathematician "should [n]ever allow himself to forget that mathematics, more than any other art or science, is a young man's game."[4] Graduate students in our early twenties, beneath the gaze of the portraits of Giants and Supergiants on the wall, we aimed for an air of cool arrogance but often reached no higher than anxiety as we trembled and tried to imagine what would become of us if inspiration failed to emerge before the passage of ten or twenty years deprived us of the qualities of youth—persistence, energy, freshness of mind—without which meaningful participation in the fellowship of mathematical research is impossible. This didn't stop us from scorning or pitying senior colleagues who were now devoting themselves to administration or history or philosophy—or undergraduate teaching—as washed-up renegades. When it began to dawn on me that some of the assumptions I had been taking for granted deserved to be questioned, the reading was buried so deeply in my past that I could not trace these beliefs back to their source. Years later I returned to *A Mathematician's Apology*, the book novelist David Foster Wallace acknowledged as one of the inspirations of his own work, "the most lucid English prose work ever on math,"[5] and there they were.

Hardy's first paragraphs are a "melancholy" and literal apology for writing about mathematics rather than "do[ing] something":

> [T]here is no scorn more profound, or on the whole more justifiable, than that of the men who make for the men who explain. Exposition, criticism, appreciation, is work for second-rate minds.
>
> . . . If then I find myself writing, not mathematics, but "about" mathematics, it is a confession of weakness, for which I may rightly be scorned or

pitied by younger and more vigorous mathematicians. I write about mathematics because, like any other mathematician who has passed sixty, I have no longer the freshness of mind, the energy, or the patience to carry on effectively with my proper job.

Even as I reproduce these lines I shudder to think how the news of this book will be greeted by whispers in the common room, the speakers perhaps not realizing that they are quoting Hardy. How old do we have to be, I wonder, before we are no longer susceptible to indoctrination?

• •

MEPHISTOPHELES: *Der ganze Strudel strebt nach oben;*
   Du glaubst zu schieben, und du wirst geschoben.*

When we wanted to proclaim our hostility to utility, we had our pick of quotations, especially from Amir Alexander's "romantic hero" period. Carl Gustav Jacob Jacobi's declaration in 1830 that "the sole objective of science is the honor of the human spirit" is a perennial favorite.[6] And as early as 1808, Carl Friedrich Gauss, "Prince of Mathematicians"— applied no less than pure—considered it to be to be "no good sign of the spirit of the times" that one keeps hearing the "petty, narrowhearted, and lazy" question—"what use is th[is] science?"—symptomatic of "coolness and lack of sense for what is great and honors mankind."

> [Q]ne thinks everything has to be related to our physical needs . . . one requires a justification, as it were, for involvement with a science, and can't conceive that there are people who study simply because studying is itself also a need.

"There are sciences," he continued, "whose study is not encouraged by the prospect of benefits to physical existence" but rather by "a pure, disinterested joy in study."[7]

Reminding his reader that "science works for evil as well as good," Hardy thought "both Gauss and less mathematicians may be justified in rejoicing that there is one science at any rate, and that their own, whose very remoteness from ordinary human activities should keep it gentle

---

* The whole whirlpool is striving upwards; you think you are pushing, but you are being pushed.

and clean." A few of us nevertheless grew up and became aware of how delicately dependent on "external goods" are the "internal goods" of Hardy's and Jacobi's idealism, and spend practically all their time (and all of us spend some of our time) keeping open the lines of communication with Powerful Beings. Alexander von Humboldt, explorer, naturalist, and royal chamberlain of Prussia, was the go-to guy in the Germany of Gauss and Jacobi for mathematicians running short of external goods. Here is a well-known excerpt from Jacobi's letter to Humboldt praising (Johann Peter Gustav) Lejeune-Dirichlet, his contemporary on the IBM timeline and, like Gauss, an early precursor of the Langlands program:

> If Gauss says he has proved something, I think it is likely. If Cauchy says it, one may bet as much in favour as against it. If Dirichlet says it, it is for sure. . . . [Details of his work follow.] Had he stayed in Paris, he would now reign there without competitor, and how different would his ostensible situation be!

"And thus," writes the historian, "Dirichlet's income was raised and he remained in Berlin."[8]

Power has many addresses nowadays, but its representatives assemble periodically so that they and the mathematicians can remind one another of the mutual benefits of their exchange of internal for external goods. Exemplary in this respect was a "gala event" at the Oberwolfach institute, in the presence of the German Federal Minister of Education and Research, to celebrate the publication of a collection of short essays by heads of major German corporations entitled *Mathematik—Motor der Wirtschaft* [*Mathematics—Motor of the Economy*]. "The list of authors in the Springer book reads like a Who's Who of German DAX companies" (Allianz, Bayer, Daimler, Lufthansa, Siemens, . . .) and their message "in a nutshell" is that "Mathematics is everywhere, and our economy would not work without it." Or, as the mathematicians who edited the volume put it, "Mathematics is not merely a fascinating science and the basis of all natural sciences and technical developments; today it has also become . . . an important factor for economic competitiveness."[9]

The CEOs used the opportunity to explain just where mathematics fits into their business models. Their explanations were straightforward, technically precise, and mercifully short on the empty generalities one expects to find in this kind of exercise. The editors put in a word for "fundamental research without concrete prospects of applications," claiming that much of the mathematics used in business was of just this type not so many years ago. Maybe so; Shell claims to be using topology and the geometry of polynomial equations along with their familiar differential equations in their tireless search for oil. But the volume's summary, written by Helmut Neunzert, an international expert in industrial mathematics, finds that the mathematics corporations use falls mainly under six headings: numerical simulation, optimization and control, modeling of risk and decisions under uncertainty, data analysis and image processing, multiscale modeling and algorithms, and high-performance and grid computing. Apart from computing, these are golden eggs mathematics has been laying for a very long time; they cover the same *industrial* mathematics specialties one finds on the careers' page of the Society for Industrial and Applied Mathematics (SIAM) in the United States.[10] The main novelty turns out to be Hardy's useless number theory: crucial for corporate applications to data storage (coding) and data security (encryption).

A 2013 report, prepared by the consulting firm Deloitte MCS Limited on behalf of the British Engineering and Physical Sciences Research Council (EPSRC) and the Council for the Mathematical Sciences (CMS), buries content very similar to Neunzert's under a mountain of management jargon and arrives at the astounding conclusion that in 2010, £ 556 billion, or 40% of total UK gross value added (GVA) that year, can be attributed to mathematical science research (MSR): £ 208 billion directly, £ 155 billion indirectly, and £ 192 billion "induced." The top two sectors for direct MSR GVA are banking/finance and computer services. "In 2010, there were over 6.9 million individuals in employment due to the wider ripple effects" of MSR in the United Kingdom. The top sector for MSR ripple employment is retail distribution, followed closely by "hotels, catering, pubs, etc."[11]

France's efforts along these lines are both more elaborate and more ambivalent. *Maths à Venir* (MAV), whose name is a pun meaning both "math to come" and "math-future," was a ceremony held in 1987 and again in 2009 in order to propitiate the Powerful Beings known in France as *décideurs*, literally "deciders," who are definitely *not* the same as elected officials but whose power to make decisions is no less univer-

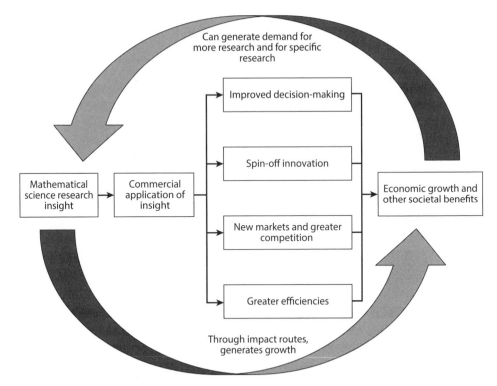

Figure 10.1. The impact of mathematics on the UK economy, or the Golden Goose as virtuous circle. (*Source*: Deloitte 2013)

sally acknowledged for being based on no discernable mandate. The 1987 MAV reportedly convinced *décideurs* to toss some welcome external goods in the direction of mathematics, and the hope in 2009 was that this gesture would be repeated. The *décideurs* to whom one needed to appeal were mostly French counterparts of the German industrialists who attended the Oberwolfach gala. But the 2009 edition of MAV, held at the newly refurbished *Maison de la Mutualité* in central Paris, was also conceived as an "*évènement exceptionnelle*,"[12] a public relations extravaganza, a celebration of mathematics' utility addressed to the *décideurs* but open to the public. The German government had sent a Minister to Oberwolfach; the French government placed MAV under the sponsorship of Prime Minister François Fillon (a no-show, but he did send a short speech). There were public conversations between mathematicians and high school students, multimedia exhibits, and lavish buffet lunches, as well as the inevitable champagne reception to which I inadvertently managed to get myself invited, all underwritten by a handful of leading French banks and industrial and defense corporations.

The Gallic pessimism on display at the *Mutualité* clashed sharply with the serenity of the Oberwolfach meeting, not to mention the self-congratulatory pomp of the Deloitte report. The president of the Société Mathématique de France hoped the public would leave convinced that "basic research (in particular mathematical research) is an excellent investment for our society," while recognizing that in an age when everyone has to "prove immediate profitability . . . it is difficult to justify . . . research . . . whose ultimate aim is the extension of human knowledge." His counterparts in the French societies of statisticians and of Industrial and Applied Mathematics felt less compelled to apologize but spoke along similar lines. Opening the meeting, Philippe Camus, a true *décideur*—president of Alcatel-Nugent, codirector of Lagardère, and head of MAV's *Comité de parrainage* (literally, "Godfathership committee")—made the stakes clear: The world outside the *Mutualité* needs to know that "mathematics is alive, it's useful, it's practical . . . and most of all, it's a strategic resource that needs to be looked after, because it will allow us in the future to preserve our social model, our way of life, and our ethic."

Mathematicians, pure as well as applied, took part in planning the two-day event, and the most charismatic spoke in plenary sessions and in panels. The organizers noted in their postmortem discussion that the "image of mathematics" presented by the meeting ("strategic resource"—in other words "merely good," a Golden Goose) "is not exactly the one mathematicians make themselves." This is consistent with the findings of Zarca's study. Comments from mathematicians in the audience betrayed distrust of corporations and *décideurs* in general. For at least some of the *décideurs* on the final panel, entitled *Mathematics, a strategic resource for the future*, the distrust was mutual. One of them had long ago decided that "I don't believe in pure research, nor do I believe . . . that direct corporate investment in research is the right way to go"; another that "corporations are not responsible for public sector salaries" (in response to a comment on the reasons for declining interest in the profession); and a third that "mathematics for the sake of mathematics is none of the business of corporations." In the aftermath of MAV, one organizer worried aloud that failing to act on the concluding recommendations "could be worse than the analytical vacuum that, for certain organizers" was their primary motivation.

• •

Readers who are not mathematicians will mainly know of Hardy, if at all, for his instrumental role in bringing the self-taught Indian mathematical genius, Srinivasa Ramanujan, to Cambridge. Years later, long after both men had died, Ramanujan returned the favor by bringing Hardy, in a straphanger role, to Broadway (in David Freeman's *A First Class Man*), to the West End (in Simon McBurney's play *A Disappearing Number*), and to four best-seller lists (David Leavitt's novel *The Indian Clerk*).

Reviewing one of the Ramanujan & Hardy plays, a *New York Times* theater critic asked "Is there a more romantic figure in popular culture than the mathematician?"[13] We may suspect him of harboring ironic intentions, but professional as well as aspiring mathematicians recognize Hardy, alone or in Ramanujan's shadow, as one of the field's most "romanticizable" personalities (if you look closely, you can see his name in the middle of figure 2.2). This is due largely to his *Apology*, dedicated from beginning to end to answering one question: "What is the proper justification of a mathematician's life?"

As graduate students in the common room, we told each other that mathematics was different from the natural sciences, that it was actually more like art. That, we've seen, is in Hardy too; he contrasted the "real mathematics" that "must be justified as arts" with "the 'crude' utility of mathematics," the kind engineers or doctors (or major German corporations) use: what usually goes by the name of *applied mathematics*, in contrast to the pure mathematics that is the primary subject of the present book as well as the *Apology*. It's not that Hardy found applied mathematics crude: "It is the dull and elementary parts of applied mathematics," Hardy argues, "as it is the dull and elementary parts of pure mathematics, that work for good or ill." The interesting parts of mathematics, pure or applied, are what Hardy compares to art. "Beauty is the first test: there is no permanent place in the world for ugly mathematics." That's one of Hardy's most quoted epigrams.[14] He allows that beauty may be "very hard to *define*," but he expects his readers to know what he means by art, and to agree that beauty is one of its criteria.

That word *beautiful* recurs regularly when one speaks of Hardy. He was past fifty and gray-haired when C. P. Snow, author of *The Two Cultures* and a close friend of Hardy's, first saw him and found his face "beautiful—high cheek bones, thin nose, spiritual and austere but capable of dissolving into convulsions of internal . . . amusement." And Leonard Woolf, like Hardy a member of the very selective secret society of Cambridge Apostles, wrote that Hardy was "one of the strangest and most charming of men"; he "had the eyes of a slightly startled fawn below the very beautiful and magnificent forehead of an infant prodigy."

The contemporary reader has outlived Hardy and has, therefore, seen art, beauty, and the relation between the two vary in meaning even more than "useful," which makes them less than ideally useful as justifications for mathematics. Nevertheless, as Herbert Mehrtens observes, the notion that mathematics has its own aesthetic, "that it is itself an 'art,' " remains "a familiar topos in mathematicians' talk about mathematics," one whose history "has yet to be written."[15] So it's an open question whether we still mean what Hardy meant when we recite his claims about beauty.[16] For Walter Pater, a precursor of the late-nineteenth-century *Aesthetic Movement*, the question would have been beside the point: "What is important . . . is not that the critic should possess a correct abstract definition of beauty for the intellect, but a certain kind of temperament, the power of being deeply moved by the presence of beautiful objects." Twice in the *Apology*, Hardy speaks of a "passion for mathematics;" Pater, comparing objects of passion, declared that "the poetic passion, the desire of beauty, the love of art for art's sake has most [wisdom]."[17] Since Hardy's *Apology* is often applauded or disparaged as a defense of mathematics as "art for art's sake," it should be remembered that this was a relatively new notion in Pater's time. John Ruskin aimed "to exhibit the moral function and end of art . . . to attach to the artist the responsibility of a preacher." For Ruskin, the utility of art was in its power of moral uplift, but Ruskin was not a utilitarian, unlike the economist William Stanley Jevons, whose prescription for improving the moral habits of workers was to expose them to music, which promoted "a general removal of the mind from its ordinary course of duties"—Russell's "lofty habit of mind," to which Hardy alludes in his *Apology*—leaving "the body in a perfect state of repose."[18]

Pater's aestheticism echoes in an extraordinary article by George Salmon "explaining" the great algebraist Arthur Cayley to the general public on the occasion of Cayley's nomination as President of the British Association. Cayley was the first occupant of the Sadleirian Chair of Pure Mathematics at Cambridge, the position Hardy held years later, when he wrote his *Apology*. In an unpublished thesis submitted in 2004, John Heard speculates that Salmon, realizing that "Cayley's work was incomprehensible" to the vast majority of his intended readers, "created a clear divide between scientists and pure mathematicians" by choosing to depict Cayley as an artist:

> My subject is the life of a great artist who has had courage to despise the allurements of avarice or ambition, and has found more happiness from a life devoted to the contemplation of beauty and truth than if he had striven to make himself richer, or otherwise push himself on in the world. . . . we can understand that it is often the lowest style of art which will attract round it the largest circle of admirers. So the fact that it is a very limited circle which is capable of appreciating the beauty of the work done by a great mathematician should not prevent men from understanding that it is like the work done by a poet or painter, work done entirely for its own sake, and capable of affording lively pleasure both to the worker himself and his admirers, without any thought of material benefit.

Heard juxtaposes the elitism of Pater and his fellow aesthetes with Salmon's rhetoric (see figure 10.2) and argues that to affirm "that the value of pure mathematics lay, not in its utility, but in the pleasure [recall Salmon's "lively pleasure"] that it gave to the elite band of pure mathematicians" did not contradict "the claim that pure mathematics was beautiful and that pure mathematicians were artists."[19]

Britain's aesthetic movement soon gave way to the promotion of beauty to ethical principle by G. E. Moore and his Bloomsbury followers, with whom Hardy was in close contact, as we'll see shortly. The "topos" of mathematics as art, however, is only one facet of a much older tradition that defines art and science in contrast to one another. Pater and Ruskin did share the habit of using the word *arts* primarily in connection with *fine arts*, and for them, as for Hardy, art and beauty were practically

Aesthetics	George Salmon
Beauty not restricted to art and literature	The world of pure mathematics is beautiful
Higher powers of perception are restricted to an elite	Only a trained elite can perceive the beauty of pure mathematics
The value of beautiful things lies in the pleasure that they give	The value of pure mathematics lies in the pleasure that it gives
To give oneself over to experiencing such pleasure is to spend one's life well	The life of a pure mathematician has been well spent in 'lively pleasure'

Figure 10.2. From Heard 2004.

synonymous. But mathematics in the *Apology* is a science as well as an art: "of all the arts and sciences, the most austere and the most remote" (ch. 28), and it is its status as science that allows Hardy to ask whether mathematics is " 'useful,' directly useful, as other sciences such as chemistry and physiology are" (ch. 6).

Early glimmerings of the "familiar topos" of the superiority of "useless" mathematics can be found in Plato and Archimedes (cf. chapter 3, note 31). That mathematics is an art has to be a much newer idea, for the simple reason that "the modern distinction between *science* and *art*, as contrasted areas of human skill and effort, with fundamentally different methods and purposes, dates effectively from [the mid-nineteenth century]," and the contemporary identification of art with the fine arts is a relatively recent development.[20] Armand Borel's 1981 lecture *Mathematik: Kunst und Wissenschaft* [Mathematics: Art and Science] at the Siemens-Stiftung cites Hardy and Poincaré but traces the theme back to the 1845 thesis defense of Leopold Kronecker. In addition to his mathematical thesis, the young Kronecker was required to argue "a number of propositions of varying degrees of seriousness, against 'opponents' chosen, partly by the Faculty, partly by the candidate among his own friends." The third of these "theses" was *Mathesis et ars et scientia dicenda* [mathematics is both art and science] and the opponent, Gotthold Eisenstein, was said to have argued that mathematics is only art.[21] A few years later, the *Bavarian Maximilian Order for Science and Art* inducted its first prestigious group of initiates, and Gauss was among them—for science, not for art.

We will return momentarily to the "beauty question" in mathematics, but we should remember that the status of mathematics as science rather than art was a matter of philosophical consensus for centuries, and beauty had nothing to do with it. Our words *science* and *art* derive from the Latin *scientia* and *ars*, the equivalents in Roman and scholastic texts of *epistêmê* and *technê*, respectively. Aristotle explained how to distinguish *epistêmê* from *technê* in his *Nicomachean Ethics* and elsewhere. For example: Science is less subject than art to doubt; the object of scientific knowledge is necessity; whereas "art is concerned with coming into being . . . how something may come into being which is capable of either being or not being."[22] On these grounds, the difference between seeing mathematics as art or science might simply map onto the difference between mathematics as invented and mathematics as discovered. But there are more decisive differences. As conceived by Plato and Aristotle, *technê* is "practical and empirical," *epistêmê* is "contemplative (*theoretikê*) and demonstrative (*apodeiktikê*) and most importantly is defined by a preexisting ontological domain." The latter means, for example, that the sciences are separate: "We cannot, for instance, prove geometrical truths by arithmetic."[23]

The Greek distinction between *epistêmê* and *technê* underwent reconsideration by medieval Arab philosophers—who called them respectively *'ilm* and *sina'a*—before reaching the hands of the scholastics, in large part as a result of al-Khwārizmī's invention of algebra (*al-jabr w'al-muqabala*). The typical object in al-Khwārizmī's "ontological domain" is a *shay*—the (still) common word for "thing" and the unknown quantity called x in our high school algebra. "When I considered what people generally want in calculating," al-Khwārizmī wrote on his first page, "I found that it always is a number." The equations look like this:

> Divide [the inheritance] between the two sons; there will be for each of them three dirhems and a half plus two-fifths of *thing*; and this is equal to one *thing*. Reduce it by subtracting two-fifths of *thing* from *thing* [italics added for emphasis].[24]

Since al-Fârâbî saw algebra as "common for number and geometry"—"a bastard by cultural as well as by disciplinary standards," in the words of Jens Høyrup[25]—it could not be a science, as we've seen, but rather a *ḥila'*, a trick or mechanism or ingenious device. One century after al-Fârâbî, ibn Sīnā (Avicenna) followed Aristotle and included geometry, astronomy, arithmetic, and music in the chapter of his *Book of Scientific Knowledge* devoted to mathematics but relegated algebra, as we've seen, to the list of "secondary parts of arithmetic."

Aristotelians will agree with al-Fârâbî that algebra can't be an authentic science because it is defined, not by its objects, but rather by its methods. Since these methods apply indifferently to objects geometric or arithmetic or of any other nature, it cannot reveal the essence of the objects and, therefore, are mere *ḥiyal*, ingenious procedures—tricks.[26] Once again, al-Khayyām dissents. "The most remarkable thing, perhaps, about Omar Khayyam's methodology is his denial of any essential disciplinary differences between geometry and algebra." Al-Khayyām opened his treatise on algebra in most unaristotelian fashion: "one of the **scientific** notions we need in the part of philosophy known as mathematics is the **art** of *al-jabr w'al-muqabala*" [my emphasis]. More explicitly:

> I say, with the help of God and His good support: The art [*sina'a*] of Algebra is a **scientific art** [*sina'a 'ilmiya*, my emphasis], whose subject is absolute numbers and measurable magnitudes *qua* unknown but connected with something known which enables one to determine them—and that thing is either a quantity or a ratio—according to a certain way in which nothing participates with them but themselves and to which you are led by examining them. And what it looks for is the accidents which are attached to its subject in that it is a subject of it with the said property. And what makes it complete is the knowledge of the scientific methods whereby one will understand this above-mentioned kind of determination of numerical and geometrical unknowns.[27]

So by the end of the middle ages—when Cardano was writing up the solution of the cubic equation in his *Ars Magna* [*The Great Art*], concluding "So progresses arithmetic subtlety the end of which, as is said, is as refined as it is useless"[28]— the clear distinction between art and science was completely mixed up, in no small part thanks to the "bastard" status of algebra, Cayley's future specialty. Then Descartes came along and turned the distinction upside down. In his *Rules for the Direction of the Mind* it's now *art* that is determined by its object "since it is

not so easy for the same hand to adapt itself both to agricultural operations and to harp-playing," but it "is certainly wrong" to say the same of
science: "all the sciences are conjoined with each other and interdepen-
dent." Twenty years later, in his conversation with Frans Burman, Descartes insisted that "one needs a mathematical mind which must then be
polished by actual practice. Now this mathematical knowledge must be
acquired from Algebra."[29] But note that "mind" in "mathematical mind"
as well as in Descartes' *Rules* . . . translates as *ingenium*. . . .

Far from pretending to shed any light on the mathematical artistry of
Eisenstein, Cayley, Hardy, or Bombieri, this pseudoarcheological detour
reminds us just how hard it is for us to be sure we know what they were
talking about. In the long view, even if our Colleges of Arts and Sciences
lack "bastard" departments of "scientific arts," no consistent meaning
can be assigned to the contrast between art and science.[30] Max Weber
thought he had put his finger on the contrast when he wrote "Scientific
work is chained to the course of progress; whereas in the realm of art
there is no progress in the same sense."[31] But progress has no constant
meaning either; certainly the "sense" of progress in mathematics is not
the same in different places and across generations.

• •

When Abraham Flexner opened his Institute for Advanced Sciences . . . in
the early 1930s, he saved the top floor—and the best offices—for mathematicians, and his first permanent position was also given to a mathematician
[Oswald Veblen] . . .. for Flexner, mathematics was the 'highest' because of
the difficult, abstract work that it involved. . . . The IAS was for thinkers—
and mathematicians were the purest of thinkers.[32]

Irving and Marilyn Lavin remind us of Flexner's "usefulness of useless
knowledge," in their reconstruction of the history of the the IAS Seal.[33]
The two allegorical figures (see figure 10.3) are labeled TRUTH (icono-
graphically correct in her nudity, as in Horace's *nuda veritas*) and
BEAUTY (draped). The fruit of the tree between them is Knowledge
presumably the usefully useless kind Flexner favored. (And the IAS is
indeed an Eden for those privileged enough to spend even a few weeks
there, but I'll let the reader speculate on what this has to do with the
scene on the Seal or, for that matter, with *Rites of Love and Math*).

Figure 10.3. Truth and Beauty: the IAS seal (Photo by Bruce White, Courtesy of the Institute for Advanced Study).

"The radical nature of Flexner's twinning of science and humanism with truth and beauty arose in part" according to the Lavins "from the radical nature of his concept for a 'modern' university by which he meant a university devoted exclusively to the pursuit of higher learning for its own sake and without regard to practical value." In the decade before Flexner had materialized his trinity of Truth and Beauty and Useful Uselessness in the form of the IAS, topologist Oswald Veblen, who later left Princeton's Mathematics Department to become the first IAS professor, had been busy tempting the snake into the garden with visions of Golden Geese. "What" he asked,[34] "do industry and engineering owe to mathematics?" Answers included "the development of locomotives, automobiles, airplanes, the telegraph, the telephone, and radio . . . overcoming 'formidable natural obstacles, bridging rivers, and damming floods'. . . applications to physics, astronomy, economics, and the social sciences." "[T]he civilization of the future will depend even more on mathematics than does the civilization of the present."

The Lavins argue that the IAS seal refers intentionally to the "Beauty is truth, truth beauty" from Keats's *Ode on a Grecian Urn*, but the IAS could just as well have adopted this motto from Wordsworth when addressing its donors:

Give all thou canst; high Heaven rejects the lore
  Of nicely-calculatedless or more;

The lines are from the poem *Inside of King's College Chapel, Cambridge* and are a plea for openhandedness in support of "immense and glorious Work of fine intelligence" albeit "for a scanty band/Of white-robed Scholars only." But did the white-robed scholars and the donors see eye to eye about useful uselessness? It's hard to find a less consensual topic; the scholars scarcely agreed among themselves from one year to the next:

What is the attitude of the academic man towards his vocation . . . ? He maintains that he engages in 'science for science's sake' and not merely because others, by exploiting science, bring about commercial or technical success and can better feed, dress, illuminate, and govern (Max Weber, 1922).

. . . an important aim of [British Science Research Council] support for long-term basic research is to enable university departments to give help to industry and Government Departments on shorter-term projects and thus bring more social and industrial relevance to their research and teaching (from the 1970–1971 SRC report)

. . . science and patents don't mix any more than oil and water (Stanford physicist D. L. Webster in 1939, in connection with funding by the Sperry Gyroscope Company),

. . . as scholars, we should not seek knowledge for its own sake (the chancellor of UCSD in 2001).

. . . between the managerial outlook and the scientific there is a basic conflict in goals (William H. Whyte, 1956).

Through closer ties to the socio-economic universe, new academic, ethical, and moral responsibilities are enriching the role and action of the university in a way that favors building the foundations of the construction of a knowledge society in harmony with a necessary sustainable development.[35]

Steven Shapin's *Scientific Life* reminds us of the stereotypical image of the scientist in the mid-twentieth century, the time of the first stirrings of Mertonian sociology of science and Hardy's *Apology*: "Scientists are internally motivated; dedicated, even called, to their work; they are selfless; resistant to convention and authority; intentionally blind to social convention and prejudice; unconcerned for fame and material reward; open. Their virtues are a pastiche of the heroic, chivalric, Stoic, and Christian. Put such people into the moral environment of corporate capitalism, and the resulting tensions are not merely mundane and contingent but ideologically essential." Karl Compton, physicist and president of MIT, drew on this stereotype when responding to a senator's question just after the war: "I don't know of any . . . group [other than pure scientists] that has less interest in monetary gain."[36] Much of Shapin's book is, in fact, devoted to deconstructing this image, which needed qualification even before today's "knowledge society." Unfortunately, Shapin's study has nothing to say about mathematicians and doesn't help us to weigh Yuri Manin's remark in a 2009 interview addressed to (Russian) mathematicians—

> "I always say, 'Why should we put ourselves on the market? We (a) don't cost anything, and (b) don't use up natural resources and don't spoil the environment.' Give us salaries, and leave us in peace."

—against this excerpt from the SIAM report:

> "Research often has a serious difficulty: too much understanding and too little transfer."[37]

Mehrtens sees mathematicians' indifference to Golden Geese as part of the modernization process, which saw the "driving out" [*verdrängen*] of questions connected with applications from "the official consciousness of mathematics.[38] But the continuing economic crisis drives them back in again. Several roundtables at that 2009 edition of *Maths à Venir*, convened during the dark early days of the crisis, addressed the role of finance mathematics. It was even proposed to draw up a charter of mathematical social responsibility, which seems to have been forgotten almost immediately, but none of the speeches at the Paris MAV noted the irony of the choice of venue: the *Mutualité*, long-time headquarters of the mutual insurance movement dedicated precisely to the ideal of *collective* solutions to the financial needs ostensibly addressed by the quants. After serving for decades as the scene of massive meetings of radical dissent, the *Mutualité* was finally handed over to a public-relations multinational and converted to a luxurious convention center, where French President Nicolas Sarkozy had planned to celebrate if he had been reelected in 2012.

Irony has not spoken its last word on the flight from utility, even when utility is understood, with Hardy, as that which "tends to accentuate the existing inequalities in the distribution of wealth." The thirteen *Creative Industries* promoted by Britain's Department of Culture, Media, and Sports were estimated in 2007 to account for 7.3% of the economy, the same proportion as financial services.[39] England has changed since Hardy's day, and it's not only because online commerce needs prime numbers, nor is it because Hardy's work in functional analysis gets applied to the Black-Scholes equation: mathematics and art turn out to be useful in much the same way.[40]

• •

It is in the abstruser sciences, particularly in the higher parts of mathematics, that the greatest and most admired exertions of human reason have been displayed. But the utility of those sciences, either to the individual or to the public, is not very obvious, and to prove it, requires a discussion which is not always very easily comprehended.[41]

I have no idea whose idea it was to invite me to the Ministère de l'Enseignement Supérieur et la Recherche on October 6, 2009, for an immersion in contemporary utilitarian public policy, in the form of a morning seminar on the French *National strategy of research and innovation: what projects for the Big National Loan* (*Grand Emprunt National*)? Entrepreneurs sat beside researchers and representatives of voluntary associations[42] in the seats and the aisles of the auditorium in the grand old École Polytechnique building in the Latin Quarter to listen to Minister Valérie Pécresse outline her hopes for the morning's four simultaneous roundtable discussions. I failed then and still fail to grasp the historic significance of the Big National Loan, piloted by a committee led by one former Prime Minister from each of the main parties.[43] Keywords pronounced by the minister—*visibilité, esprit d'entreprise* [entrepreneurial mindset], *valorisation, gouvernance*—resurfaced throughout the morning's events. In her greetings and again at her concluding speech three hours later, she insisted that basic research not subject to *retour sur investissement* was "at the heart of our national research and innovation strategy." My notes have the minister saying that "progress of science must not be neglected" and stressing the parallel between *quête de rente* [search for profit] and *quête de sens* [search for meaning].[44] There is an "urgency to renew and rationalize research" to confront its "major defect"—the absence of *fluidité* between research and innovation.

Otherwise basic research was hardly mentioned at all, certainly not at the roundtable I chose to attend, entitled *The Big Loan to reinforce the innovation ecosystem*. Finally a key concept—ecosystem —that might help me figure out how I fit into the grand national scheme.[45] A big-enough fish to merit invitation to the seminar, I am nevertheless too small to be a predator—most likely a bottom-feeder. The minister situated the innovation ecosystem in a "new research landscape" characterized by three more key concepts: *autonomie*, *décloisonnement* [removing barriers], and the *université au coeur de l'innovation* [university at the heart of innovation] as a "post-crisis springboard" [*tremplin d'après-crise*].

Facing a panel too numerous for the space at the table, the innovation ecologists in the audience were a comparatively young and hungry crowd, overflowing the seating in the assigned seminar room ("I'm sorry about these material conditions") and compulsively consulting their portable communications devices. The session was led by the president of something called the *pôle de compétitivité* SYSTEM@TIC, who made a distinction I couldn't quite follow between *problèmes de moyens* and *problèmes soft* and introduced the four themes of the presentation, namely,

1. Getting from *emergence de l'idée* to *création d'un startup*;
2. Technology transfer;
3. *Problèmes de financement*;
4. *Décloisonnement de l'écosystème*.

Of these, only the fourth theme seemed at all relevant to my situation—the *décideurs* want to break down the barriers and let me out to swim among the sharks, and vice versa.[46] Naturally this was the theme dropped from the program for lack of time. But I did learn from the speakers, as well as from *décideurs* in the audience, that

1. It is important to focus on people rather than (just) on patents—has the researcher set loose from the *cloison* made millions or tens of millions of euros?

2. Research turns money into ideas; innovation turns ideas into money.[47]

Two recurrent themes—the *knowledge economy* and *competitiveness of the French economy*—were repeatedly attributed to then-President Sarkozy. As for my own responsibility in all this, "It's up to scientists to leave the labs to meet citizens and explain how what they are doing contributes to their well-being."* Not the scientist's well-being, of course, but that of the citizen, 49% of whom, according to a 2006 opinion poll, thought that the first responsibility of researchers is to "improve the well-being of humanity," far ahead of "knowledge," "French competitiveness," and "improving everyday life."[48] No figure was given for the percentage of French citizens who referred to anything that looked like the "relaxed field" or "internal goods" or *telos* of a "tradition-based practice." But we don't know whether or not mathematics was mentioned in the survey.

Will scientists really meet ordinary citizens, rather than *décideurs*, on the other side of the *cloison*? In their concluding remarks (just before the champagne reception), the speakers insisted that, while it was understood that basic research brings no *retour sur investissement*—in other words, the research laboratory cannot help to repay the Big Loan—basic research must be "associated" with *laboratoires d'innovation*. "All our research will evolve in the direction of encouraging [public-private] exchanges." At one point during the *ecosystème* panel, I was tempted to speak up and point out that, actually, the "entrepreneurial mindset" was not unfamiliar to scientists, and if most of us have chosen to follow a not especially easy path to a not particularly lucrative career, it was because we decided early in our lives that the *esprit d'entreprise* was a mindset to be avoided at all costs! This is just another way of underscoring our attachment to our "relaxed field," and you might think the *décideurs* should hear how the *chercheurs* feel about the *décloisonnement* in store for us. But how much time would a marine ecologist have for a soft-shelled creature showing up at the lab, fresh from the sludge, with

---

* "*Il reste aux scientifiques de sortir du labo à la rencontre du citoyen et d'expliquer comment ce qu'ils font contribuera à son bien-être.*"

a point of view? Would the *décideurs* have understood why Neal Koblitz, coinventor of elliptic curve cryptography, "would have felt queasy" about accepting the $1000 per month consultancy he was offered in 1997 and gave the money away "since there was little that I did . . . during my five years as consultant that I wouldn't have been glad to do anyway free of charge," given that the trillion dollar e-commerce industry has spun off tens if not hundreds of millions for the data security techniques based on Koblitz's ideas? Or would they have simply disregarded Koblitz's hopelessly romantic unentrepreneurial mindset?[49]

"The dichotomy between truth and interest is one of the standard topoi of the logic of scientific authorship," writes Mario Biagioli, who adds that "truth . . . has to be priceless because it cannot belong to the logic of interest and its ubiquitous unit of measure—money." Anyway, you were right to be skeptical about Sarkozy's claim to authorship of the themes cited earlier. Along with three of Pécresse's keywords, those themes can be found on the first two pages of the report on the *First European Forum on Cooperation between Higher Education and the Business Community*, held in Brussels in February 2008 "in line with the European Commission's initiative of May 2006 on modernizing Higher Education in Europe, a key element of the Lisbon strategy"; Valérie Pécresse had given one of the opening speeches. These same themes and keywords reappear in item 9 of the 2009 *European Commission Legislative and Work Program List of Strategic and Priority Initiatives*, where one reads that

> [t]he Communication of the Commission "Delivering on the Modernisation Agenda for Universities: Education, Research and Innovation" (COM(2006) 208 final) highlights that Universities have to recognize *"that their relationship with the business community is of strategic importance and forms part of their commitment to serving the public interest."*[50]

Neither of these documents refers even once to mathematics, but at least one French *décideur* has seen the connection: Philippe Camus, MAV Godfather, had written on the MAV 2009 Web site that "Mathematics are at the heart of most key technological and economic issues. . . . It's more crucial than ever to promote closer ties between the world of mathematical research and the corporate world."

Back in the United States, the 2012 SIAM Report concurs: "Our economy and that of the developed world is in the midst of a transition from a product-based economy to a knowledge-based economy. . . . We are convinced that the mathematical and computational sciences have contributed and will continue to contribute to the nation's economy . . . Universities will continue to play a key role." This is true enough and hardly surprising, because the concept of the "knowledge economy" was popularized by U.S. economists long before it became fashionable in Europe. SIAM then goes on to channel the voices of Powerful Beings, recognizable by their use of the word *must*:

> But this will not happen by itself; university faculty must actively encourage students to consider careers in industry and prepare those students for the very different world they will encounter upon graduation.[51]

• •

> I don't know why we are here, but I'm pretty sure that it is not in order to enjoy ourselves.

This unsourced quotation, probably apocryphal but attributed to Wittgenstein on dozens of websites,[52] may well be in keeping with the European Commission's Strategic and Priority Initiatives: the *écosystème* panelists worried that professors and researchers suffer from "soft emotional problems," such as a reluctance to work on projects that are not their own. But it is belied by mathematicians' accounts of their experience. Helmut Neunzert and his colleagues published a study in 2004 entitled "Mathematics Dream Job" [*Traumjob Mathematik*], in which they report that about 80% of their subjects had "fulfilled their occupational dreams." Neunzert stressed that "this is an unusually high percentage," but we have seen that an even higher 91% of pure mathematicians in France placed "the pleasure of doing mathematics" at the top of their list of motivations. A 2011 survey in Abu Dhabi reports that mathematics (and chemistry) teachers are the most satisfied of the emirate's (generally highly satisfied) teaching staff.[53]

One of the good things the Simons Foundation has done is to record interviews with mathematicians, by mathematicians, and make them freely available on the Internet. Their value as sociology is questionable—the selection is mainly limited to Giants—but you may want to believe that mathematicians speak with greater candor to their peers than to even the most sensitive social scientist. So we watch Paul Sally of Chicago reminisce about his 1968 collaboration with Joe Shalika at IAS: "we would work there eight or ten hours a day, filling every blackboard in the place, just having the time of our lives." IAS professor Robert MacPherson quotes Kolmogorov's opinion that "the only way somebody can be a scientist is that somehow their personality gets frozen at an early age . . . at the playful stage." Whenever he's talking with Pierre Deligne, MacPherson feels like "I'm the eight-year old and he's the kindergartener." Deligne himself recalls how pleased he was to learn as a student in Brussels that "one could earn one's living by playing, i.e., by doing research in mathematics."[54]

I hope my distinguished but playful colleagues won't take it as a mark of disrespect if I remind them at this point that, asked about the utility of his discovery of the magnetic field, Michael Faraday supposedly answered, not that we would someday be using it for activities as diverse as data mining and shopping, but rather: "Of what use is a newborn babe?" If you grant the legitimacy of that question, you may be ready to join Alison Gopnik in contemplating the "philosophical baby" in its relaxed field. Note the uncanny resemblance of the "useful uselessness of immaturity" to life at the School of Mathematics at the IAS:

> Play is the signature of childhood. It's a living, visible manifestation of imagination and learning in action. It's also the most visible sign of the paradoxically useful uselessness of immaturity. By definition, play . . . has no obvious point or goal or function. It does nothing to advance the basic evolutionary goals of mating and predation, fleeing and fighting. And yet these useless actions—and the adult equivalents we squeeze into our work-day—are distinctively, characteristically human and deeply valuable.[55]

Outside this relaxed field, it's considered poor form to admit that we are motivated by pleasure. Aesthetics is a way of reconciling this motivation with the "lofty habit of mind," and it has been that way at least since Plato's *Philebus* listed the constituents of "the good," placing pleasure in fifth and last place, with beauty second, followed by "mind and wisdom" and "sciences and arts and true opinions."[56] Nevertheless, the suspicion grows that not only beauty, but goodness (or utility) and truth serve as excuses to talk about pleasure on the sly. Truth's pleasures are relished more widely than you might think. Henri Cartan, a founder of Bourbaki, confessed that "I greatly enjoyed discovering what was true." Benedict Gross, coauthor of one of the most influential papers on the Birch-Swinnerton-Dyer conjecture (and part-time installation artist): "When you discover a mathematical truth, everything immediately becomes clear. . . . The beauty of mathematics is just a pleasure to behold." Ingrid Daubechies, current president of the IMU, opines that "We like logical thinking as an activity—figuring things out gives us pleasure."[57]

The pleasure of "figuring things out" is not so different in mathematics and theoretical physics—Daubechies was originally trained as a physicist—and if we are about to quote Richard Feynman at length, it's because he was never at a loss for words. In a televised interview, he explained that he "used to enjoy physics and mathematical things because I used to play with it. It was never very important, but I used to do things for the fun of it." When he was a young physicist at Cornell, Feynman showed a calculation involving the rotation of a Cornell cafeteria dish to his senior colleague Hans Bethe, who told him, "that's very amusing, but what is the use of it?" No use, Feynman replied, just "the fun of it."

> I **relaxed** and started to play . . . with rotation . . . and that just led me back into quantum electrodynamics . . . and I continued to play with it in the **relaxed** fashion . . . and in very short order I worked out the things for which I later won the Nobel prize[58] [my emphasis].

David Hume's *Treatise of Human Nature* takes for granted that the reasons for "curiosity, or the love of truth," specifically the "discovery of the proportions of ideas, consider'd as such"—and draws his examples from "mathematics and algebra"—are bound up with pleasure. The relation is not as simple as cause and effect: "'tis not the justness of our conclusions, which alone gives the pleasure." Indeed, "to fix our attention or exert our genius" is "of all other exercises of the mind . . . the most pleasant and agreeable."

But that's not all: "The truth we discover must also be of some importance." Later in this section, Hume uses "importance" and "utility" interchangeably.[59] When Neunzert was asked "what it is that fascinates me about mathematics," he summarized his answer in a few words: "That one can do something useful with its playfulness." Industrial and applied mathematicians in the United States are reported to feel the same way: "In many cases, academic mathematicians derive considerable satisfaction from seeing the technologies they know and love being applied in productive and interesting ways to real-world problems."[60]

At Gradgrind's utilitarian school in Dickens's *Hard Times*, pleasure is a world away, more precisely the world of Sleary's circus. In reality, the historical utilitarians and their immediate followers held pleasure in such high regard that some of them attempted to quantify it as a surrogate for utility. The "exact utilitarianism" of Francis Ysidro Edgeworth's *Mathematical Psychics* hypothesized that "Pleasure is the concomitant of Energy;" to understand emigration, for example, Edgeworth thought it would help to write things such as

The happiness of the present generation may be symbolized

$$\int_{x_0}^{x_1} n[F(xy) - cy]dx + cd$$

"Inexact" utilitarians, such as Jevons, John Stuart Mill, or Jeremy Bentham, were less keen on mathematical formalism, but all emphasized the centrality of pleasure in their systems—Bentham invented a "hedonic," or "felicific, calculus"—to the point that it has recently been proposed to define historical "utilitarianism" as "an acceptance of two postulates," the first of which is " 'Social utility' is an aggregate (estimated or precise) of individual utilities (reflecting individuals' pleasure, happiness, etc)."[61] In this respect, at least, mathematicians line up with the utilitarians, while the European Commission, with no mention of either pleasure or happiness on any of the 117 pages of their 2009 road map, sides with the Gradgrinds.[62]

I have already hinted that mathematicians refer to "beauty" when they want to talk about pleasure while maintaining a "lofty habit of mind"; talk about art is loftier still. But let no one be fooled. For Edmund Burke, "the beautiful is founded on mere positive pleasure." Hume thought it "an universal rule, that their beauty [of every work of art] is chiefly derived from their utility, and from their fitness for that purpose, to which they are destined," but we have already seen that utility for Hume is also bound up with pleasure. As for Kant, "the feeling of pleasure or displeasure . . . is precisely the riddle" whose solution required him to write the *Critique of Judgment*.[63]

Much as Edgeworth understood pleasure in his *Mathematical Psychics* on the model of physics, Burke sought physiological grounds for beauty; he was convinced that "beauty acts by relaxing the solids of the whole system. . . . a relaxation somewhat below the natural tone seems to me to be the cause of all positive pleasure." The last chapters of part IV of Burke's *On the Sublime and Beautiful* are devoted to showing that "the genuine constituents of beauty, have each of them . . . a natural tendency to relax the fibres." Neuroimaging now maps Burke's "solids" and "fibres" in order to trace emotion through the activation of brain regions. Researchers at London's Wellcome Laboratory report that activation of the *medial orbito-frontal cortex* (mOFC) is associated with a reported experience of beauty, with the intensity of the experience reflected in the strength of the activation. T. Ishizu and S. Zeki claim to have demonstrated that this correspondence is similar for auditory and visual beauty and that, therefore, "there is a faculty of beauty that is not dependent on the modality through which it is conveyed." More intriguing, for our present purposes, is the following passage:

> [A]ctivity in mOFC correlates with the experience of pleasure and reward, whether real or imagined, and its expectation. This naturally raises, at a neurobiological level, an issue long discussed in the humanities, namely the relationship of aesthetic experience to pleasure.[64]

Now if it turns out that *mathematical* beauty is also hardwired in the mOFC of the beholder,[65] it stands to reason that one would want to equip automatic theorem provers with mOFCs. We seem to be back in Steven Pinker's cheesecake bakery, an uncomfortable spot for mathematicians seeking to frame their apologies. Artist and art critic Roger Fry, a leading member of Bloomsbury, posits a "pure moralist" for whom the "life of the imagination"—the description applies to pure mathematics, though Fry is naturally thinking of the arts—can be justified only if it is "shown not only not to hinder but actually to forward right action, otherwise it is not only useless but, since it absorbs our energies, positively harmful." What Fry calls "the Puritanical view," a version of which appears to reign in today's European Commission, "regards the life of the imagination as no better or worse than a life of sensual pleasure, and therefore entirely reprehensible." Fry dismisses as "special pleading" a second view, that of "moralists like Ruskin" who argue "that the imaginative life does subserve morality." Fry's own "Apology" rests on different tenets:

> [M]ost people would, I think, say that the pleasures derived from art were of an altogether different character and more fundamental than merely sensual pleasures, that they did exercise some faculties which are felt to belong to whatever part of us there may be which is not entirely ephemeral and material.

Note that Fry seeks the eternal, not in the foundations of mathematical truth but rather in the "fundamental . . . pleasures" of art. Like pure mathematics, the imaginative life that art expresses "is separated from actual life by the absence of responsive action." And, therefore, art isn't faced with the "moral responsibility" implied by the life of action; instead, "it presents a life freed from the binding necessities of our actual existence."[66] Art, in other words, is a *relaxed field*, as is the "life of the imagination" more generally. We don't have to choose between Pinker's evolutionary determinism and "whatever part of us . . . is not entirely ephemeral"; we might even choose to see the latter as the focus of a tradition-based aesthetics—and here Fry's aesthetics shades into ethics, specifically the ethics of the philosopher G. E. Moore, whose *Principia Ethica* has been called Bloomsbury's Bible.[67]

Irving Lavin argues in his article on the IAS seal that the theme of the relation of truth and beauty to science first "became explicit and central" in Fry's essay "Art and Science," which we quote shortly. "It may be that Fry and Flexner were acquainted," Lavin speculates. Fry lived a few blocks away from Abraham Flexner while he was Curator of Painting at the Metropolitan Museum in New York, and Fry certainly knew Simon Flexner, Abraham's older brother. It is known that Flexner was advised—by mathematicians, not by Fry—to recruit Hardy to the IAS but was "readily convinced that there was no way to lure [him] away from Cambridge," where they met in 1932.[68]

• •

Grothendieck insists throughout *Récoltes et Sémailles* on the importance of beauty in his life, in mathematics and beyond. This quotation is typical:

> The deepest and most fruitful work is the one that attests to the most liberated sensitivity to apprehending the hidden beauty of things [and a footnote adds: such delicate sensitivity to beauty seems to me intimately linked to something to which I've had occasion to refer under the name of making demands (on oneself), or of "rigor" (in the full sense of the term), which I described as "attention to something delicate in ourselves," an attention to a quality of comprehension of the thing that is probed. This quality of **comprehension** of a mathematical thing cannot be separated from a more or less intimate, more ore less perfect perception of the particular "beauty" of the thing.].[69]

Langlands, invited by Notre Dame's philosophy department to address the question of "beauty in mathematical theories," was, as usual, more circumspect. After reminding his audience that Jacobi declared his devotion to the "honor of the human spirit" in response to the opinion he attributed to Fourier, that "the main goal of mathematics is public utility and the explanation of natural phenomena," he continued:

> I am not sure it is so easy. I have given a great deal of my life to matters closely related to the theory of numbers, but the honor of the human spirit is, perhaps, too doubtful and too suspect a notion to serve as vindication. The mathematics that Jacobi undoubtedly had in mind when writing the letter, the division of elliptic integrals, remains, nevertheless, to this day unsurpassed in its intrinsic beauty and in its intellectual influence.

Acknowledging that referring as Fourier did to "public utility" is "if not then at least now, often abusive," Langlands concluded that "it is not easy to find an apology for a life in mathematics." "Nevertheless," though we are "only animals," human beings

> have also created—and destroyed—a great deal of beauty, some small, some large, some immediate, some of enormous complexity and fully accessible to no-one. It, even in the form of pure mathematics, partakes a little of our very essence, namely its existence is, like ours, like that of the universe, in the end inexplicable.[70]

Before the *Apology* shifts to a polemic on utility, Hardy concludes his aesthetic chapters with "a few disjointed remarks" on the nature of mathematical beauty, summarized in a single sentence: "In both theorems [of Euclid and the Pythagoreans, together with their proofs] there is a very high degree of *unexpectedness*, combined with *inevitability* and *economy*."[71] Louis J. Mordell, Hardy's successor as Sadleirian Professor at Cambridge (and the first professional mathematician I ever met), later added "simplicity of enunciation." With that, it's only a slight exaggeration to say that the theorizing of mathematical beauty—on internal, structural grounds—comes to an end.[72]

Aesthetic judgment in mathematics[73] is hampered by its meager lexicon; it doesn't inspire "lofty" habits in the use of language. *Pretty, appealing, attractive*, and the like, carry less weight than the all-purpose *beautiful* and the perennial favorite *elegant*, and in practice they overlap with words like *clever* or *ingenious* that reflect a different set of concerns. *Harmonious* or *symmetric* seem to me to beg the question. So in principle Hardy's trio of *unexpectedness, inevitability*, and *economy* should provide a welcome alternative. André Weil's topological insight, the one that converted me to number theory, qualifies on all three counts; and I recognized these qualities in the dream strategy that lingered with me in Paris in 1992. But after a moment's thought, we realize that we have made no progress at all toward elucidating what beauty in mathematics has to do with beauty elsewhere. The problem is that *unexpectedness, inevitability*, and *economy* are of little help in grasping the beauty of Hardy's "forehead" or "high cheekbones" or Shakespeare's "After life's fitful fever he sleeps well," quoted in the *Apology* and a particular favorite of Hardy's.

Even if we suspect that the word *beautiful* in mathematics is used mainly in order to express personal approval, we are entitled to ask what kind of approval it expresses and how, if at all, it differs from positive value judgments expressed by words like *true* or *good*. Roger Fry had studied mathematics at Cambridge and once attempted to distinguish "esthetic response . . . from the responses made by us to certain abstract mental constructions such as those of pure mathematics. . . . " His conclusion is reminiscent of Russell's "stern perfection": "Perhaps the distinction lies in this, that in the case of works of art the whole end and purpose is found in the exact quality of the emotional state, whereas in the case of mathematics the purpose is the constatation of the universal validity of the relations without regard to the quality of the emotion accompanying apprehension."[74]

All three of Hardy's criteria can be found, with some effort, in Fry. In his *Art and Science*, for example, Fry returned to the theme of emotion in science: "the recognition of inevitability in thought is normally accompanied by a pleasurable emotion, and . . . the desire for this mental pleasure is the motive force which impels to the making of scientific theory." Elsewhere he wrote that "a certain *quality of surprise, or at least unexpectedness*, is essential to keep our contemplation at full stretch"; and he drew his readers' attention to Rembrandt's (and Shakespeare's) "economy."[75] But while each of these qualities meets with Fry's approval, he mentions them only briefly; they are not central to his aesthetic vision.

Fry, like Hardy, was a Cambridge Apostle, as were Leonard Woolf, Russell, John Maynard Keynes, E. M. Forster, and Lytton Strachey. G. E. Moore was Hardy's Apostolic sponsor, and you could read the *Apology*'s aestheticism and cult of uselessness as an expression of the ideal of Moore's *Principia Ethica*, which was "that personal affections and aesthetic enjoyments include *all* the greatest, and *by far* the greatest, goods we can imagine."[76] Thus Moore subordinates aesthetics to ethics. He makes it possible to find an ethical justification for something as useless as pure mathematics, provided it can be framed in terms of "aesthetic enjoyments." In the *Apology*, Hardy attempted to do just that.

Clive Bell, Bloomsbury's other major art critic, agreed with Moore. "To pronounce anything a work of art is . . . to make a momentous moral judgment," he wrote. Echoing Moore, he insisted, "The starting point for all systems of aesthetics must be the personal experience of a peculiar emotion." Bell lists mathematicians alongside mystics and artists as capable of attaining "ecstasy":

> The pure mathematician rapt in his studies . . . feels an emotion for his speculations which arises from no perceived relation between them and the lives of men, but springs, inhuman or super-human, from the heart of an abstract science. . . . Before we feel an aesthetic emotion for a combination of forms, do we not perceive intellectually the rightness and necessity of the combination?[77]

Bertrand Russell, rather than Hardy, may have been the mathematician the Bloomsbury critics had in mind. Bell exchanged letters with Russell, and Fry painted Russell's portrait. My cursory exploration of the available source material, on the other hand, has unfortunately yielded no evidence that Hardy ever talked about art or beauty—or about anything at all, for that matter—with Bell or Fry. All I have been able to establish is an affinity at the level of ideas. Hardy shared with Bell and Fry, as well as with Moore, the readiness to recognize the aesthetic as an intrinsic good; if pressed to categorize Hardy's three-word aesthetic theory, we would be tempted to call it "formalist," in the manner of the Bloomsbury critics.

There is also a difference. The emotional response central to Bloomsbury's aesthetic is absent from Hardy's discussion of mathematical beauty. "[O]bsessive math is not done by geeks, but by passionate people who feel as much as they think." This is the lesson David Auburn, author of *Proof*, a play about fictional mathematicians, wanted to share with his audience after learning in the *Apology* "about the pleasure, passion, and joy of doing" mathematics. Auburn was right to recognize that Hardy's book is a work of deeply felt emotion, not limited to the "melancholy" of the *Apology*'s first lines. But it is telling that the words *passion* and *pleasure* hardly figure in the *Apology*, and *joy*, not at all. Only once does Hardy's aesthetic emotion approach ecstasy: "Greek mathematics . . . is eternal because the best of it may, like the best literature, continue to cause intense emotional satisfaction to thousands of people after thousands of years."

Hardy did call his research "the one great permanent happiness of my life," and when we read this we feel the "haunting sadness" that struck Snow and most subsequent readers, because the sixty-three-year-old Hardy told us at the very beginning of his *Apology* that mathematics is "a young man's game" and he is no longer a player. The word *permanent* occurs ten times in the *Apology*, *emotion*, three times. *Beauty*, which recurs throughout the text, is the word Hardy chose to allude to his emotional reaction to mathematics. The Powerful Beings to whom he thought he owed allegiance, he may have felt, would find it more "serious"—the word occurs in nearly every chapter of the *Apology*—than a direct appeal to emotion. Most mathematicians who have written on the topic have made the same choice, because they also want their aestheticism to be taken seriously. It is, therefore, ironic that aestheticism as a movement, far from permanent, belongs to an increasingly irretrievable past. Even Fry, who could write "[i]f by some miracle beauty could be generated without effort, the whole world would be the richer," used the word *beauty* sparingly in connection with art, while Bell preferred to avoid it altogether, opting instead for *significant form*. More than a century separates us from the time when serious people who were not mathematicians could speak of beauty unfiltered, without apologies. Is it any wonder that, in popular culture's serious precincts, the mathematician has become the romantic figure of choice?

chapter δ

How to Explain Number Theory at a Dinner Party

Fourth Session: Order and Randomness

In 1801 Gauss published the *Disquisitiones Arithmeticae*, considered the founding text of modern number theory. Much of the *Disquisitiones* is concerned with the study of solutions to equations of degree 2 (quadratic equations) in two variables, such as $x^2 + y^2 = 3$ or $x^2 + xy + y^2 = 11$: equations of ellipses or hyperbolas. For our purposes here, quadratic equations (especially equations of ellipses) with integer coefficients are completely understood, thanks to Gauss. This does not mean that one cannot still ask questions about quadratic equations in two variables whose answers are unknown. In this as in all areas of mathematics, our state of knowledge can be determined only relative to the questions we have chosen to ask. Nevertheless, we are here only looking for a way to classify quadratic equations in two variables into two kinds: those that have infinitely many solutions and those that have none at all. Legendre and Gauss explained exactly this dichotomy more than two hundred years ago for quadratic equations.[1] When there is at least one solution, there is a *formula* for writing down all the solutions. So, for the purposes of a dinner party, I can consider this problem to be closed.

Matters are quite different when the degree of the equation is allowed to increase to 3, as with the two equations we considered last time.

$$y^2 = x^3 - x, \tag{E1}$$

$$y^2 = x^3 - 25x. \tag{E5}$$

These *cubic equations* in two variables are two examples of *elliptic curves*.

Figure δ.1 shows a picture of the curve with equation (E5), which should not look at all like an ellipse; the curves do have a distant con-

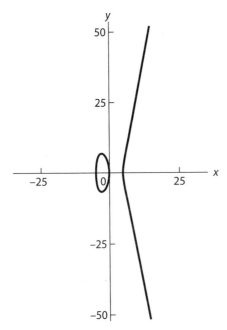

Figure δ.1. Graph of the cubic equation $y^2 = x^3 - 25x$ [formula (E5)].

nection with ellipses, but for number theorists the connection survives only in the name.

It is a famous theorem of Gerd Faltings, proved in 1983, that equations in two variables of degree 4 and higher almost always have only finitely many solutions in rational numbers, as predicted by (Hardy's future successor) L. J. Mordell in 1922; when they have infinitely many, it is always for an uninteresting reason, basically because they are the wrong sort of equations—the graph will cross itself like figure δ.2a) or have a sharp cusp like figure δ.2b) rather than be smooth like the one in figure δ.1.

For example, Faltings's theorem implied that the equation

$$x^n + y^n = 1, \qquad (Cn)$$

where n is a positive integer, has only finitely many rational solutions as soon as n is at least 4. If we write

$$x = \frac{a}{c}, y = \frac{b}{c},$$

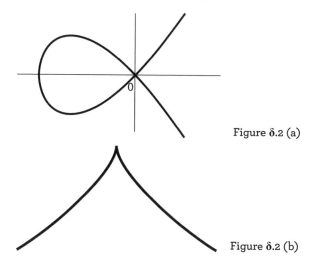

Figure δ.2 (a)

Figure δ.2 (b)

where a, b, and c are whole numbers, then equation (Cn) becomes the more familiar *Fermat equation*,

$$a^n + b^n = c^n, \qquad\qquad (Fn)$$

and Fermat's Last Theorem, proved by Andrew Wiles, says that (when n is at least 3) this equation has *no* solutions when a, b, and c are all positive. At one point in Wiles's proof of Fermat's Last Theorem, he makes direct reference to a theorem of Langlands (and a version due to Jerrold Tunnell, whom we already encountered in chapter 9) on automorphic forms; at one point in Faltings's proof, he also uses automorphic forms (not in the Langlands style); and both Wiles and Faltings were utterly dependent on Grothendieck's algebraic geometry.

So equations of degree 1 or 2 are easy to understand; equations of degree 4 or more may be complicated, but at least one knows not to look for infinitely many solutions. The degree that continues to obsess number theorists is 3.

Equations (E1) and (E5) are very special examples because they have a direct connection with triangles, as do all equations of the form

$$y^2 = x^3 - N^2x, \qquad\qquad (EN)$$

where N is any positive whole number [so (E7) has equation $y^2 = x^3 - 49x$, and so on]. It turns out that equation (EN) has infinitely many solutions in rational numbers precisely when there is a right triangle with *area N*, all three of whose sides—what we called *a*, *b*, and *h* in session 2—are *rational* numbers.

If there is such a right triangle, then *N* is called a *congruent number*, and so the problem of determining whether or not (EN) has infinitely many solutions is called the *congruent number problem*. There are many right triangles with area 1, but it can be checked that none of them has all three sides rational (for example if $a = 1$ and $b = 2$, then $h = \sqrt{1^2 + 2^2} = \sqrt{5^2}$, which we have already seen is irrational). Thus 1 is not a congruent number, and, therefore, equation (E1) has only finitely many rational solutions. On the other hand, (E5) has infinitely many rational solutions. The first interesting one is $x = -4$, $y = 6$. And you can check that the right triangle with sides $a = 9/6$, $b = 40/6$, $h = 41/6$ (you have seen these numbers before) has area

$$\frac{1}{2}ab = \frac{1}{2} \times \frac{9}{6} \times \frac{40}{6} = \frac{9 \times 40}{2 \times 6 \times 6} = 5,$$

as promised.

I have already made the point that a problem in number theory is interesting when it is the subject of one or more stories. You are right to notice that there is no triangle visible in equation (EN) and that I have not explained why you might think of the *N* that indexes the equation as the area of anything. It takes a leap of the imagination to see the equation $a^2 + b^2 = h^2$ in the right triangle, but the leap was taken for us so many years ago that we no longer notice that we are leaping every time we write down the Pythagorean theorem. The story that relates equations (EN) to right triangles is not so well known but is also quite old; it goes back at least to Arab algebraists of the tenth century, and is fascinating in itself. It has been retold many times, most recently in the 1980s by Jerrold Tunnell in a particularly engaging way, one that revealed a close connection with the most contemporary concerns of number theorists. There are infinitely many ways to write equations for elliptic curves, and most (for example $y^2 + y = x^3 + x$, which has the obvious solution $x = y = 1$) cannot be attached to such immediately appealing stories. But a web of stories has grown up in the last fifty years or so around the problem of determining whether or not a given elliptic curve has infinitely many points.

The starting point is the method of congruences. In the last session, we introduced the notion of approximate solutions relative to a prime number p, and we used congruences to count the numbers of approximate solutions $S(p)$ to equations (E1) and (E5). One would need to perform a great many calculations (easy with a computer program) and test a great many cubic equations in two variables to appreciate the following theorem.

Theorem (Hasse) For any elliptic curve, and for any prime number p, the number $S(p)$ as just defined is very close to p in the following sense: except for a finite set of exceptional primes p that depends on the elliptic curve in a simple way, the difference between $S(p)$ and p is at most equal to $2\sqrt{p}$.

We have been alluding indirectly to Hasse's theorem practically since the beginning of the book. It was the inspiration for Weil's idea to count solutions to polynomial equations using topology, which in turn inspired Grothendieck's new foundations for geometry, which was in turn the basis for Deligne's solution—using a trick, in Grothendieck's opinion— to the last of Weil's conjectures. But Hasse's theorem is quite remarkable on its own. For example, there are 25 possible pairs x, y for $p = 5$. Hasse's theorem says that for most cubic equations in two variables, the number $S(5)$ of approximate solutions modulo 5 is more than $5 - 2\sqrt{5} = 0.527$ and less than $5 + 2\sqrt{5} = 9.47$. Since $S(5)$ is always a whole number, this means, in other words, that there is always at least one approximate solution modulo 5 [unless 5 is one of the exceptional primes, which is usually not the case but is the case for (E5)], and there are also always at least 16 pairs x, y that are *not* solutions. Similarly, the number $S(7)$ is always at least 2 and at most 12 out of a possible 49 pairs. After calculating the numbers $S(p)$ for a variety of equations, one can easily understand the statement of Hasse's theorem, but it was the height of sophistication at the time and still requires a good deal of advanced mathematics to prove.

Hasse's theorem provides a glimpse of order into the mysterious problem of determining whether or not a cubic equation in two variables has finitely or infinitely many rational solutions. If you are heuristically inclined, you might guess that if the equation has infinitely many rational solutions, then $S(p)$ would have a tendency to be at the high end of the allowed range (i.e., closer to $p + 2\sqrt{p}$ than to $p - 2\sqrt{p}$). If you try to justify the heuristics, you will soon admit that this is quite a wild guess, but the really tricky part is to make sense of what it means for something like $S(p)$ to "tend" in one way or another. It was the two British mathematicians Bryan Birch and Sir Peter Swinnerton-Dyer who gave a rigorous expression to this wild heuristic guess, providing a very precise notion of how the size of the expressions $S(p)$ can be used to predict whether or not an elliptic curve has infinitely many solutions. This notion is defined in the framework of the prime number theorem mentioned in the first session; it can be considered a substitute for a *formula* for the number of solutions.

You will remember from chapter 2 that the *Birch-Swinnerton-Dyer* (BSD) *conjecture*[2] was the guiding problem for the first part of my career and that it has the distinction of being one of the seven Clay Millenium Prize Problems, like the Riemann hypothesis, whose solution carries a million dollar price tag. Jeremy Gray's essay, "A History of Prizes in Mathematics," written to accompany the Clay Mathematics Institute's official presentation of the Millennium Problems, traces the tradition of awarding prizes for solving mathematical problems back to the sixteenth century. Not surprisingly, the earliest prize Gray records was given for solving a problem that has already come up during our dinner party. It was awarded in 1535 by Antonio Fior to Niccolò Tartaglia, for (re)discovering Scipione del Ferro's formula for the three solutions to the cubic equation, and consisted in "thirty dinners to be enjoyed by [Tartaglia] and his friends."

The BSD conjecture is an attempt to discern order in the apparently unpredictable nature of solutions to elliptic curves, following Hasse's theorem, which places strict limits on the numbers of approximate solutions for varying p. The "tendency" (also known in the literature as "bias") to which Birch and Swinnerton-Dyer drew attention is a rather subtle phenomenon. There are coarser ways to measure the difference between p and $S(p)$. In the few examples given before, $S(p)$ was always greater than p, but it is easy to find examples where $S(p)$ is less than p; Hasse's theorem shows only that there are no examples where $S(p)$ is less than $p - 2\sqrt{p}$. In fact, on the coarse measure it can be seen that it is equally likely that $S(p) - p$ is positive or negative.

You will have noticed that for the curve (E1) we have calculated $S(3) = 3$ and $S(7) = 7$. This is not a coincidence. Remember that 3 and 7 are the first two prime numbers that cannot be written as sums of two squares. One can prove that for equation (EN) (any N), we have $S(p) = p$ whenever p is a prime in the congruence class of 3 modulo 4—in other words, whenever p cannot be written as the sum of two squares. There is a reason for this: each of equations (EN) has an extra symmetry. Suppose (x, y) is a solution to equation (EN). If you multiply x by -1 and y by $\sqrt{-1}$, then you get another solution, as you can verify if you like algebra:

$$(\sqrt{-1} \times y)^2 = -y^2 = -(x^3 - N^2x) = (-x)^3 - N^2(-x)$$
$$= (-1 \times x)^3 - N^2(-1 \times x).$$

Now you will recall that $\sqrt{-1}$, also written i, is an *imaginary number*, and so the new solution will be in imaginary numbers if the original solution consisted of rational numbers, but this should not obscure the main point, which is the presence of an extra symmetry. Although this is by no means obvious, it turns out that the additional symmetry of the equation imposes a predictable additional structure on the numbers $S(p)$, a *reason* that permits calculation of some of the $S(p)$ without any additional effort.

Most elliptic curves do not have extra symmetries. For those that do not (for example, the elliptic curve with equation $y^2 + y = x^3 + x$), it was conjectured by Mikio Sato and John Tate that the only predictable feature of $S(p)$ is that imposed by Hasse's theorem. In other words, the difference between $S(p)$ and p is at most $2\sqrt{p}$, but within that range the difference is as random as possible. What it means to be as random as possible is again framed by the prime number theorem and cannot be explained in a few words. The histogram in figure δ.3 plots the difference $p - S(p)$ for large collections of primes p.[3]

The *Sato-Tate conjecture* was proved by Laurent Clozel, Nick Shepherd-Barron, Richard Taylor, and me in a series of papers completed in 2005 and 2006, for a large class of elliptic curves (including the example $y^2 + y = x^3 + x$ and infinitely many others), and then for all elliptic curves in my joint paper with Tom Barnet-Lamb, David Geraghty, and Taylor. Number theorists are interested in detecting patterns.

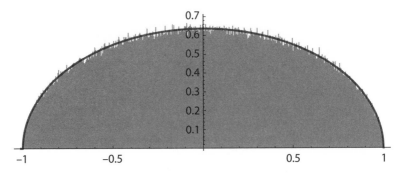

Figure δ.3. Numerical confirmation of the Sato-Tate conjecture (William Stein).

A perceptible pattern demands explanation. This explanation typically takes the form of a story, expressed in the language of proof, that provides a convincing reason for the existence of the perceived pattern. Where no reason can be found, one expects there to be no pattern—or, rather, as random a pattern as possible. But randomness is also a pattern.

> P. A.: I had been thinking of getting a copy of the author's book for my future in-laws' coffee table, but I'm finally convinced it would be pointless.
>
> N. T.: Is it because of the formulas? Have they joined the latest exodus of 10,000 potential readers?
>
> P. A.: Formulas are not the problem. On the contrary, it was thanks to a formula that Tartaglia and his friends enjoyed all those dinners. Abel, on the other hand, went hungry after he showed there was no formula and died at 26. Maybe he would have led a happier life if he had found a formula.
>
> N. T.: A formula like Scipione del Ferro's, like Ed Frenkel's love formula, gives an immediate impression of control: not only do you know the problem has a solution, you can actually write it down. But mathematicians are equally happy to know there is no formula, provided an alternative structure is available. Once the work of Abel and Galois shifted attention to structural features of solutions, formulas were no longer the goal, and the ones that worked up to degree 4 were consigned to the realm of the routine and predictable. The same thing happened with polynomial equations in two variables: the degree 2 polynomials, like

those we saw in chapter β, were central to Gauss' systematic approach to number theory 200 years ago, but their formulas have long been understood. Now attention is focused on degree 3 polynomials in two variables, where the Birch-Swinnerton-Dyer conjecture offers a very different kind of formula.

P. A.: You are talking as if you agree with André Weil's comment about achieving "knowledge and indifference at the same time."

N. T.: I found the part about "furtive caresses" more eloquent.

P. A.: And about the "majestic beauty" that "can no longer excite us."

N. T.: Jeremy Gray writes in his essay that "A good problem is one that defies existing methods . . . but whose solution promises a real advance in our knowledge."[4]

P. A.: Gray could just as well have defined a good problem as one whose solution promises a real advance in our indifference. You are reminding me that the author has given my in-laws excellent reasons for not wanting to read his book, and formulas have nothing to do with it.

N. T.: The author did try to explain that the problem of understanding solutions to equations of degree 3 in two variables is the first really hard problem one encounters in number theory.

P. A.: I'm sure every problem looked hard until it was solved, at which point it didn't look hard any more; instead, the solution became "routine and predictable." That's what I understand by "achieving knowledge and indifference at the same time." My in-laws have every reason to assume solution of the Birch-Swinnerton-Dyer conjecture will lead inexorably to indifference, in exactly the same way.[5]

N. T.: It's the first really hard problem those of us encounter who are living in the present, which is when the Clay Millennium Problems were chosen, and it stands to reason that if one can't solve the first problem, one isn't going to have much luck with the ones that follow.

P. A.: It's the first of a series of doors you will need to unlock, and the series is endless, as the author himself wrote. For those of us who stand by and watch your indifference grow after you unlock them, all those doors look alike.

N. T.: Let me try an analogy. You don't really care about Nora, do you?

P. A.: Of course I care about Nora! If I didn't care about her, why would anyone in the audience care about her?

N. T.: And yet Maurice Maeterlinck, at the turn of the twentieth century, wrote of his "disgust" at the presence of living human performers on stage, which leads to "the abolition of the symbol, the annihilation of the dream and, consequently, the destruction of art." He once wrote an essay calling for a "theater of androids" like the figures in a wax museum.[6]

P. A.: There are now theaters like that, in Disneyland, for example. And cartoon characters can also provoke catharsis. Performing artists, on the other hand, perform, which is different from what mechanical figures or cartoon characters do, and caring about the characters has something to do with that.

N. T.: Maybe the audience comes to watch you care about Nora. It's no concern of theirs whether you will still be caring about her after they leave the theater or whether you will have moved on to some other problem.

P. A.: But the audience wants to care about Nora, wants to feel pity and fear, at least for the duration of the performance, not just to experience a vicarious catharsis through watching some performer's caring. Are you saying that the author expects readers to pick up the book just in order to watch mathematicians care about the Birch-Swinnerton-Dyer conjecture?

N. T.: That's not exactly how I would put it, but it's not a small matter to be able to care about something like that. Some readers, if not your future in-laws, will have wondered how that's possible.

P. A.: From the vantage point of page 320, it's hard to disagree. But what makes you think anyone will read that far?

The Veil of Maya

*The attraction of everything problematic, the joy of X, how-
ever, is too great in such more spiritual, more spiritualized
people, for this joy not to flare up again and again like a
bright blaze over all the distress of what is problematic,
over all the danger of uncertainty, even over the jealousy of
the lover. We know a new happiness.*[1]

Scholarship seems not to have considered which metaphorical veil, or
Schleier, Hilbert was inviting his listeners to lift, among the many on
offer to an educated German in the late nineteenth century. Was it the
"veil of poetry" [*der Dichtung Schleier*] a radiant goddess handed to
Goethe, revealing herself as Truth; or was it one of the seven veils of
Oscar Wilde's *Salomé*, which Richard Strauss orchestrated, a few years
after Hilbert's lecture, in the *Schleiertanz* that concludes his opera?

Rather than assuming Hilbert sought to awaken the prurient inclina-
tions of coming generations, it is safest to assume that he was thinking
of of the "veil of Isis" that Schiller introduced into German poetry, that
Kant identified with the barrier separating human understanding from
the *Ding an sich*, that Schlegel enjoined his reader to tear—"Whoever
cannot bear the view of the goddess, let him flee or perish." The piercing
of this veil has enjoyed a long career as a metaphor for the pursuit of the
scientific knowledge of nature, not least in the title—*Isis*—of what is
still one of the most prestigious international journals of history of
science.

Hilbert's listeners in Paris knew that lifting veils is a risky business.
In Schiller's poem *The Veiled Statue at Saïs* [*Das verschleierte Bild zu
Sais*], a young seeker of Truth learns that it is hidden behind the veil of
the goddess. Ignoring a priest's warning, he raises Isis's veil and falls

"senseless and pale." "For all time, the serenity had gone from his life. A deep melancholy carried him off to an early grave." Schlegel's comment was a direct retort to Schiller's poem, and Novalis agreed: "If it is true that no mortal can lift the veil . . . then we will just have to try to become immortal." [2]

Hilbert spoke in Paris in the spirit of Novalis. He famously affirmed his turn-of-the-century optimism in another sentence from the same lecture: "Here is the problem, seek the solution. You can find it through pure thought, because in mathematics there is no *Ignorabimus*."* Mehrtens sees Hilbert's near-contemporary Felix Hausdorff as a partisan of a different strand of modernism, one less confident that mathematics was destined to unveil an unequivocal Truth. Writing in a literary journal two years after Hilbert's lecture, under the pseudonym Paul Mongré, Hausdorff imagined a Truth-seeker in his garret, staring at his navel and thinking "Everything is thus, but must it be thus? . . . Could there not be a crystal space, where one could see around the corner and sense one's way into another I [*in ein anderes Ich hineinempfinden*]? . . . Or why not three genders . . . Men, Middlers, Mothers. . . ." Mongré's essay—historians translate the pseudonym as "Paul to-my-liking"—is entitled "The Veil of Maya" [*Der Schleier der Maja*],[3] unquestionably a reference to Schopenhauer, by way of Nietzsche, philosophers traditionally seen as more sympathetic to the arts than to the sciences.

Classic accounts of the modernist turn in mathematics frame it as a confrontation between camps led by Hilbert, the optimist formalist, and by L.E.J. Brouwer, founder of intuitionism and critic of Cantor's abstract set theory. Mehrtens brackets Hausdorff between Hilbert and Brouwer—a modernist (like Hilbert) and "working mathematician," who devoted considerable attention to philosophy (like Brouwer). Hilbert's is by far the biggest name in the word cloud of figure 2.2, and Brouwer's looms large beside his, but Hausdorff's name is absent, which doesn't seem fair, because it was he as much as anyone who made Cantor's set theory speak the language of geometry—not the Euclidean geometry of circles, triangles, and straight lines, but the geometry of our intuition of continuous space.[4] It was Hausdorff who invented the term *topological space*,

* *"Da ist das Problem, suche die Lösung. Du kannst sie durch reines Denken finden; denn in der Mathematik giebt es, kein Ignorabimus!"*

now indispensable, for the spaces of his geometry. The spaces Hausdorff called "topological" are now known as *Hausdorff spaces*; they fit marvelously with our intuitions of geometry freed of notions of length, but more flexible axioms have also proved useful, notably in Grothendieck's geometry. A direct line, or ladder, can be drawn from Hausdorff's conception of space to Grothendieck's, passing through Brouwer and Weil, among others, and on past Grothendieck to ∞-categories, as well as the claim I imagined I heard Beilinson make, that geometry is an illusion, a Veil of Maya, that serves to hold fixed points together.

The (small-*f*) foundations of the topology studied today were largely established by Hausdorff, together with contemporaries in France and among the students of the name worshippers Luzin and Egorov in Moscow. Hausdorff lends his name to the mnemonic math majors use to distinguish the spaces he defined among all topological spaces: "a Hausdorff space is one in which any two points can be *housed off* from one another." This is the kind of sophomoric humor that would appeal to the author of *Against the Day*'s mathematical jokes, and it's possible Hausdorff himself would have appreciated it. A contemporary reviewer of his topological masterpiece, the *Grundzüge der Mengenlehre*, published in 1914, noted its "occasional glimmer of humor." Paul Mongré allowed his humor much freer reign; his greatest literary success, a one-act satirical comedy, was performed more than three hundred times throughout the German-speaking world in the years leading up to World War I.[5]

Mongré's principal philosophical text, published when Hausdorff was thirty, was called *Das Chaos in kosmischer Auslese* [Chaos in cosmic selection]. It "aimed at nothing less than to destroy permanently *every* type of metaphysics." Mongré called this position "transcendent nihilism." The argument, derived from a "kind of metaphorical use of [set-theoretic] mathematics," was summarized by a contemporary reviewer: "it is impossible to attribute to the system of limitations and syntheses which define *our* reality any ulterior objective precedence above other systems, apart from its simple relation to us."[6] This is the theme of *Der Schleier der Maja*'s "Truth-seeker," but Mongré was not dogmatically attached to Truth; in his *Sant' Ilario*, he had written that "[i]f not truth itself, then surely the belief in holding truth is, to a dangerous degree, antagonistic to life and murderous for the future [*lebensfeindlich und zukunftsmörderisch*]."[7]

Mongré's transcendent nihilism was complemented by what Hausdorff called "considered empiricism." Two years before Liebmann was to call mathematics a "free, creative art," Hausdorff, also giving an inaugural address in Leipzig as an unpaid *Professor extraordinarius*, spoke of the "creative freedom of [mathematicians'] concept formation [*Gedankenbildung*]," especially in their treatment of space. Hausdorff's lecture was devoted to the "Problem of Space" [*Raumproblem*], and he identified three *Spielräume*—literally, "play spaces"—in the " 'Freedom of Choice' among hypotheses on the formulation of the mathematical concept of space,"—the *Spielraum* of thought, of intuition, and of experience [*den Spielraum des Denkens, den Spielraum der Anschauung, den Spielraum der Erfahrung*].[8] It was with respect to the *Spielraum* of experience as it bears on the study of physical phenomena that Hausdorff saw the need for a considered empiricism.[9] The *Spielraum* of thought, in contrast, was "very large indeed"; it was the source of the "creative freedom" that, Hausdorff added, mathematics had acquired "not without struggle against philosophical attempts at suppression."[10]

We encountered that word "freedom" in our very first chapter, in Cantor's famous dictum that "the essence of mathematics lies in its freedom." We saw it again in Liebmann's characterization of mathematics as a "free, creative art" and in Brouwer's interpretation of real numbers as "free choice sequences"; it can be found in Hilbert's writings as well.[11] In 1942, as a seventy-four-year-old Jew in Nazi Germany, Hausdorff was faced with the prospect of being deported to the Theresienstadt concentration camp from the Endenich cloister in Bonn, where he was about to be interned. Hausdorff, his wife, and his sister-in-law, chose— one would like to say freely, within the very narrow *Spielraum* still available to them[12]—to take an overdose of barbiturates and thereby escape the fate that had been ordained for them.

A team of eminent scholars is completing a definitive edition of Hausdorff's collected works—"unique . . . in the annals of mathematical publishing"—with the care befitting the literary figure he undoubtedly was. This scholarship has confirmed the immortality of the "Dionysian mathematician"[13] in the land of mathematical Giants. A few years ago the prestigious Hausdorff Research Institute for Mathematics was created in Bonn, where Hausdorff spent the greater part of his career, and which, since the war, has been Germany's principal mathematical center. Here he is honored as, perhaps, the first modern mathematician to give a name to what we have been calling the "relaxed field"—he called it the "*Spielraum* of thought"—and as a mathematician who never lost his sensitivity to his chosen field's problematic attractions while remaining fully aware that every veil lifted only reveals another veil.

Chapter 1 Introduction: The Veil

1. (Hilbert 1900).
2. See Gillies (1995). Although Kuhn's work has been criticized by professionals, I cite him because his vocabulary is so familiar and lends itself to the picture imagined in the previous paragraph.
3. (Kuhn, 1996, p. 33).
4. *mit der reicheren Entfaltung einer Wissenschaft die Notwendigkeit auftritt, die ihr zu Grunde liegenden Begriffe und Prinzipien zu verändern. Es geht auch in diesem Punkte der Mathematik nicht anders wie den Naturwissenschaften : neue Erscheinungen stürzen die alten Hypothesen um und setzen andere an ihre Stelle* (Kronecker 2001, p. 233). Subsequent chapters, especially chapter 7, will return to Cantor's set theory and the Foundations Crisis.
5. Siegel was sympathizing with his contemporary, Louis J. Mordell, who had just written a famously savage review of Serge Lang's book *Diophantine Geometry*. Mordell and Siegel make brief appearances in the next chapter (Lang 1995).
6. Gray (2010, p. 37) refrains from using Kuhn's framework, undoubtedly for good reasons.
7. Mumford, whose career is evenly divided between the purest branches of pure mathematics and the most challenging questions of artificial pattern recognition, has been awarded most of the highest distinctions in the field; he taught at Harvard for many years before moving to Brown. The quotation is from Casazza (2011). Frank Quinn (2012, p. 33) reports that Paul Halmos claimed to be in "roofing and siding." Humanists see things differently. Alex Csiszar quotes Bruno Latour's claim that "[i]t is hard to popularize science because it is designed to force out most people in the first place," and concludes that "the particular difficulty that mathematicians have in talking to nonmathematicians about what they do, in fact even in talking to fellow mathematicians who are not quite a part of the same subdiscipline, is a telling indicator of the discipline's tendency to exclude all but the very few" (Csiszar 2003).

Chapter 2 How I Acquired Charisma

1. That the quest for certainty can awaken a vocation is a staple cliché of mathematical reminiscences (which this and similar paragraphs in this chapter exemplify only by way of pastiche). Thus, Helmut Hasse chose mathematics "because even in my earliest youth the impulse toward knowledge of objective, irrefutably valid truths was so strong that it overshadowed every other interest" (Hasse 1952, my translation). It's tempting to claim that the certainty of mathematics appealed to me as well, as a response to the turbulence of 1968, but it didn't really happen that way. It was only after comparing the biochemistry and mathematics programs during my first year in Princeton that I opted

for the latter. Certainty, in the philosophical sense, was not an issue for me at the time. The stability of the hierarchy set out in the IBM poster was appealing, but not more so than the apparently endless string of Nobel Prizes in molecular biology. There are many paths to mathematics and just as many mathematical personalities.

2. The Eames couple was and still is famous for the lounge chair they designed in 1956. Through the Eames Studio, Redheffer produced a series of portraits of more than a hundred distinguished mathematicians, many of whom were still alive when I first saw the IBM poster. The portraits are posted in the corridors in the heart of the UCLA Mathematics Department (Gamelin 2005).

Other noteworthy collections include the portraits of academicians on the top floor of the Steklov Institute of Mathematicians in Moscow, the Michael and Lily Atiyah Portrait Gallery at the University of Edinburgh (a mix of historical figures and contemporaries and collaborators of Sir Michael Atiyah), and *Faces of Mathematics* at Herriot-Watt University in Edinburgh, whose mission is to "present [. . .] the human side of this most austere and challenging area of modern science." The book (Cook 2009) pairs portraits of contemporary mathematicians with one-page interviews.

3. Here E. T. Bell is notoriously inaccurate. Caroline Ehrhardt (2011, p. 196) calls his book a "biographical novel." See also the discussion of Amir Alexander's (2011) book in chapter 6.

4. On April 5, 2012, IBM released a digitized version as a free iPad app called "Minds of Modern Mathematics" (still including only one female mind), expressing the hope that "classes and students will use the app, provoking more people to pursue math, science or technology-related educations and jobs." Large-scale versions of the IBM timeline are on display at the New York Hall of Science in Queens and at the Museum of Science in Boston. (http://www.wired.com/gadgetlab/2012/04/new-ibm-app-presents-nearly-1000-years-of-math-history/). The word *giants* was inadvertently suggested by one of the Princeton University Press readers of my book proposal.

Many future mathematicians are still initiated into the field at high school summer programs like the one I attended at Temple. The PROMYS program at Boston University is one of the most successful and influential; 100 of its graduates are currently professors, 70% of them in mathematics (Clay Mathematics Institute 2013). It is much more mathematically challenging than the Temple program, devotes less attention to the "romantic" chapters in the field's history, and is wary of promoting a "great man" version of history of mathematics. Still, the Giants and Supergiants are inevitably part of the story. Glenn Stevens, who has been leading PROMYS since 1989, explained (in a private communication) that "the point is not to see the pioneers as the 'great masters' so much as to have an internal understanding of how they might have been motivated to develop these theories in the first place. We hope the students will begin to identify personally with the 'masters.' "

5. Neil Chriss, in (Lindsey and Schachter 2007, p. 111).

6. (MacIntyre 1981, p. 181). Also, "insofar as the virtues sustain the relationships required for practices, they have to sustain relationships to the past—and to the future—as well as in the present" (p. 206).

7. (Smith 1759). Smith adds: "It is not always the same case with poets, or with those who value themselves upon what is called fine writing."

8. (Krantz 2002, p. 186). Weil remained a snob even when he was apparently being

generous: "It is also necessary not to yield to the temptation (a natural one to the mathematician) of concentrating upon the greatest among past mathematicians and neglecting work of only subsidiary value. Even from the point of view of esthetic enjoyment one stands to lose a great deal by such an attitude, as every art-lover knows . . . " (Weil 1978). This is the published version of a lecture I attended in which Weil explained that what "one stands to lose" is precisely a feeling for the gulf separating genius from "subsidiary" talent.

I hasten to add that Weil's vanity was remarkable because it was atypical—humility is still expected and can in fact be read as a sign of prestige, since the mathematicians who know the most are also those who realize how much more there is to know. And I shouldn't be too hard on Weil, tempting target that he is, not only because his name is on many top-ten lists, but specifically because of his overwhelming influence on number theory, in particular on the kind of number theory to which this book devotes much of its attention.

C. L. Siegel was last seen in chapter 1 complaining about a pig in a beautiful garden.

9. (Weber 1922).

10. (Menand 2001, pp. 153-54). Benjamin Peirce has given his name to the coveted mathematics instructorships ("BPs") at Harvard; he is better known as the father of philosopher C. S. Peirce.

11. See (Zarca 2012, p. 318) and the surrounding discussion. Maybe this is typical of "high-consensus fields." "Physicists often describe how they are able to rank-order the achievements of fellow physicists" (Hermanowicz 2007).

12. http://mathoverflow.net/questions/10103/great-mathematicians-born-1850 -1920-et-bells-book-x-fields-medalists. One of the lists granted Kolmogorov only an "honorable mention." It would be interesting to compare the list to Redheffer's UCLA portrait gallery, for which I was unable to find a catalogue online.

13. No less prestigious are the *Nemmers Prize* (since 1994), the *Kyoto Prize* (every four years since 1985), and the *Crafoord Prize* (since 1982), the latter only occasionally granted to mathematicians. The monetary value of three of the four IMU prizes—the Fields Medal, Nevanlinna Prize, and Gauss Medal—is modest, but the *Chern Medal*, created in 2010, is worth $250,000, and some of the other prizes mentioned bring close to $1 million. And this year (2014), billionaire Yuri Milner and his Silicon Valley friends will start to "transform [mathematicians] into rockstars" with the first of the new $3 million *Breakthrough Prizes* in mathematics. Milner thinks scientists "should make at least a fraction of what some Wall Street trader makes" (Walker 2013).

14. For Kovalevskaia, see (Ann Hibner Koblitz 1983) and (Audin 2008). For Schwartz, the best source is his autobiography (Schwartz 2001). Several complete biographies of Grothendieck have long been in the works; the most complete is (Scharlau 2010) and its two companion volumes. There's also Grothendieck's own 900+-page *Récoltes et Sémailles*, as well as the two-part report (Jackson 2004). Perelman is the subject of the modestly titled (Gessen 2009). Two trade books were written about Erdős shortly after his death [in this book I cite (Hoffman 1998)], and there is also a sixty-minute documentary.

15. In hindsight I realize he was spaced out rather than distant, focusing his inner eyes on his research rather than his visible eyes on the person to whom he was speaking.

This characteristic trait, familiar to anyone who has lived with a mathematician, was already well known in antiquity (see chapter 6).

16. For example, the polynomial equation $x^2 + y^2 = 1$ in two variables defines a circle; the equation $x^2 + 2y^2 = 5$ defines an ellipse; the equation $x^2 + y^2 = z^2$ defines a cone in three dimensions. There are pictures in chapters β and δ for those who don't remember the x- and y-axes of high school algebra. Once there are more than three variables, the figure defined by the equation can no longer be represented by a picture, but it can be represented in many other ways, which together make up the subject of algebraic geometry.

17. See (Rashed 2005). The IBM chart informs us that al-Khayyam is best known for his Persian poetry cycle, the *Rubaiyat*, and Wikipedia says the same thing; but (Rashed and Vahabzadeh 1999), the definitive edition, in Arabic and French, of al-Khayyam's mathematical works claim (on p. 5) that no contemporary documents prove that al-Khayyam the poet and al-Khayyam the mathematician (and philosopher) were, in fact, the same individual.

18. More precisely, what Weil counts are solutions to *congruences*, where the number of solutions is always finite: see chapter γ. Tim Gowers (2013) has given a very readable account of Weil's topological idea in the presentation of Pierre Deligne's work that accompanies the announcement that Deligne was awarded the 2013 Abel Prize, most notably for his proof of Weil's conjectures.

19. Take a swimming pool (any shape you like as long as there are no islands in the middle), cover it completely with dandelion fuzz, and set up a smooth current on its surface. After one hour (or any other lapse of time you choose), at least one piece of fuzz will be back where it started. This is an example of *Brouwer's fixed point theorem*, which, of course, is not about swimming pools and currents but about flat surfaces— specifically flat surfaces without *holes*, in the technical sense discussed in chapter 7— that are subject to an internal motion that is continuous in the sense that nothing is allowed to be torn. (More precisely, the theorem deals with continuous mappings— using the surface as a map of itself that may be distorted but is not torn.) For more complicated surfaces, or configurations of higher dimension, there is the more elaborate Lefschetz fixed point theorem that tells you how many points are likely to be left in place after a continuous motion.

Mathematical economists have used versions of this theorem to convince themselves that economic systems tend to equilibrium, in spite of considerable evidence to the contrary, but that's another story.

20. Henry Brougham, *Lives of the Men of Letters and Science, who Flourished in the Time of George III*, quoted in (Shapin 2008, p. 8). Shapin explains that Brougham was including "natural philosophers"—scientists, in other words—among philosophers.

21. (Weber 1978, section 10).

22. The term *research program* is associated with Lakatos, especially (Lakatos 1978). In this book I use the term more informally, as in (Mazur 1997, especially section 6).

23. B. H. Gustin uses the word in much the same way as I do: "Much as religious charisma must be routinized . . . the charisma of science is regularized, attenuated, and embodied in the institution of publication, as well as, of course, in prizes and prestigious professorships" (Gustin 1973, p. 1131). Gustin does not stress the symbolic importance

of mathematical charisma. An anonymous reader suggested that I was using the word *charisma* to designate "an emergent property of one's structural position" rather than an intrinsic (and magical) aspect of one's character, and that what I mean by charisma corresponds rather to "what sociologists mean by the term 'status,' which is to say the prestige accorded to individuals because of the abstract positions they occupy rather than because of immediately observable behavior" (Gould 2002). I thank the same reader for bringing to my attention the following quotation from (John Levi Martin 2009): "The very charisma that makes this person appear extraordinary may better be seen as a property of the pattern of attributions of charisma, reversing cause and effect."

The Martin quotation certainly corresponds to my application of the term to my own situation; but I wouldn't necessarily characterize the charisma of the Giants on the wall in this way. One feature that distinguishes them—and also initiators of research programs, like Grothendieck or Langlands—is that they are remembered for *creating* the "structural positions" to which their names are subsequently attached, rather than for fitting into preexisting positions—including the positions that preexist in association with features of the natural world (like elementary particles, or aromatic hydrocarbons, or DNA) and that often have no counterparts in mathematics. It is common to use words like "visionary" and "prophetic" when speaking of their work, and the reader will recall from the preface my observation that the way we talk about value in mathematics borrows heavily from the discourses associated with religion. It therefore doesn't seem at all inappropriate to adapt Weber's terminology of charisma in connection with such Giants, nor to suggest that this charisma rubs off, as in Weber's model of "routinization of charisma," on those of us who make even modest contributions to their research programs.

24. (Grothendieck 1988, p. 50).

25. (Shils 1968; Max Weber 1978, Part one, III, p. 242). For "seriousness" see chapter 8.

26. Felix Browder (1989) cites Weber's distinction between "rational-bureaucratic" and "charismatic leadership" to explain how M. H. Stone "transformed" the University of Chicago mathematics department "into the strongest mathematics department in the U.S. (and at that point probably in the world)." Stone's charismatic leadership was organizational and maybe personal, but not scientific in the sense I have in mind; he transformed the department but not the discipline.

27. The *Matthew Effect* consists "in the accruing of greater increments of recognition for particular scientific contributions to scientists of considerable repute and the withholding of such recognition from scientists who have not yet made their mark" (Merton 1968), quoted in (Cole 1970). The name is an allusion to Matthew 25:29: "For unto every one that hath shall be given, and he shall have abundance: but from him that hath not shall be taken even that which he hath." Merton also introduced the expression "role model" into sociology. The present chapter attempts to demonstrate that the Giants and Supergiants of the IBM poster are much more than role models in Merton's sense.

28. See the text that accompanies Serre's photo in (Cook 2009, p. 144), or the article (Huet 2003) for samples of his charm.

29. (Max Weber 1922). Analyzing the Grothendieck-Serre correspondence, Leila Schneps observes (following Grothendieck) that "self-analysis in any form strikes Serre

as a pursuit fraught with the danger of involuntarily expressing a self-love which to him appears in the poorest of taste" (Schneps 2014b).

30. This position was expressed with memorable violence by G. H. Hardy (2012), as we'll see in chapter 10, and this may account in part for the suspicion greeting mathematical public intellectuals in the English-speaking world. The situation is rather different in France, where Fields medalists regularly participate in public debates. Attitudes may be changing, in part thanks to the success of Gowers's blog and similar Web-based platforms.

31. Three well-known incidents hint at a different model—the growth of the Moscow mathematical school in the twentieth century (Graham and Kantor 2009), the creation of the Bourbaki movement in Paris in the 1930s, and the self-education of G. Shimura and his contemporaries in Tokyo in the 1950s. Kuhn's model of paradigm shift is as artificial as class struggle in explaining these developments. In each case a new generation, inspired by foreign practices, created a new and original approach. For Bourbaki it was Göttingen, especially the abstract algebra of Emmy Noether, the one woman on the IBM poster. Grothendieck's program was mainly built on the Bourbaki model, whereas Langlands was more eclectic, incorporating important insights of the Bourbakist Séminaire Cartan, the novel ideas of Shimura and Taniyama in Japan, and the branch of the Moscow school around Gelfand, as well as many others. The Frye quotation is from (Frye 1957, p. 186).

32. (Shils 1968).

33. (Bourdieu 1984, p. 208). Nor am I thinking of passages like this one: "The privileged classes find in what one can call charismatic ideology . . . a legitimation of their cultural privileges . . . " (Bourdieu and Passeron 1964, 1985, p. 106). This recent quotation from (Clerval 2013), also in the Bourdieu spirit, is more appropriate. "The intellectual petite bourgeoisie as a whole is in a paradoxical social position . . . dominated by the bourgeoisie in being exploited in its work and in suffering the degradation of working conditions . . . but it also receives . . . a kind of partial retrocession of capitalist surplus value that bears witness to its participation in the exploitation of popular classes." Exploitation by mathematicians is primarily *symbolic* and inadvertent; a good example is the subject of chapter 4.

34. (Joyal 2008, p. 154). Joyal is a mathematician, not a sociologist, and the quotation is from a recent textbook. Reuben Hersh, also a mathematician, has promoted a social constructivist approach to mathematics in a series of books (sometimes with Philip Davis), notably (Hersh 1997). But his primary interest is the social nature of mathematical truth. Truth is important, but it's only one of the values that motivates mathematicians.

35. (Ehrhardt 2011).

36. "The advent of 'league tables' of university excellence, first produced in 2003 by Shanghai Jiaotung University, was perhaps the inevitable consequence of the convergence during the 1990s of liberalisation of international markets, enabled by new communications technologies, and the shift of the global economy towards one based on information and knowledge. . . . universities, as sources of innovative ideas and highly skilled manpower, have come to be seen as vital agents in maintaining national competitiveness. . . . Amongst governments, the rankings indicated the extent to which

their universities were achieving the excellence presumed to be needed to drive and support national economic prowess" (Boulton 2010).

37. (Boulton 2010), *Times Higher Education Supplement*, June 24, 2010.

38. (Cole and Cole, 1967; 1968). See also (McCain 2010) for some of the history of citation indices. In 1980, sociologists could write that "it is not logically necessary that . . . bibliometric centrality is equivalent to intellectual centrality in the development of a research area . . ." while claiming that it was legitimate to use bibliometric indices to analyze "eminence hierarchies" in the case under study (Hargens et al. 1980).

39. (Hirsch 2005).

40. (Hagstrom 1964).

41. ". . . status honor is normally expressed by the fact that above all else a specific *style of life* is expected from all those who wish to belong to the circle" (Weber 1978, vol. II, section IX, p. 932). Val Burris proposes in (Burris 2004) to explain the persistence of status rankings of university departments on the basis of a Weberian model of social exchange, which of course is not inconsistent with authentic merit. Burris studies reciprocal hiring practices of leading sociology departments, but I have no doubt that mathematics departments in the United States, United Kingdom, and France behave similarly. Mathematics departments are ranked by *U.S. News and World Report* and by the *Times Higher Education Supplement*, for example; the lists are not consistent.

42. (Cole and Cole 1973).

43. Langlands (2013), expressed with characteristically understated acerbity. Mario Biagioli (2003, p. 266) writes more generally that "The [scientific] author is the producer of the work, but he or she is also 'produced' (i.e., recognized and rewarded as such) by his or her peers.

44. (Foucault 1984, p. 112).

45. The topic is explored at length in (Mazur 1997) and from a different point of view in chapter 7.

46. The lectures presented during the seven installments of SGA were published in 12 volumes that come to a total of 6209 pages. There are very few pictures.

47. (Shapin 2008, pp. 266–267).

48. Langlands's fairy tale article is (Langlands 1979a). The analogy of motives or automorphic forms with elementary particles should not be taken literally; and it should be understood that both Langlands's "speculation" and Grothendieck's "imagining" are rooted in the kind of formally rigorous reasoning that characterizes mathematics.

49. Formulated in the early 1960s by the English number theorists Bryan Birch and Sir Peter Swinnerton-Dyer, now emeritus professors at Oxford and Cambridge, respectively. See (Ash and Gross, 2011). I discuss some implications of the conjecture at length in chapter δ.

As for the other Clay problems related to number theory: the third of Weil's conjectures, solved by Deligne on the basis of Grothendieck's program, was a geometric version of the *Riemann Hypothesis*, and it is widely believed that the completion of Langlands' program is intimately connected with the solution of Riemann's original version, the first of the Clay problems (see chapter α for a brief description). The *Hodge conjecture* predates Grothendieck's work—though early in his career Grothendieck corrected Hodge's original prediction—but its most satisfying interpretation is as a precursor (or *avatar*, see chapter 7) of Grothendieck's theory of motives, which, in turn, is the starting point for Langlands's *Märchen*.

50. "Seen as" is conceptual shorthand. "Seen as an avatar of an automorphic form," in the sense to be explored in chapter 7, is more accurate. The existence of an automorphic form connected to an elliptic curve was already conjectured at the time—it's a long and complex story, described in (Ash and Gross 2011)—and had long been known for equations (E1) and (E5) but was proved in general only in 2000 by C. Breuil, B. Conrad, F. Diamond, and R. Taylor, using the methods introduced by Wiles in his proof of Fermat's Last Theorem.

51. (Mulkay 1976, pp 448, 454).

52. The articles were J. Coates and A. Wiles, "On the conjecture of Birch and Swinnerton-Dyer," *Inventiones Math* **39** (1977), 223-251; B. H. Gross and D. B. Zagier, "Heegner points and derivatives of L-series," *Inventiones Math.* **84** (1986), 225–320; K. Rubin, "Tate-Shafarevich groups and L-functions of elliptic curves with complex multiplication," *Inventiones Math.* **89** (1987) 527–559; B. Perrin-Riou, "Points de Heegner et dérivées de fonctions L p-adiques," *Inventiones Math.* **89** (1987) 455–510; V. A. Kolyvagin, "Finiteness of $E(\mathbf{Q})$ and ш$(E,\mathbf{Q})$ for a subclass of Weil curves," *Izv. Akad. Nauk SSSR Ser. Mat.* **52** (1988), 522–540; K. Kato, "p-adic Hodge theory and values of zeta functions of modular forms," *Astérisque* **295** (2004): 117–290. The reader will have guessed that *Inventiones Math.* is a "great journal." A few of the authors were a bit older and already had "great jobs" at the time of their breakthrough, but most belong roughly to my age cohort. If this chapter were autobiographical, I would let the reader know how it felt to enter professional life at the same time as these colleagues, especially before I acquired tenure.

53. "Research scientists gain in professional repute to the extent to which they are seen to have contributed intellectually to areas of scientific consensus" (Mulkay 1976, p 448). On mentoring in science, see (Hooker et al. 2003).

54. In *Truth and Truthfulness*, Bernard Williams, who describes himself as a genealogist, refers to genealogists who talk this way as "deniers." The Persian scholar al-Biruni had already met such colleagues in the eleventh century:

"[I]f he is shown that arithmetic and geometry are impossible to understand unless one proceeds systematically from first principles, unlike other sciences in which he may be acquainted with something of their middle (parts) or their ends without knowledge of their beginnings, he thinks that this is intended to [turn him away] from his appreciation and to confuse him. This, he imagines, is similar to the ignorance into which (non-initiate) members of (secret) sects (are led) with regard to the doctrines of their sects until they had taken the oaths, entered into the covenants, and made a long practise and training. This adds to his revulsion, so that the stopping of his ears with his fingers becomes his most potent recourse, and the raising of his voice in shouts his most powerful equipment."

From al-Biruni's preface to (al-Biruni 1976, p. 6). Other references here and later are to (Corfield 2012).

55. (Bloor 1991, p. 86). The *strong programme* in the Sociology of Scientific Knowledge is the genealogical approach to sociology of science pioneered by Bloor, in the book just cited, and Barry Barnes.

A recent article by Barany and MacKenzie (2014), which takes as its starting point the central role of chalk and blackboard (or their substitutes) in mathematical communication, is to my mind an unusually successful and insightful example of a "critical and naturalistic" study within the genealogical approach to sociology of mathematics. Its examination of the practice of "following the argument" at seminar talks is one of a number of fruitful contributions to a future tradition-based sociology of mathematics. For my purposes, its focus is nevertheless excessively epistemological, as the concluding sentences suggest: "Mathematical ideas are not pre-given as the universal entities they typically appear to be. The most important features of mathematics can be as ephemeral as dust on a blackboard." When the authors write, "Given the stereotype of the lone mathematician and the importance of breakthrough stories in post-facto accounts of mathematical innovation . . . the predominance of project-work in mathematics is . . . surprising," I'm surprised to learn this is surprising. The view of mathematics as a "science of ideals" to which the authors refer may be widely shared by mainstream sociologists of science, less so by mathematicians—unless the "ideals" are "naturalized" in connection with the values explored in the present chapter.

56. Exceptions are (Hagstrom 1964) and (Stern 1978). Mathematics is frequently considered along with other sciences by sociologists of both tendencies, for example, in (Hagstrom and Hargens 1982).

(Restivo 2001) was not quite, as the author claimed, the first book by a sociologist on the sociology of mathematics; it was preceded by (Heintz 2000) and followed by (MacKenzie 2001) and (Rosental 2003). All four books, while very different, are genealogical in inspiration, though Heintz makes some gestures in the tradition-based direction.

A much more recent study, closer to the encyclopedist tendency than to the other two, is (Zarca 2012). Zarca's is the only study I've seen based on an extensive survey of values, practices, and social conditions of mathematicians and (for comparative purposes) scientists in related fields. The survey, conducted anonymously by Internet, included responses of roughly 1000 to 2000 individuals (some responses were more complete than others). The author restricted attention to scientists working in France (I responded to the survey) but quotes other studies and biographical material to supplement his data. The book appeared after I had completed a first draft of this chapter, and I could not incorporate Zarca's conclusions at length, but I note that he and I share many of the same concerns. In particular, he devotes a subsection of twenty pages to "l'élitisme des mathématiciens" (in which he quotes Weil extensively!) but does not treat the hierarchy as a constitutive factor in mathematics.

57. The statement deserves to be qualified: most of the encyclopedist texts I cite in sociology are empirical studies (like Zarca's) and (unlike Zarca's) have very little to say about philosophical questions, whereas the Strong Programme seems more interested in philosophy of Mathematics than in mathematics; see note 60. Here I refer primarily to English-language philosophy. For continental philosophy, the subjective experience of the mathematician is often central; this was especially true of the phenomenological approach initiated by Husserl and still very much a factor in France (but, interestingly, totally absent from Zarca's discussion).

58. If philosophers of Mathematics truly monopolized the certification of mathematical legitimacy, in the way that psychiatrists monopolize the dispensing of anxiety medicine, then logicians, working on behalf of philosophers of Mathematics, would

have to be viewed as mathematics' "puppet government." Most mathematicians, however, would disregard the latters' directives even if most logicians hadn't ceased issuing them long ago, and most philosophers know it. For example, Marcus Giaquinto's philosophical treatment of the Foundations Crisis—see chapter 3—is entitled *The Search for Certainty* (New York: The Clarendon Press, Oxford University Press, 2002), but it concludes with the reminder that "[m]athematics is definitely not just logic." Pointing to the variety of modes of typical mathematical reasoning, "reasonings that involve modalities (contingency, necessity, possibility), or that are affected by the passage of time, or that concern individuals, or that require iconic displays, or that link heterogeneous acts of thought or speech or kinds of things.," Emily Grosholz concludes that "it is no wonder that logic all by itself cannot express mathematics" (Grosholz 2007, p. 284).

59. The distinction, which overlaps my classification, between *mainstream* and *mavericks* in philosophy of mathematics, goes back to (Kitcher 1984). What I call *philosophy of mathematics* is often called the *philosophy of mathematical practice*; I will avoid this expression, because it is long-winded and because it implies that what deserves to be called "philosophy of mathematics" is concerned with something other than what mathematicians actually do.

I've always thought it peculiar that the sociologist Bloor attempted to account for "The Standard Experience of Mathematics"—the title of the chapter from which the preceding quotation is taken—without reference to any sociological data whatsoever. (He describes a game invented by mathematical educator Z. Dienes but not the people playing the game.) Such a priorism is typical of the philosophy of Mathematics, and I agree that sociology cannot be merely empirical, but the notion of a sociology without people challenges the imagination.

60. (Greiffenhagen 2008). In her review of (MacKenzie 2001), Bettina Heintz (2003) makes a similar claim: "The few [sociologists] who have dealt with mathematics did it in a rather programmatic way and with reference to sometimes quite arbitrary examples from the history of mathematics. The empirical reality of modern mathematics, however, was hardly touched." One of the sociologists Heintz had in mind was Sal Restivo, who claimed in 1999 that sociology of mathematics—by which he apparently also means sociology in the SSK mold—"has attracted very few practitioners and seems to be a citation orphan in core STS." See also (Greiffenhagen, in press) and (Greiffenhagen and Sharrock 2011), a helpful antidote to the exclusive focus on epistemic consensus—on Mathematics rather than mathematics—that, as the authors demonstrate, led to a misreading of the relation between public and private aspects of mathematical research. Greiffenhagen and Sharrock find that even Reuben Hersh (see note 34), who for thirty years has taken the lead in promoting a humanistic alternative to the philosophy of Mathematics, exaggerates the contrast between the (formalist) "front" and (constructivist) "back" of mathematics in (Hersh 1988; 1991).

It bears mentioning that, as with most contemporary work in philosophy of Mathematics, Greiffenhagen's examples are taken from logic, whose organization and practices are far from representative of mathematics as a whole; but the phenomena to which his video analysis draws attention are familiar in more central branches as well. See also (Greiffenhagen and Sharrock 2009a, 2009b).

61. (Corfield 2012, pp. 253, 255). This is in line with the frequently repeated observation that what matters in mathematics is not truth as much as interest; see the discus-

sion of heuristics in the next chapter. (C. S. Fisher 1973) claims that mathematicians' "methods of attack are not rationally chosen," that "They are derived from a combination of training, taste, and personal predilection."

62. The preferred technique, the *Arthur-Selberg trace formula*, is a close relative of the fixed point theorem mentioned in note 19. Langlands' unpublished letters are magnificently preserved at http://publications.ias.edu/rpl/, along with his published articles. For Act 1, see especially the letters in sections 5 and 6 that record the evolution of Langlands' ideas during the crucial early years; the goals of Act 2 are described in the articles in sections 8 and 10, especially those published in the 1970s.

I have repeatedly been invited to contribute a chapter (after paying a hefty publication fee) to a book project on endoscopic surgery; however, this has nothing to do with endoscopy in Langlands' sense.

63. His first work on (FL1) was in collaboration with G. Laumon; his complete solution was published in (Ngô 2010). Objects $\mathcal{G}$ and $\mathcal{H}$ are examples of what are called *perverse sheaves*, a notion developed by a number of mathematicians around 1980 to reduce the study of complicated geometric objects to a more familiar geometry. Grothendieck has written in *Récoltes et Sémailles* that he dislikes the terminology.

64. (Sinclair and Pimm 2010). Membership in the tradition is in principle not class-based. In his *Vie de M. Descartes*, Adrien Baillet tells the story of the Dutch peasant and shoemaker Dirck Rembrantsz van Nierop, who, not satisfied with the mathematics he had found in books written in the vernacular, left his village in the hope of meeting Descartes. The first time he showed up at the philosopher's door, he was sent away. The second time, he was offered money, which he refused. Finally, on the fourth visit, he was admitted to the presence of Descartes, who "immediately recognized his talent and merit" and soon counted him among his friends. Rembrantsz is remembered as "one of the leading astronomers of his century" as well as a cartographer and published a considerable number of books in mathematics. (Baillet 1691, Tome II, pp. 555–57).

65. For Jevons, see note 18 to chapter 10. For mass culture, see chapter 8, especially the comments of F. R. Leavis and Pierre Boulez.

66. Respectively, the American Mathematical Society, the Society for Industrial and Applied Mathematics, and the National Science Foundation.

67. See the list near the end of the chapter and in note 75.

68. Gowers's announcement was at http://gowers.wordpress.com/2012/01/21 /elsevier-my-part-in-its-downfall/. The signatures were collected at http://www .thecostofknowledge.com/, which also hosts the Statement of Purpose.

69. (Clarke 2010) lists "five traditional functions of journals: Dissemination, Registration, Validation, Filtration, and Designation." Henry Cohn, a signatory to the Statement of Purpose, added the function of "archiving . . . particularly crucial in mathematics, because the half-life of math papers is much greater than in other scientific fields. . . ." Merton, of course, was a functionalist. For the record, my own position is that a scientific journal is not reducible to a collection of functionalist functions but is rather an expression of a *social relation*. For more of this kind of talk, see (Marx 1858). Tao's and Cohn's comments were at http://publishing.mathforge.org/discussion/7 /whats-the-point-of-journals/#Item_0.

70. (Gustin 1973, p. 1128).

71. Many of the "greatest journals" are in fact published at relatively low cost by

university presses or professional societies, but there are apparently not enough of them to fill a CV, especially in certain branches of mathematics.

72. (Bellah 2011, p. 633, note 54).

73. http://publishing.mathforge.org/discussions/?CategoryID=0. Mathematical physicist John Baez, whose "This Week's Finds in Mathematical Physics" was one of the first successful mathematics blogs, opened a Math 2.0 discussion with a proposal to add a simple 1- to 5-point rating system to existing preprint servers. Baez emphasized that his goal was not to replace the current peer review system; others have made more radical proposals (for example, at www.papercritic.com).

At lunch one day in Paris, a colleague who had signed the SoP described a less radical vision of a future without commercial publishing, of journal editorial boards directly vetting articles that had been posted on dedicated internet file servers, Tao's skepticism notwithstanding; maybe a single file server, like arxiv.org, would suffice. "What about the journal hierarchy?" gasped a publisher's representative. No problem; editorial boards would sort that out among themselves.

74. For an academic mathematician, these duties include but are not limited to publication in peer-reviewed journals, speaking at national and international conferences, training PhDs, and applying for research grants. Not all mathematicians are academics, of course; but most professional pure mathematicians do teach at universities.

75. Also the Research Institute for Mathematical Sciences (RIMS) in Kyoto, the Isaac Newton Institute near Cambridge University, institutes in Bonn and Münster, Moscow and St. Petersburg, Barcelona, Vienna, Montreal, and Beijing; as well as the even longer list of institutes I haven't yet visited, notably IMPA (Instituto de Matemática Pura e Aplicada) in Rio de Janeiro. Weekly conference centers on the Oberwolfach model include the Centre International de Rencontres Mathématiques (CIRM) on the limestone cliffs overlooking the Mediterranean coast near Marseille and the spectacular Banff International Research Station (BIRS) in the middle of Banff National Park in Canada.

76. See also note 43. Grigori Perelman reportedly turned down his 2006 Fields Medal and the $1,000,000 prize awarded by the Clay Mathematics Institute in 2010 for his solution to the Poincaré Conjecture, in part because he felt the award overlooked the fundamental earlier contributions of Columbia University professor Richard Hamilton. Perelman's unusual attentiveness to what is ostensibly one of the guiding ideals of science—*disinterestedness*, third on Merton's list of norms—is discussed in (Zarca 2012, pp. 20–21). See also chapter 6, note 36.

77. (Bourdieu and de Saint Martin. 1996, p. 19).

CHAPTER α HOW TO EXPLAIN NUMBER THEORY (PRIMES)

1. Debbie Epstein, Heather Mendick, and Marie-Pierre Moreau , "Imagining the mathematician: young people talking about popular representations of maths," *Discourse: Studies in the Cultural Politics of Education* 31, No. 1 (February 2010): 45–60. The Eilenberg Lectures mentioned in this chapter are named after the late Samuel Eilenberg, professor for many years at Columbia, who had an impressive gray beard.

2. (Gowers 2008).

3. (Hasse 1952).

4. At least since (Hardy 2012), which included brief proofs of the irrationality of the square root of 2 and the infinity of primes to illustrate his mathematical aesthetic. See chapter 10.

5. This is also a recurring theme in Thomas Pynchon's novel *Against the Day*; see bonus chapter 5. Every number theorist knows, by the way, that the Riemann hypothesis doesn't quite claim that $N/\log(N)$ is statistically optimal; but I don't advise you to aim for a more accurate statement at your next dinner party. I also recommend that you avoid mentioning, as I have up to now, that $\log(N)$ stands for the *natural logarithm of N*; this year (2014) marks the 400th anniversary of John Napier's publication of his *Description of the Wonderful Canon of Logarithms* [*Mirifici Logarithmorum Canonis Descriptio*], and their wonders could fill the dinner hour all by themselves.

6. (Doxiadis 1992). Highly recommended. Although the twin primes conjecture still seems a long way from being settled, a breakthrough in that direction, due to the previously obscure (but now famous) Yitang Zhang, gave number theorists around the world an unexpected reason to celebrate during the spring of 2013. At about the same time, Harald Helfgott announced the solution of a problem closely akin to the Goldbach conjecture: that every odd number greater than 5 is the sum of three primes.

7. Frege was the first to define the natural numbers in this way: in (Frege 1884), the number 1 is defined (in section 77) as the concept (*Begriff*) "equal to 0," where 0 has been defined (in section 74) as the concept "not equal to itself." (Rotman 1987) places the discovery of zero at the heart of several fundamental transformations in painting and money as well as arithmetic.

8. P. A. is mistaken: Kronecker's considerations are metaphysical, not necessarily theological. But the mistake is understandable.

9. This is what the logicians don't understand. It's perfectly legitimate for them to want to post rules, like traffic signs, to keep the explanations from crashing into one another. The problem is when they convince themselves that they were the ones who built the roads.

> P. A.: Please don't count on my support if you're trying to drum up sympathy for some obscure and tedious feud with logicians.
> N. T. (Crestfallen): As you wish.

10. Both of these ideal types were encountered briefly in chapter 1; we will see more of them in the next chapter.

CHAPTER 3 NOT MERELY GOOD, TRUE, AND BEAUTIFUL

1. (Collini 2011).

2. Quotations from the CBC TV program "Power and Politics," June 28, 2011, was still online two years later at http://podcast.cbc.ca/mp3/podcasts/powerandpolitics_20110628_18370.mp3, but the podcast has since been taken down. I thank Vladimir Tasic for transcribing the sentences quoted.

3. Quoted in (Donoghue 2008, p. 5) and by Joan Wallach Scott in her March 30, 2011, lecture on academic freedom at the IAS.

4. (Scott 2009, pp. 452–453). Compare (Fischer 2007): "From the point of view of system theory, the politically intended penetration of research by social, economic and political goals and standards . . . is pathological because science [is] called upon or even forced to make values and codes foreign to its system into its primary ones. . . . only such research as "pays off" is socially "relevant" or politically correct is to be financed by society." Wagner (2007) finds "this diagnosis . . . of special relevance, since the contemporary reform of the European science system" requires that "science itself . . . be organized according to market-economic principles."

5. The titles speak for themselves: *Science Mart* (P. Mirowski); *Lowering Higher Education: The Rise of Corporate Universities and the Fall of Liberal Education* (J. Cote and A. Allahar); *A vos marques, prêts . . . cherchez!* (I. Bruno).

6. Here is a recent French version of this cliché, published as the conclusion of the conference *Maths à Venir* in December 2009, to which we return in chapter 10. "[*Les mathématiques] interviennent de manière cruciale dans de nombreuses sciences naturelles, humaines ou sociales, dans la technologie moderne, et dans la vie de tous les jours, même si on n'en a pas toujours conscience. Elles sont utilisées pour l'imagerie médicale, les jeux vidéo, les moteurs de recherche sur internet, la téléphonie mobile, dans les modèles climatiques, dans la finance, pour ne citer que quelques applications. La vitalité et la santé de l'école mathématique française sont donc devenues un enjeu stratégique.*"

7. Quotation from M. J. Nye, *Michael Polanyi and His Generation: Origins of the Social Construction of Science*, reviewed in (Shapin 2011).

8. (Flexner 1939). Among other contributions, Flexner mentioned electricity, radio transmission, relativity theory, insurance, and bacteriology. We return to this article, to Flexner, founder of the Institute for Advanced Study, and to *Maths à Venir*, in chapter 10.

9. See for example the Mathematics and Climate Research Network, http://www .mathclimate.org/.

10. "Houses of Refuge and Entertainment: Why are Institutes for Advanced Study Proliferating?" an after-hours lecture by Peter Goddard, February 17, 2011.

11. The *four color theorem* is the claim that it is possible to color any map with a maximum of four colors so that no two contiguous countries have the same color. For a social science perspective, see (MacKenzie 1999).

12. The *Kepler conjecture* asserts that the "grocer's" stacking of identical balls (oranges) in three-dimensional space is the most efficient. Like the four color theorem, Kepler's conjecture has now been proved with the help of computers. The *classification of finite simple groups* is a basic problem in algebra: it asks for a complete list of all the distinct finite groups—like the Galois groups we will encounter in chapter β—that are simple in the sense that they are not put together from smaller groups.

13. For all this, see http://code.google.com/p/flyspeck/wiki/FlyspeckFactSheet

14. (Gray 2004, quotations on pp. 27, 37).

15. See www.math.ethz.ch/news_events/bombieri_math_truth.pdf.

16. Mentioned in (Ruelle 1991, pp. 3–4); emphasis added.

17. Transcribed from (Voevodsky 2010), quotations from 8'50" and 43'30". Gödel's second incompleteness theorem denies the possibility of proving the foundations of the interesting parts of mathematics consistent. I note in passing that the "proscience" side

of the "Science Wars" of the 1990s explicitly discounted the possibility of a reaction like Voevodsky's to Gödel's theorems.

18. Still the most prestigious honor in mathematics; see chapter 2. The IAS has long had the world's largest concentration of Fields Medalists. Among the mathematicians mentioned in this chapter, Bombieri, Deligne, Atiyah, and Tao, as well as Voevodsky, have all been Fields Medalists; all but Tao either are or have been IAS professors as well.

19. As does "the knowledge that material things exist" and "all evident reasoning about material things" (Descartes 1991, Part 4, paragraph 206).

20. (Heidegger 1962, paragraph 21).

21. Gödel's more famous first incompleteness theorem put an end to another goal of Hilbert's program: to show that every true theorem in the system admits a proof based on the rules. See note 25.

22. From the announcement of a workshop held in the Netherlands in late 2011, at Lorentz Center (2011).

23. See note 15. Bombieri adds that these considerations "may change with time."

24. (Frege 1879, Preface, p. 5). quoted at http://plato.stanford.edu/entries/frege -logic/.

25. The most sustained attempt to mechanize the verification of mathematics is associated with David Hilbert, whose goal was to read off the provability of a proposition from its syntax. Gödel's incompleteness theorems are conventionally said to have put an end to Hilbert's program and to have ended the Foundations Crisis by declaring it incurable. Frege himself insisted in his *Grundlagen der Arithmetik* that, although "It is possible . . . to operate with figures mechanically, just as it is possible to speak like a parrot, that hardly deserves the name of thought. It only becomes possible at all after the mathematical notation has, as a result of genuine thought, been so developed that it does the thinking for us, so to speak." So is Frege a mechanizer? Edward Kanterian (in a private communication) quotes from Frege's article on Boole: "'our thinking as a whole can never be coped with by a machine or replaced by purely mechanical activity.' But he immediately adds: 'It is true that the syllogism can be cast in the form of a computation'. Overall, he seems undecided about the relation between inferring as a 'live' and as a mechanical process." Thanks to David Corfield for forwarding Kanterian's message.

26. Blog post March 11, 2011, on the *n-Category Café*, http://golem.ph.utexas.edu /category/.

27. See (MacKenzie 2001) for a review of these arguments.

28. (Lanier 2010, p. 10).

29. The question opening this paragraph is not merely rhetorical and is one of the main themes of (MacKenzie 2001). On pages 326–327, MacKenzie quotes a computer scientist who argued that mathematicians' failure to perform proofs formally, "by manipulating uninterpreted formulae accordingly [*sic*] to explicitly stated rules" proves that "mathematics today is still a discipline with a sizeable pre-scientific component, in which the spirit of the Middle Ages is allowed to linger on" (Dijkstra 1988); a philosopher (Peter Nidditch) who claimed that "in the whole literature of mathematics there is not a single valid proof in the logical sense"; and another philosopher (Paul Teller) who doubts "that mathematics is an essentially human activity." Already in

1829, Thomas Carlyle could complain of "the intellectual bias of our time," that "what cannot be investigated and understood mechanically, cannot be investigated and understood at all" and that "Intellect, the power man has of knowing and believing, is now nearly synonymous with Logic, or the mere power of arranging and communicating" (Carlyle 1829).

30. Examples can be found in (Daston-Galison 2007) and (Alexander 2011).

31. The meaning of this passage, of fundamental importance in understanding the role of mathematics in Western culture, is obscured by modern translations. In this 1917 translation by Bernadotte Perrin of chapter 14 of Plutarch's *Life of Marcellus*, as in Dryden's seventeenth-century translation, the first two occurrences of "mechanics" correspond to the word *organikos*, the third to *mekhanikos*. We return to the distinction between *mekhanikos* and geometry in chapter 8.

32. (Rashed and Vahebzadeh 2000, p. 112).

33. (Cziszar 2003).

34. In "Science as Vocation," Weber, writing before Hardy's *Apology*, asserts that "academic man . . . maintains that he engages in 'science for science's sake.'" Many mathematicians must feel this way, but it's not so easy to defend this position in public. Weber himself goes on to ask "what . . . does science actually and positively contribute to practical and personal 'life'?" In addition to technological applications and "tools and . . . training for thought," Weber adds a third objective: "to gain *clarity*." Weber's own discipline of sociology contributes in exemplary fashion to this objective. So, I would like to say, does mathematics. The so-called A-G requirements for admission to the University of California seem to support this contention: mathematics is item c, and the A-G guide expects prospective students to arrive with "awareness of special goals of mathematics, such as clarity and brevity . . . parsimony . . . universality . . . and objectivity." But very few students graduating from the UC system will devote their lives to research in pure mathematics. It takes a book at least the length of this one to explain what kind of clarity is to be gained from that particular choice.

35. (Tymoczko 1993).

36. "*Cela semble si naturel aux mathématiciens, tant l'aspect artistique de notre discipline, plus encore que des autres sciences, est évident, que nous ne voyons pas où est le problème. . . . ce qui fait généralement avancer un mathématicien, c'est le désir de produire quelque chose de beau. . . . quand nous prenons connaissance d'un résultat ou d'un théorème, notre premier souci est de juger de sa beauté.*" [Comptes Rendus 2010].

37. (Muntadas 2010, pp. 47–48, 58, 82). Muntadas interviewed dealers, collectors, gallery owners, museum directors and curators, "docents," critics, media representatives, and a few artists.

38. His article, renamed "Narrative and the Rationality of Mathematical Practice" is (Corfield 2012).

39. The first quotation is from (Hardy 1929), the second, due to Atiyah, is reproduced on page 256 of (Corfield 2012). Thanks to David Corfield for reminding me of the latter quotation. Proof also plays an important role, not to be explored here, in institutional validation. For Bourbaki, while logic is "extremely useful," it is but one aspect of their "axiomatic method . . . indeed the least interesting one" (Bourbaki 1950, p 223).

40. Atiyah is hardly the only mathematician to see proofs in this way. Compare

"The object of mathematical rigor is to sanction and legitimize the conquests of intuition, and there was never any other object for it" (Pólya 1962, p. 127).

Ian Hacking's new book (2014) makes an illuminating distinction between *Cartesian* and *Leibnizian* mathematicians. The Atiyah quotation would be representative of the former attitude, while Voevodsky's project is typically Leibnizian insofar as it aims at mechanization. Hacking's consistency in treating the two attitudes in parallel, if pursued by other philosophers, could go a long way toward reconciling the philosophies of Mathematics and mathematics.

41. (Huizinga 1950, p. 49).

42. Thus, Ursula Martin defines mathematical practice descriptively as " . . . producing conjectural knowledge by means of speculation, heuristic arguments, examples and experiments, which may then be confirmed as theorems by producing proofs in accordance with a community standard of rigour" (Martin 1998). Compare the following quotation from G. Cantor's 1867 doctoral thesis: "In mathematics, the art of proposing a question must be held of higher value than solving it." For epistemology and heuristics in the work of (philosopher and mathematician) Bernhard Bolzano, see (Konzelmann Ziv 2009).

43. http://www.math.ias.edu/~vladimir/Site3/Univalent_Foundations.html, http://ncatlab.org/nlab/show/homotopy+type+theory#introductions_31.

44. Urs Schreiber (2012) offers an example from one of his own papers on the border between topology and string theory. "[A] certain kind of problem that poses itself in the context of string theory, which . . . was generally regarded to be among the more subtle problems in a field rich in subtle mathematical effects . . . finds an elegant and simple solution once you regard it from the perspective of homotopy type theory." In technical language: "homotopy type theory . . . automatically produces the correct answer, the 'E 8-moduli 3-stack of the supergravity C-field in M-theory.' A solution that looks subtle to the eye of classical logic becomes self-evident from the point of view of homotopy logic/homotopy type theory." "It is one of those cases," Schreiber adds, "where a simple change of perspective leads with great ease to a solution of what seemed to be a difficult technical problem." A new mathematical formalism—new foundations in Manin's sense—is adopted, at least by the affected subculture, when there is an accumulation of similar examples.

45. (Manin 2004). In (Garfield 2010), Jay Garfield points out a second distinction between foundationalism of *content*—"certain sentences or cognitive episodes are taken to be self-warranting and to serve as the foundation for all other knowledge" and foundationalism of *method*—"certain faculties or methods of knowing are taken to be self-warranting and foundational". Garfield mentions Descartes' method of clear and distinct perceptions as an example of the latter and argues that the early Buddhist philosopher Nāgārjuna (see chapter 7) is specifically opposed to this kind of foundationalism. In the epistemology of mathematics the two—axioms and rules of argument—tend to be considered part of one system of foundations.

46. (Weyl 2009, p. 188; Benacerraf 1973; Grosholz 2007).

47. (Kreisel 1967; Gray 2010, p. 203, p. 273 for the Enriques quotation).

48. Mehrtens (1990, p. 12) is describing objections to the set-theoretic *axiom of choice* and alludes to Brouwer's account of the difference between formalists and intuitionists (p. 178).

49. (Bloor 1991, p. 155).

50. J. Avigad, "Understanding Proofs," in (Mancosu 2008, 317–353).

51. (Pimm and Sinclair 2006, chapter ω).

52. (Zarca 2012, Tableau 17, pp 278–279 and surrounding discussion). More precisely, respondents were asked to rate "components of the socioprofessional and psychosocial dimensions of the professional ethos." Teaching came next after pleasure on the scale of importance; at the very bottom were applications to science and the "social world" (20% and 8%, respectively, for pure mathematicians), along with administration of research (12%), concern for France's international standing (17%), and "the glory of establishing an important result" (18%).

Zarca does not indicate whether or not his respondents explained what they meant by *plaisir*, but he does cite Weil's comparison of mathematical to sexual pleasure, noting that the former typically lasts longer. None of the items on the list looked at all like a Golden Goose, but this may just have been an oversight on Zarca's part.

53. (Zarca 2012, pp. 259–66, especially tableau 14). Some readers may be wondering whether or not the mathematicians who chose to accept Zarca's e-mail invitation to participate in his Internet survey, rather than to ignore it, form a representative sample of the profession. Zarca addresses just this question in an appendix (pp. 343–34). Information identifying the respondents was sealed off from their responses, and Zarca was thus able to ascertain that his sample is indeed representative as far as age, position, and sex are concerned. Conversations after the survey was closed revealed that a higher than average proportion of "high-level" mathematicians did not bother to fill out the questionnaire. By including the opinions expressed in these conversations in his discussion and in autobiographical writings, Zarca feels he managed to compensate for their relatively low level of participation. There remains the question of whether the sample might be "psychologically unrepresentative" of the profession precisely because it gives excessive weight to those who were eager to respond when there was no obligation to do so. (For example, this might imply that the high proportion of respondents for whom pleasure is a primary motivation is merely an artefact of the methodology.) Zarca admits there is no way to know whether or not this was the case but concludes that to disqualify the results on such grounds would amount to denying the possibility of "any survey that aims to establish [statistics] for relatively concentrated populations."

54. A handful of amateurs have left their mark on mathematics, including the French jurist Pierre de Fermat, Supergiant of the seventeenth century, when the status of professional mathematicians had not yet been defined, and (in the mid-twentieth century) the German Kurt Heegner, whose ideas continue to be used in work on the Birch-Swinnerton-Dyer conjecture. But research in pure mathematics is overwhelmingly reserved for professionals.

55. (Beebee 2010). In the mid-1930s, the British journal Philosophy published an series of letters on the topic "The Present Need of a Philosophy." The responses covered much the same utilitarian ground as the BPA letter (no computers, of course) but were more carefully argued. I thank Brendon Larvor and David Corfield for these references.

56. (Zarca 2012, p. 274; Huizinga 1950, p. 132). Quotations in the next two paragraphs are from pp. 160–61 and p. 203; the Aristotle quotation is from *Politics*, viii 1339 A, 29.

57. Of course Huizinga did not use Kuhn's language of paradigms. The Bellah quotation is from (Bellah 2011, p. 594).

58. (Freud 1958), translation slightly modified.

59. (Lindsay and Schachter 2007, pp. 108–109). The complex relations between mathematicians and the reality of "the real world" is explored in chapter 7. For the comparison of formalized mathematics to chess in the writings of Weyl and Heinrich Behmann, see Mancosu (1999).

60. *Der Meister der Moderne ... bestimmt sich als "freier Mathematiker," als "Schöpfer," und die Gegenmoderne wirft ihm "Schöpferwillkür" vor"* (Mehrtens 1990, p. 10).

61. (Huizinga 1950, p. 3, p. 2; Netz 2009, p. x).

62. Hermann Weyl also saw mathematics as *"schöpferish,"* but (as usual) his version of the argument was idiosyncratic. "Mathematics plays a central role in the construction of the world of the mind. Mathematizing, like myth, speech, or music is one of the primary human creative attitudes. . . . " And again: "Mathematics is not the rigid and petrifying schema, as the layman so much likes to view it; with it, we rather stand precisely at the point of intersection of restraint and freedom that makes up the essence of man [. . .] Theoretical creation is something different from intuitive insight; its aim is no less problematic than that of artistic creation." (Weyl 1968, Vol III, p. 293, my translation; Vol. II, p. 533, translation in (Mancosu 1998, p. 136)).

63. (Burghardt 2005, p. 77–78). Burghardt attributes the term "relaxed field" to G. Bally (*entspannten Feld*) and K. Lorenz. Bachem (2012) mentions that "bluethroats and blackbirds sing the most artful songs when they have no objective"—here "artful" (*kunstvoll*) refers to "songs of greater complexity and subtlety."

64. (Dewey 1987, p. 283).

65. The European Research Council asks reviewers to assign objective grades on a scale of 1–4 when answering the following questions:

> Does the proposed research address important challenges at the frontiers of the field(s) addressed? Does it have suitably ambitious objectives, which go substantially beyond the current state of the art (e.g. including inter- and transdisciplinary developments and novel or unconventional concepts and/or approaches)? . . . Does the proposed research involve highly novel and/or unconventional methodologies, whose high risk is justified by the possibility of a major breakthrough with an impact beyond a specific research domain/discipline?

66. (MacIntyre 1981, p. 175).

67. (Muntadas 2011), pp. 38, 66 et passim.

68. (MacIntyre 1981, p. 207). Corfield (2012) also cites these passages.

69. (Wright 2011). Wright goes on to say that "the full value of philosophy—as both activity and research discipline—does depend upon the existence of the equivalent of a 'grass roots,' a culture of philosophical education and awareness. . . . So of course philosophy should be funded, but only because it belongs to the kind of value that the subject has that the style of thinking it involves, and its preoccupations, is capable of communication to, and beneficial impact upon, the lives of the population in general.

In essence, the reasons why philosophy should be funded are more or less the same as those (non-instrumental) reasons why education generally should be funded."

The same holds for mathematics. The chapters labeled by Greek letters are meant to be an experiment in communication. If I had space to explain the structure of the experiment, I would devote it to talk about the creation of ways of thinking. But the premise of this book is that the "style of thinking" as well as the "preoccupations" of mathematics are grounded in a practice that is the book's proper subject.

70. Jacques Rancière writes that "Art exists as an autonomous sphere of production and experience since History exists as a concept of collective life" and dates this existence back to the mid-eighteenth century (see also note 20 in chapter 10). Replacing Art by (small-*m*) mathematics in the preceding sentence, it says that the existence of mathematics as a self-conscious tradition-based practice is tied up with its projection in history, which is consistent with the themes of chapter 2. The timing for mathematics may be different.

Some differences between mathematics and the arts also have bearing on the purpose of this book. Those involved with the arts exhibit much more cultural confidence—you're not likely to read a book called *An Artist's Apology* unless the artist did something genuinely and specifically awful; artists tend to have better parties than mathematicians; and art, unlike mathematics, is a vehicle for large-scale individual and institutional investment. These three differences may be linked.

71. (Bellah 2011, p. 200). Preceding quotations from pp. 139–44, 188. Capitalization of "Powerful Beings" is my addition.

CHAPTER 4 *MEGALOPREPEIA*

1. "It is difficult to live in a world whose productive medium is money." Nicole El Karoui, French mathematician, specialist in Mathematical Finance, speaking at the Académie des Sciences, quoted in *Les Echos*, November 20, 2008.

2. Gertrude Stein, *Dr. Faustus Lights the Lights*.

3. That is, ethics made quantitative, as opposed to the ethics of mathematics, which is the guiding theme of this book. Quotations are from (Boulding 1969).

4. (My[confined]space 2008; Fisher et al. 2009; Ellerson 2009).

5. The rebellion was crushed by eight Roman legions under the command of Crassus (played by Laurence Olivier in the 1960 Stanley Kubrick film). "A top quant farm" is from (Patersson 2010, p. 140).

6. (Rocard 2008).

7. (Legge 2013; Ferguson 2012, pp. 3, 206; Parliamentary Commission 2013, Vol. 1, p. 7).

8. At the Hague, where he would have liked to see Milton Friedman in the dock: see (Besson 2008). In (Jouve 2011), a colleague asks *"Devra-t-on un jour traduire messieurs Black et Scholes devant le tribunal pénal international de La Haye?"*

9. (Kahane et al. 2009). *Le Monde* chose not to print the response, which appeared instead in the online mathematics journal of the CNRS, the French national research agency.

10. Both the SMAI (Société de Mathématiques Appliquées et Industrielles) and the SMF (Société Mathématique de France) published special dossiers on mathematics and finance in the wake of the crash: *Matapli*, **86, 87** (juin, novembre 2008) and *Gazette des Mathématiciens*, **119** (janvier 2009). The Yor quotation is on pp. 75–76; Karoui's is on p. 24 of *Matapli* **86**.

11. (Lambert-Mazliak 2009; Rogalski 2010; Guedj 2008).

12. (OMI 2012).

13. (G. Cohen 2012). Addressing the French Senate, Gauss Prize winner Yves Meyer reported that "when Nicole El Karoui was teaching there, 70 % of École Polytechnique graduates went to work for banks" (Comptes Rendus 2010). Meyer may have been referring to mathematics students; in his opening speech at *Maths à Venir* 2009 (see chapter 10), Philippe Camus, quoting the president of EADS, claimed the figure was 25% of all graduates (Roy 2010, p. 55).

14. Quoted in (Halimi 2012).

15. See (Taylor 2007, pp. 176ff) for the history of "The economy as objectified reality."

16. Writers about quants seem to have been issued identical stocks of clichés. Bridge metaphors abound, alongside allusions to Faust and the Devil. Just after Lehman Brothers collapsed, an article entitled "Don't blame the quants" protested that "[w]hen a bridge collapses, no one demands the abolition of civil engineering" (Shreve 2008). A quant quoted anonymously by the *New York Times* in 2009 rejected the analogy: her work is "not like building a bridge. If you're right more than half the time you're winning the game." In his fine review of Patterson (2010])' for the *Notices of the American Mathematical Society*, David Steinsaltz (2011) writes "one can hardly imagine top civil engineering academics reacting with nonchalance if their best graduates were not building bridges but finding bridges prone to collapse so they can cash in buying insurance on them."

17. (Smith 1776, chapter VI). Interestingly enough, this is the sole use of the word *derivative* in *The Wealth of Nations*.

18. (Braudel 1981, p. 471). Brian Rotman links these innovations to the semiotic opportunities and perils created by the introduction of the number zero: "If the xeno part of xenomoney threatened collapse of the world money system from the past, from unsupportable debt . . . the money part of it threatens . . . collapse from the future, from an unsustainable mutability of money signs created by the financial futures markets." "Xenomoney" is Rotman's (1987, pp. 96, 93) term for a certain kind of fictitious capital, supplanting paper money and viewed as "a sign able to signify its own future." Writing in 1987, Rotman predicted the collapse of the options market on purely semiotic grounds; was anyone paying attention?

19. Mephistopheles:

> *Ein solch Papier, an Gold und Perlen Statt,*
> *Ist so bequem, man weiß doch, was man hat;*
> *. . .Und das Papier, sogleich amortisiert,*
> *Beschämt den Zweifler, der uns frech verhöhnt.*
> *Man will nichts anders, ist daran gewöhnt.*
> *. . .*

Kaiser:

> *Das hohe Wohl verdankt euch unser Reich;*
> *Wo möglich sei der Lohn dem Dienste gleich.*
> *Vertraut sei euch des Reiches innrer Boden,*
> *Ihr seid der Schätze würdigste Kustoden.*
> *. . . Wo mit der obern sich die Unterwelt,*
> *In Einigkeit beglückt, zusammenstellt.*

20. From the Benjamin Jowett translation of *Politics*, Book 1, 1259a, online at classics.mit.edu. Thales was traditionally considered to be one of the founders of Greek mathematics as well as philosophy.

21. (Compétences 2005).

22. (Patterson 2010, p. 38). Black became ineligible for the Nobel Prize when he died in 1995. Estimates for total losses due to the collapses depend on sources and timelines, but they tend to be in the low billions of dollars for LTCM and in the trillions for the 2008 crisis. See, for example, FCIC (2011) for losses just within the United States.

23. MacKenzie's (2003, p. 835) article is unsurpassed for the clarity of its account of the Black-Scholes equation. In its emphasis on "performativity"—roughly, how the mathematical model brings its own reality into being —it is a strikingly successful example of what I called the "genealogical" approach to the sociology of mathematics. But it touches only briefly on questions of mathematical practice.

24. (Wilmott 2000) was written nearly a decade before the 2008 crash!

25. Maybe not so many after all. Columbia doesn't list its students online, but NYU's comparable Masters in Mathematical Finance graduates only 30–35 students per year.

26. Such equations also provide one class of *avatars*, one *théorie cohomologique*, for Grothendieck's hypothetical motives. This is the avatar relevant to the *Hodge conjecture*—one of the six remaining million dollar Clay Millennium Problems. See chapter 2 for the Clay problems and chapter 7 for Grothendieck avatars.

27. Compare the Galois symmetry of chapter β. Much more general equations of this kind, with no special symmetry, are treated in Chung and Berenstein (2005), whose bibliography refers to a paper Langlands wrote on this topic by with F.R.K. Chung.

Although the financial data themselves are assumed to vary stochastically, figure 4.1 expresses a deterministic relation between the option price and the price of the underlying security. MacKenzie (2003) includes an account of the authors' initial difficulty in finding its solution. Black wrote that he "had never spent much time on differential equations, so I didn't know the standard methods used to solve problems like that." Ultimately, they obtained the solution by financial reasoning. A mathematician—not necessarily a specialist in differential equations—would recognize that the equation in figure 4.1 is a slightly modified version of the heat equation whose discrete form underlies the *Stable Equilibrium* game. Compared to the usual form of the heat equation, the arrow of time is reversed, in the sense that instead of initial conditions, the solution is determined by terminal conditions. I thank Ivar Ekeland for this explanation.

28. This requires qualification. Deterministic systems are not always predictable;

probabilistic methods were introduced in physics to study deterministic systems, like the movement of an ideal gas, or the Brownian motion of particles in a liquid suspension, with too many parameters to calculate deterministically. The equations of Brownian motion provide the model for the evolution of stock prices and, thus, for the Black-Scholes equation. Even when the number of parameters is small, initial conditions cannot, in general, be known with sufficient accuracy to permit reliable prediction; this is the insight at the root of what is popularly known as *chaos theory*. Probabilistic methods can also be used to provide qualitative (and quantitative) information about chaotic dynamical systems.

29. The attempts of Rudolf Carnap and Ian Hacking have been particularly influential. For the sociologist Elena Esposito, "the numbers calculated on the basis of stochastic models represent a kind of doubling of reality, a fictive reality, that does not compete with real reality but forms an alternative description, the increases the available complexity" (Esposito 2007, p. 30).

30. (Thorp 1966, p. 182). This book, which explains Ed Thorp's "winning strategy" for blackjack—Wikipedia considers Thorp "the father of card counting"—was followed by *Beat the Market*, written with Sheen Kassouf, which applies similar probabilistic methods to stock options. In *The Quants*, Patterson calls Thorp "The Godfather."

31. (Steinsaltz 2011; Ekeland 2010; Wilmott 2000). The *Notices of the AMS* is the American Mathematical Society's house journal, and is undoubtedly the most widely-read of all general-interest professional journals for mathematicians.

32. (Tao 2008a), posted on Tao's blog shortly before the Lehman collapse, lists the following (not very plausible) assumptions, without which the Black-Scholes model is unrealistic: infinite liquidity, infinite depth, no transaction costs, no arbitrage, infinite credit, infinite divisibility, (unlimited) short selling, no storage costs. Most of these assumptions need no explanation, and most of Tao's account is accessible to a reader who doesn't know what a differential equation is. At the end it becomes technical: Tao actually derives the Black-Scholes differential equation using a discrete (difference equation) stochastic model not so different from the probabilistic hybrid between *Stable Equilibrium* and *Matthew Effect* described previously.

33. (Derman and Willmott 2009). "As more and more complex securities . . . were downgraded, banks experienced . . . losses and write-downs which reached US$700 billion in November 2008 . . . far in excess of what pricing models, rating models and risk models would have predicted" (Crouhy 2009). Hélyette Geman, a specialist in financial mathematics, was quoted in the Swiss *Le Temps* on November 12, 2008: "Our reputation is now tarnished because mathematical models have made the risks of toxic investments opaque . . . From 1986 to 1996 the contribution of probability was very positive" but there followed "a phase of excessive mathematization where the beauty of the results" was distorted by confusing the model with the real world.

34. (Shreve 2008; Chemillier-Gendreau and Jouini 2008).

35. (Krugman and Wells 2011; J. Macdonald, "Elephant Tears," review of *Money and Power: How Goldman Sachs Came to Rule the World*, London Review of Books, 3 November 3, 2011).

36. More precisely, "by 2007 the trade in derivatives worldwide was one quadrillion (thousand million million) US dollars—this is 10 times the total production of

goods on the planet over its entire history," says Stewart. "OK, we're talking about the totals in a two-way trade, people are buying and people are selling and you're adding it all up as if it doesn't cancel out, but it was a huge trade" (Harford 2012). See also (BIS 2013).

37. (El Karoui and Pagès 2009).

38. (Chance and Brooks 2012). Quotations are from pp. 17–18, exercises are below p. 25.

39. The speaker is Neil Chriss, from Lindsay and Schachter (2007, p. 134); see chapter 2, note 5. Academia was not for Chriss—recall what he said about Hesse's *Glass Bead Game*. Not everyone who moved from academia to Wall Street had Chriss' options: there were more mathematics and physics PhDs than jobs at the time. Admittedly, starting pay on Wall Street even then was often higher than a full professor's salary. But who cared about money?

40. (O'Neil 2008).

41. (Harkinson 2011).

42. (Patterson 2010, p. 295).

43. (Jouve 2011, Overbye 2009). To be fair, God sometimes gets into the act as well, for example, in this sentence from Georg Simmel's *Psychology of Money*: "Just as God in the form of belief, so money in the form of the concrete is the highest abstraction to which practical reason has attained." Simmel also quotes the Meistersinger Hans Sachs: "Money is the secular God of the world." [Quoted on page 238 of (Simmel 1978).]

44. (Lindsay and Schachter 2007, p. 202).

45. If you don't know how much you need to qualify as an "Ultra High Net Worth Individual" (UHNWI), then you probably are not one, which would explain why you have not been approached by "[b]rands in various sectors, such as Bentley, Maybach, and Rolls-Royce in motoring, . . . to sell their products. Figures gathered by Rolls-Royce suggest there are 80,000 people around the world with disposable income of more than $20 million. They have, on average, eight cars and three or four homes. Three-quarters own a jet aircraft and most have a yacht." (Wikipedia 2013). More precise information can be found in Capgemini and RBC Wealth Management (2013).

46. The Rand quotation is unsourced and perhaps apocryphal but is often attributed to her on the Internet and, as such, deserves to count as part of her influence. For Rand's influence on Greenspan, see (Greenspan 2007, pp. 51–53).

47. This is from the translation by H. Rackham, online at http://www.perseus.tufts .edu. To translate *epistemon* as "scientist" might be more appropriate—see chapter 10—and certainly more in line with the objectivist view of wealth creation—or the neoliberal view of Baroin or Thatcher's TINA, which come to much the same conclusions.

48. (Smith 1759); the second quotation is in (Bowles and Gintis 2011).

49. (Alvey 2000).

50. (Stephenson 1999, p. 29).

51. (Higgins 2001, pp. 168–69; Patterson 2010, pp 82–85; Nocera 2009).

52. A study by Olivier Godechot (2011, p. 14) determined that "finance constitute[d] 37% of the headcount of the top 0.01%" of salaries in France in 2007. On p. 2, he cites an earlier British study estimating that "70% of the recent increase of the share of the

top 1% in the United Kingdom was captured by workers of the financial industry." In 2013, HNWIs account for roughly 1% of the U.S. population and about .6% of the population of France and the United Kingdom.

53. Quotation from (Lindsay and Schachter 2007, p. 200). Goldman Sachs has changed since 1994. Greg Smith, explaining "Why I Am Leaving Goldman Sachs" in the *New York Times* in March 2012, caused something of a sensation when he wrote that "five different managing directors refer to their own clients as 'muppets,'" and "These days, the most common question I get from junior analysts about derivatives is, 'How much money did we make off the client?'"

54. Quotation from (Lindsay and Schachter 2007, p. 46). Some slivers of my grandparents' lifestyle are still on display at the Lower East Side Tenement Museum.

55. http://www.urbandictionary.com defines *fuck-you money* as "An amount of wealth that enables an individual to reject traditional social behavior and niceties of conduct without fear of consequences." The quotation is from (Stephenson 1999, p. 26).

56. (MacIntyre 1981, p. 149). MacIntyre also mentions *megalopsuchia*, "magnanimity."

57. (Guerzoni 1999, p. 338).

58. Crassus, who "acquire[d] wealth dishonestly, and then . . . squander[ed] it uselessly" was "the most covetous man in the world" according to Dryden's translation of Plutarch's *Life of Crassus*. Plutarch explains that Crassus made his money in real estate rather than finance: "making advantage of the public calamities," by buying houses on fire "and those in the neighbourhood" at bargain prices. Plutarch uses the words *philoploutia* and *philokerdes* rather than *pleonexia*.

59. "Since kings and princes have a lot of property and wealth, they should give bigger recompenses and they should spend with more pleasure and promptness." From Romanus Aegidius's *De regimine principum*, written between 1277 and 1279; quoted in Guerzoni (1999, pp. 355–356). Guerzoni traces the cult of what Latin authors called *magnificentia* from Plato and Aristotle through Cicero and Aquinas to the Italian sixteenth century, when "aristocratic magnificence . . . became a heroic princely attribute" and "every action or aristocratic consumption could be morally justified by the desire to pursue full magnificence."

60. [Muntadas 2011], quotations on pp. 18, 102, 32, 31.

61. (MacIntyre 1981, pp. 181, 178).

62. This is (painter) Allan Kaprow, quoted in (Muntadas 2011, p. 102).

63. (Cook 2009, p. 46; Ferguson 2012, p. 230; Simons 2010, starting at 48′20). Krugman (2009) also questions the "social value" of high-speed trading. Patterson (2010, p. 116) mentions an allegation of fraud directed at Simons's Renaissance Technologies but gives no reason to think it is credible.

Ferguson's opinion doesn't sound especially flattering, but if you've read his book, you'll know that Simons, almost uniquely among Ferguson's headliners, is accused neither of committing nor conniving with nor even facilitating criminal behavior. At worst he is "a pure drag on the economy, like spam e-mail," "providing no social benefit"—but hasn't the same been said (albeit on a much more modest scale) of pure mathematics?

64. (Patterson 2012, pp. ii, 234).

65. Quoted in (Overbye 2009).

66. The Clay Mathematics Institute, already mentioned in chapter 3 in connection with the million dollar Millenium Prize Problems, was founded in 1998 by Landon T. Clay III, who, according to the CMI Web site, has "had a distinguished career as a successful businessman and in finance and science-based venture capital funding." CMI supports many research and educational initiatives, among them the PROMYS program for high school students mentioned in note 4 to chapter 2. AIM was created by John Fry, of Fry's Electronics, the chain of big-box electronics stores based in California, and is (temporarily) located in the Palo Alto big box; they receive funding from the NSF as well as the Fry family.

Eric Weinstein imagines a "Nobel Prize for Backers"—HNWI individuals who have experience with risk management. See http://streamer.perimeterinstitute.ca/Flash /47c17555-ac4e-4143-aa10-b3d4ad499b40/viewer.html (around 15′).

67. (Pieper 2007).

68. (Mehrtens 1990, p. 377–385; following quotations from p. 385). For the IHP, see http://www.ihp.fr/fr/presentation/histoire.

69. In this respect mathematicians are not alone. A recent *New York Times* article entitled "Billionaires With Big Ideas Are Privatizing American Science" described "a cottage industry . . . offering workshops, personal coaching, role-playing exercises and the production of video appeals" in order "to help scientists bond quickly with potential benefactors." "Today, federal funding of basic research is on the decline. . . The best hope for near-term change lies with American philanthropy."

The author of this book feels obliged to disclose that he has personally benefited from the generosity of several of the philanthropic foundations mentioned in this section.

70. (Lafforgue 2008), my translation.

71. (Skovsmose 2010). "The good" and "the true" converge when responsibility is mechanized.

72. (Lave 1988, pp. 125–126).

73. (Steinsaltz 2011, p. 703). In (Mirowski 2004, p. 6), we read that "Politically pugnacious economists . . . threw their weight behind the supposed 'positive/normative distinction' in their theories. . . . Philosophers, together with statisticians and economists, began to pretend that elaborate statistical algorithms . . . perhaps fortified with game theory . . . could somehow provide solutions to the value relativity of measurement and quantification. . . . All . . . tended to treat some generic thing called 'mathematics' as if it were capable by itself of cutting the knot binding science and the economy."

74. (Einstein 1949).

CHAPTER β HOW TO EXPLAIN NUMBER THEORY (EQUATIONS)

1. Ignoring degenerate linear equations, like $x = x$, that are tautologically true for all numbers x, and inconsistent linear equations, like $x = x + 1$, that have no solutions.

2. This is the celebrated *quadratic formula,* known since antiquity but which only gradually took on this standard form, starting with Christoff Rudolff's introduction of the radical sign in the sixteenth century. At present count, YouTube features 1290

"quadratic formula song" videos, in which high school students and their teachers set $B^2 - 4AC$ and company to tunes by the likes of Justin Bieber and Lady Gaga as well as to "Pop Goes the Weasel." German students instead study the equation $x^2 + px + q$—the "pq-Formel"—and if you want to see the solution set to music you should check out the rap *Die Lösungsformel*, http://www.youtube.com/watch?v=tRblwTsX6hQ. There are (in 2013) videos but no songs about the formula in French and neither videos nor songs in Dutch. A colleague has alerted me to efforts to eliminate the quadratic formula from the British curriculum, jeopardizing future dinner-party explanations of number theory to British Performing Artists.

3. This is a manner of speaking. As a general rule, the problems of finding integral solutions and finding rational solutions are both interesting. It's the latter that drives the plot at this point in the story.

4. This was the tenth problem on Hilbert's list from the 1900 Paris Congress. The (negative) solution was completed by Yuri Matiasevich in 1970 on the basis of earlier work by Martin Davis, Hilary Putnam, and Julia Robinson. See Putnam's forthcoming *Intellectual Autobiography*.

5. The graphs of quadratic equations are the *conic sections* that we will meet again in the next chapter.

6. The complete version, for even numbers or for odd numbers that are not primes, is just a little more complicated to state: any positive integer n can be written as the sum of two squares if and only if each of its prime factors of the form $4k + 3$ occurs an *even* number of times in its prime factorization.

7. It would be more accurate to call it *Brahmagupta's equation*, since Brahmagupta, working in Rajasthan, analyzed such equations in the seventh century; and Bhaskara II solved them completely by the twelfth century—500 years before John Pell.

8. You can find it in any elementary book on number theory—for example, in (Stark 1987, pp. 239–45).

9. *Posterior Analytics*, Book I, 7. The author is indebted to Marwan Rashed for this and all subsequent discussions of ancient Greek and Islamic philosophy of mathematics.

10. Quoted in (Fisch 1999, p. 145).

11. In his commentary on Euclid, al-Khayyam "argues that a geometrically obtained irrational ratio can be understood only by accepting irrationals as numbers—and in so doing he emerges as the first mathematician to admit irrationals to the status of numbers" (Özdural 1998). For al-Khwarizmi's ontology of numbers, see chapter 10.

In Hamilton's quaternions, $i^2 = j^2 = k^2 = -1$ and $ij = k$ but $ji = -k$. They make an appearance in chapter 5 also in number theory, but not in this book.

12. (MacIntyre 1981, p. 196).

13. (Cook 2009, p. 7).

14. Suppose we label the four solutions:

$$A = \sqrt{2} + 3i \; B = \sqrt{2} - 3i \; C = -\sqrt{2} + 3i \; D = -\sqrt{2} - 3i$$

There are 24 ways to permute the solutions, but only 4 of them are in the Galois group: these are

$$A \leftrightarrow B, C \leftrightarrow D \quad \text{(exchange } A \text{ and } B, \text{ exchange } C \text{ and } D;$$
$$\text{in other words permute } 3i \text{ and } -3i);$$

$A \leftrightarrow C, B \leftrightarrow D$ (in other words, permute $\sqrt{2}$ and $-\sqrt{2}$);

$A \leftrightarrow D, B \leftrightarrow C$ (in other words, permute the pairs $3i$ and $-3i$ and $\sqrt{2}$ and $-\sqrt{2}$);

The other 20 permutations fail to preserve the rules of arithmetic. For example, we see that $A + D = 0$. If we switch A with B but leave C and D unchanged, then $A + D$ is replaced by $B + D$, which is not equal to 0. Thus this permutation is not allowed in the Galois group. In the Hamlet analogy, say A is Gertrude, B is Claudius, D is Hamlet, and the equation $A + D = 0$ corresponds to the sentence "Gertrude is Hamlet's mother." Exchanging Gertrude with Claudius while leaving Hamlet alone gives the sentence "Claudius is Hamlet's mother," which belongs to a different play.

15. In the version emphasized by John Tate.

16. For example, pp. 184–186, pp. 200–202.

Chapter 5: An Automorphic Reading of Thomas Pynchon's *Against the Day*

1. (Dalsgaard 2012, p. 165).

2. From the review by Martin Paul Eve at http://www.berfrois.com/2012/02 /pynchonite-generosity-martin-eve/.

3. On December 7, 2008, it gave one, which is still there and is still the only (real) one, as you can check for yourself.

4. (Sante 2007; Menand 2006). Menand is being too modest; his book *The Metaphysical Club* (Menand 2001) includes a perfectly creditable account of the history and philosophy of probability and statistics in its chapters on C. S. Peirce.

5. Compare http://quarterlyconversation.com/the-christophe-claro-interview.

6. So does Wikipedia: "The paths of Stencil and Profane through the novel form a sort of metaphorical V." Also see also the next note.

7. In his 1990 review of *Vineland*, Rushdie wrote, "His novel 'V.' was actually V-shaped, two narratives zeroing in on a point, and "Gravity's Rainbow' was the flight path of a V-2 rocket, a deadly parabola that could also be described as an inverted V." Dalsgaard (2012) alludes to the "difference in propulsion between the ballistic-rocket and narrative parabolas in *Gravity's Rainbow*."

More meticulous analyses point to a morally charged dialectic in *Gravity's Rainbow* between deterministic (meta)physics, symbolized by the parabola, represented by the Pavlovian Ned Pointsman, and pointing at death; and statistical or stochastic (meta) physics, personified by the sensitive Roger Mexico and associated in the text with ellipses and the richness and unpredictability of life: the "ellipse of uncertainty" of the rocket's landing spot, an ellipse of blood left after delivery of a child, an ellipse of light illuminating an anarchist encampment. The scene of a particularly convoluted intertwining of dynamics of life and death is illuminated, literally, by a flashlight beam alternating between elliptical and parabolic outline. It's nevertheless clear that *Gravity's Rainbow* is written under the sign of the parabola.

8. But compare with Wikipedia: " . . . the line approximates a segment of a small circle upon the surface of the (also approximately) spherical Earth."

9. Like a Keplerian orbit, rather than a Copernican circular orbit—so it is worse than Copernicus after all. . ..

10. In late 2013 Roberto Natalini, an applied mathematician in Rome, sent me a copy of (Natalini 2013), a chapter on the role of mathematics in the work of the late writer David Foster Wallace, in a book devoted to the latter. Wallace had an avowed interest in mathematics; he was a great admirer of G. H. Hardy's *Apology* (see chapter 10) and wrote a full-length book on Cantor's set theory. Natalini finds several mathematical structures in Wallace's *Infinite Jest*, so precisely calibrated with the (fragmented) plot that it is impossible to believe that their presence is merely incidental. Here I just want to mention that, on Natalini's analysis, the trajectories of the two main characters of *Infinite Jest* follow the two arcs of a hyperbola that "are close to merging in the middle of the novel, when [the characters] live a few hundred meters apart."

Natalini and his colleague Lucia Caporaso coordinate the Italian Web site MaddMaths, dedicated to popularizing mathematics and guided by two principles: that "mathematics is more entertaining and less frightening than what is usually believed" and that "mathematics is not only the basis of innovation and technological development but also of the modern idea of citizenship" (Caporaso and Natalini 2013).

11. (Engelhardt 2013). Real and imaginary axes are present both within Pynchon's narrative and outside it in our reading of it, yet another doubling or bilocation, just as the quaternions are a doubling of the original complex numbers. The Joyce quotation is from (Ellmann 1983, p. 573).

12. Pynchon chose the twenty-fifth anniversary of C. P. Snow's "Two Cultures" lectures for one of his rare departures from public silence: (Pynchon 1984).

13. (Dalsgaard 2012, p. 157).

14. This MH, not the author of (Harris 2003).

15. (Harris 2003).

16. (Slade 1990). I have no idea what he means by this. Slade thanks the mathematician John S. Lew for bringing the parallels between Atman and Galois to his attention. See also (A.-G. Weber and Albrecht 2011).

CHAPTER 6 FURTHER INVESTIGATIONS OF THE MIND-BODY PROBLEM

1. Quotation from Has translated from French subtitles; the translation of Plato is from (Nightingale 2004, p. 115), with slight modifications suggested by Marwan Rashed.

2. *The aesthetics and psychology of the cinema*, trans. Christopher King.

3. (Lévi-Strauss 1992, 99–100).

4. "[A]n immense musical accelerator, built for the sole purpose of detecting the Higgs boson of the universe of love" (Spice 2013). See especially (Dreyfus 2010).

5. Actually an authentic equation in the theory of vertex algebras, reproduced in figure 6.8 from the article (Frenkel el al. 2011).

6. See http://videos.tf1.fr/infos/2010/mathematiques-le-genie-russe-refuse-1 -million-de-dollars-5787192.html.

7. (Spinney 2010; Mazzucato 2010).

8. "It's worth noting that almost all films involving mathematicians depict them as clever, whether they are also crazy, schizophrenic, or ruthless murderers." referring to Robert Connolly's *The Bank*, from (Emmer 2003). The first of these mathematical psychopaths may well have been Sherlock Holmes's nemesis Professor Moriarty who, after writing a "treatise upon the Binomial Theorem which has had a European vogue" gave up a "most brilliant career" in pure mathematics in order to pursue a life as the "Napoleon of crime." Quotations are from *The Final Problem* by Arthur Conan Doyle.

9. (Alexander 2011, p. 3). The next quotation is from p. 161.

10. It's not true, for example, that Galois first wrote up his main discoveries the night before his fatal duel; nor was he ignored and rejected by the mathematical establishment; and there is no reason to believe his duel over Stéphanie was arranged by his political enemies. Alexander was not the first to bring these facts to our attention—see (Rothman 1982)—but to my knowledge he was the first to theorize why the fabrication was and remains so successful.

11. Or Thales, who, in his cameo in Plato's *Theaeteus*, falls in a well while studying the stars.

12. The quotation under the entry for *wrangler* dates from 1751. Toby Gee, whom we will be meeting in chapter 8, was Senior Wrangler his year.

13 (Diderot and Grimm 1765). Diderot's 1748 novel *Les bijoux indiscrets* is dubious about the allure of mathematics. The novel concerns a magic ring that endows a woman's *bijou* with the gift of speech, and the secrets revealed in this way. The *bijou* of a woman who had studied geometry spoke incomprehensibly: "*Ce n'était que lignes droites, surfaces concaves, quantités données, longueur, largeur, profondeur, solides, forces vives, forces mortes, cône, cylindre, sections coniques, courbes, courbes élastiques, courbe rentrant en elle-même, avec son point conjugué . . .*"

14. (Clairaut 1745), reprinted and placed in context in (Boudine and Courcelle 1999). My thanks to Olivier Courcelle and Jean-Pierre Boudine for this reference. See also (Boudine 2000), quoted in (Justens, 2010). Diagram from books.google.com.

15. Maupertuis was a friend of the Marquis d'Argens, author of the novel *Thérèse philosophe*, which features a chapter entitled *Thérèse se procure machinalement des plaisirs charnels*. The quotation is from Mauperthuis' *La Vénus Physique*.

16. With respect to the sixty-six-day version of the book, no longer considered reliable. The quotations can be found in the authoritative 1804 version, edited by François Rosset and Dominique Triaire. In my edition the flirtation begins on page 231 and the marriage is announced on page 607.

17. (Alexander 2011, p. 74; Brown 1997). The French were hardly the first to acknowledge the erotic potential of mathematics. Medieval Hindu exercise manuals included problems like this one from the *Manoranjana*, a commentary on Bhaskara's *Lilavati* written by one Ramakrishna Deva (Brahmagupta and Colebrook 1817):

> The third part of a necklace of pearls, broken in an amorous struggle, fell to the ground: its fifth part rested on the couch; the sixth part was saved by the wench; and the tenth part was taken up by her lover: six pearls remained strung. Say, of how many pearls the necklace was composed.

18. Byron wrote in his diary that "Lady B. would have made an excellent wrangler at Cambridge" (see note 12).

19. {Zamyatin, 1924, Record Twenty-seven; Edgeworth 1881), my emphasis. Edgeworth continues: "but we seem to be capable of observing that there is here a greater, there a less, multitude of pleasure-units, mass of happiness; and that is enough." See also chapter 10, note 61.

20. (Jung 1969, p. 59).

21. Bourbaki's structuralism is treated at length in the next chapter. The catastrophe theory of (the non-Bourbakist) René Thom, like structuralism, is a twentieth-century revival of Leibniz' *mathesis universalis*, but arising within mathematics itself. At one time or another, catastrophe theory, like the structuralist double binary opposition scheme illustrated here, has been applied to everything. For a catastrophe-theoretic approach to sex, see (Hubey 1991). Hubey's abstract begins "A nonlinear differential equation model and its associated catastrophe is shown to model the simplest version of the sexual response of humans. The mathematical model is derived via well-known and non-controversial aspects of sexual orgasm as can be found in the literature." Hubey's love equation is

$$\Psi + 2\omega_n\xi\Psi + \omega_n^2\Psi + \mu\Psi^3 = A \sin(\Omega t)$$

where "Ψ represents 'sexual tension,', Ω is the frequency of the 'excitation,' and μ is some non-linearity parameter."

22. From his letter to Simone Weil, 1940. The same imagery, culminating in an allusion to the Bhagavad-Gita, reappears in Weil's 1960 article quoted in chapter 7. Both of Weil's texts derive pleasure from witnessing an intimate coupling between two branches of mathematics; a text by Grothendieck from the 1980s, quoted less frequently, evokes the "deep kinship between the two passions that had dominated my adult life" by casting the mathematician and *la mathématique* as the erotic partners:

> It's surely no accident if in French as well as in German, the word that designates [mathematics] is of the female gender, as is "la science" which encompasses it, or the still vaster term "la connaissance" [knowledge] or again "la substance." For the genuine mathematician, by which I mean the one who "makes mathematics" (as he would "make love") there is indeed no ambiguity regarding the distribution of roles in his relation to *la mathématique*, to the unknown substance, then, of which he acquires knowledge, which he knows by penetrating her. Mathematics is then no less "woman" than any woman he has known or merely desired—of whom he has sensed the mysterious power, attracting him into her, with that force at once very sweet, and unanswerable (Grothendieck 1988, p 496).

Frenkel, at two generations' distance from Grothendieck, would never express himself in this way, but some mathematicians in the United States—men as well as women—found the imagery of *Rites*'s trailer uncomfortably reminiscent of the "distribution of roles" Grothendieck found so natural and protested when Berkeley's MSRI announced it was cosponsoring a special showing of *Rites* in a downtown Berkeley theater. The protestors worried that spectators viewing *Rites* would get the message that the "genuine mathematician" is a man and convinced the MSRI to withdraw its sponsorship (the showing was nevertheless sold out, with more than 100 left waiting on the pavement and many standing in the back of the theater). Some of the most interesting comments were recorded at http://www.math.columbia.edu/~woit/wordpress/?p=3318 on Peter Woit's always stimulating blog "Not Even Wrong."

It's funny to note that the only mainstream film to have represented the mathematician's partner unequivocally as a sex object is the avowedly feminist *Conceiving Ada*, whose "computer genius" protagonist shares her private moments with a husky bearded boyfriend, who divides his on-screen time between (a) watching adoringly while she interacts with phantoms behind the computer screen, (b) getting her pregnant, and (c) making coffee.

23. http://aejcpp.free.fr/lacan/1969-12-03.htm, reproduced from *Magazine Littéraire* Spécial Lacan, number 121 de Février 1977. See also http://www.psychanalyse -paris.com/Il-eusse-phallus.html for a discussion of Lacan's analysis of the castration complex in terms of the golden ratio.

24. Elisabeth Roudinesco, quoted in French Wikipedia. More recently, French philosopher Alain Badiou has attempted to place Lacan's theory of love on a sound theoretical footing. I am indebted to Vladimir Tasić for drawing my attention to this excerpt from a discussion of Badiou's *La Scene du Deux*. UC Irvine professor Juliet Flower MacCannell writes that "Badiou has succeeded. He has formalized—to an unparalleled degree—the work of the [Lacanian] object a' [the upper-right corner of Figure 6.5, MH] in structuring the nonrelation of the sexes and opening the way to the relation that is in excess over them: the supplement of Love." MacCannell quotes Badiou (her translation) in (MacCannell 2005):

> An amorous encounter is that which attributes eventmentally to the intersection— atomic and unanalyzable—of the two sexed positions a double function. That of the object, where a desire finds its cause, and that of a point from which the two is counted, thereby initiating a shared investigation of the universe. Everything depends, at bottom, on this $u \leq M$ and $u \leq W$ being read in a double fashion: either that one is assembling there the inaugural non-relation of M and W, inasmuch it affects the non-analyzable u with being only what circulates in the non-relation. The positions M and W are then only in this misunderstanding over the atom u, cause of their common desire, a misunderstanding that nothing sayable can lift off, since u is unanalyzable. This the first reading. Or one reads it in the other direction: starting from u, either $(W - u)$ and $(M - u)$, two positions like those the atom u supports by subtracting itself from it (pp 186–87 of Badiou's original text).

Tasić made the following attempt at an interpretation:

> The sexes (M and W) are not in a relation but both dominate the same atom (u) in some poset. Apparently the successful solution of Lacan's riddle ("there is no sexual relation") is a love formula having something to do with the notion of a partially (but not totally) ordered set.

25. However, see http://www.cambridge-news.co.uk/Whats-on-leisure/Art/Cambridge -don-Dr-Victoria-Bateman-dares-to-bare-in-nude-portrait-20140515065433 .htm. *Rites* was in competition at the June 2010 Sexy International Paris Film Festival (SIPFF) but was not listed among the prize winners. See www.sexyfilmfestparis.fr.

26. She has instead been consecrated, like her contemporary Galois or her father, as

a romantic martyr; see the film *Conceiving Ada* mentioned before, as well as William Gibson's novel *The Difference Engine* and several books for a general audience.

27. Quotations from (Alexander 2011, pp. 86–89).

28. (Alexander 2011, p. 13). This may not be unrelated to their poor reputation as lovers: " . . . it's maybe because of math's absolute, wholly abstract Truth that so many people still view the discipline as dry or passionless and its practitioners as asocial dweebs" (Wallace 2000).

29. *Rites of Love and Math*, opening frames.

30. (Alexander 2011, p. 49). After reading this paragraph, Alexander pointed out a third possible archetype, see following text.

31. *Чем глубже мы погружаемся в материальный мир, тем дальше мы от него отдаляемся в направлении мира идеального. Light and Word* (*Свет и Слово*), in [Parshin 2002].

32. From *Proof*, by David Auburn; cited in (Hofmann 2002).

33. All quotes from (Ikeda and Bonnet 2008). Gross, a number theorist, is especially famous for his work with Don Zagier on the Birch-Swinnerton-Dyer conjecture; see chapter 9 and endnote 52 in chapter 2.

34. (Menkes 2010).

35. http://abcnews.go.com/Entertainment/wirestory?id=10017982&page=2.

36. http://www.youtube.com/watch?v=YZXLLNbi76o.

37. Comments and quotation at http://www.youtube.com/watch?v=eQAuSGvQjN0&feature=related.

38. See http://www.claymath.org/poincare/laudations.html. Mikhail Gromov, also on hand at the Perelman ceremony in Paris, had been quoted a year earlier to the same effect: Perelman's "main peculiarity is that he acts decently. He follows ideals that are tacitly accepted in science" (Gessen 2009, p. 111). Gessen herself decided while writing her book that Perelman "has an internally consistent view of the world that is entirely different from the view most people consider normal"—her definition of "crazy" (on the book's amazon.com page).

39. No coincidence: there's always a champagne reception.

40. Two films, actually, including one still in production in collaboration with my colleague Pierre Schapira.

41. (Alexander 2011, pp. 253–255).

42. Much less that she anticipated Kepler's theory of planetary orbits by a good twelve centuries.

43. (Aczel 2001).

44. (Metz 1977, p 10, p. 68).

45. We will encounter the very different figure of naked truth, Horace's *nuda veritas*, in chapter 10.

46. From http://en.allexperts.com/q/Russian-Language-2985/truth.htm. I thank Yuri Tschinkel for this link. A. N. Parshin, on the other hand, points out an old Russian proverb where the emotional content of the two terms seems to be reversed. The proverb is based on Psalm 85:11, which reads "Truth shall spring out of the earth; and righteousness shall look down from heaven" in the King James version. In the Russian Bible, where the verse is numbered 84:12, "Truth" and "righteousness" are translated *istina* and *pravda,* respectively, and so the proverb becomes *Истина от земли, а правда*

с небес: *istina* from the earth, and *pravda* from the heavens. The Hebrew words are *emeth* and *tsedek*.

47. Among the Chinese characters composing Mariko's name is one meaning "real" or "true."

48. (Florensky 1996). The original text can be found online: ... *истинный художник хочет не своего во что бы то ни стало, а прекрасного, объективно-прекрасного, то есть художественно воплощенной истины вещей. . . . Лишь бы это была истина—и тогда ценность произведения сама собою установится.*

49. http://www.thomasfarber.org/b_body.php.

50. In which a police inspector screams, "Did you fucking know about this? You and your fuckin' equations?"

51. To forestall possible misunderstandings, I remind readers that the equation in question was printed in an article having nothing to do with love or film several years before Frenkel began his collaboration with Reine Graves. It might also be mentioned that the algorithm for the online dating service OK Cupid, created by four Harvard math majors in 2004, was sold in 2011 for $50 million, making it by some measure the most successful love formula in history—unless you count Facebook, created at Harvard that same year.

52. *Точно узнать продолжительность жизни, по мысли Френкеля, вполне реально, если иметь в распоряжении статистику половых актов конкретного человека.* http://woman-today.ru/themes/world/9942/. No one knows where they got this idea.

53. As a teenager, he may have seen this movie on Soviet TV: http://www.imdb.com/title/tt0216755/.

54. A. N. Parshin points out that the name-worshippers' postulate continues "but God is not His name."

55. Graham and Kantor (2009) give a detailed account of Florensky's martyrdom. Florensky has been repeatedly accused of antisemitism, especially in connection with the Mendel Beilis blood libel of 1913. Pyman (2010) examines the evidence for this accusation and finds it inconclusive.

56. From *Light and Word* in (Parshin 2002) and *Staircase of Reflections* (Parshin 2007), respectively.

57. Erwin Panofsky claims Suger wrote this verse under the influence of Pseudo-Dionysius the Areopagite, author of *On the Divine Names* and *The Celestial Hierarchy*, among other works. According to Parshin, Pseudo-D. insists in *The Celestial Hierarchy* on the fact that "every angel is a mirror: he accepts light in order to reflect it." (Parshin *2005*). I found this translation at http://www.esoteric.msu.edu/VolumeII/Celestial Hierarchy.html:

> [The Hierarchy] moulds and perfects its participants in the holy image of God like bright and spotless mirrors which receive the Ray of the Supreme Deity—which is the Source of Light; and being mystically filled with the Gift of Light, it pours it forth again abundantly, according to the Divine Law, upon those below itself.

58. Florensky is quoted in (Parshin 2007) in the Russian original: *Тело есть наиболее одухотворенное вещество и наименее деятельный дух, [. . .], но лишь в первом приближении; Тело есть осуществленный порог сознания; Тело есть*

пленка, отделяющее область феноменов от области ноуменов. The translations are mine. The Russian word *пленка*, which traditionally designates a very thin boundary between two media, is also the contemporary term for the material on which the images of a film are recorded. Note that for Florensky, the body is a film separating two regions of the mind.

59. *То, что за телом, по ту сторону кожи, есть то же самое стремление само-обнаружиться, но сознанию сокрытое; то, что по сю сторону кожи есть непосредственная данность духа, а потому не выдвинутая вне его. Сознавая, мы облачаем, и переставая сознавать - разоблачаем себя самих.* (The English version is my translation, with Parshin's help.)

60. In (Parshin 2008). The concept of a mirror that exchanges the images on either side is paradoxical. The reader is encouraged to try to draw a diagram, which may or may not resemble that drawn by Florensky in his *Limits of Knowledge*; for Florensky, there is an infinite staircase of steps between the two worlds. I thank A. N. Parshin for this reference.

61. See *Rig Veda*, 10:129.

62. *Ведь любовь возможна к лицу, а вожделение—к вещи; рационалистическое же жизнепонимание решительно не различает, да и не способно различить лицо и вещь, или, точнее говоря, оно владеет только одною категорией, категорией вещности, и потому все, что ни есть, включая сюда и лицо, овеществляется им и берется как вещь, как res* (Florensky 1997).

63. (Zinoman 2010).

64. Faust, Part I, lines 1856–59; translation http://www.online-literature.com /goethe/faust-part-1/5/; Part II, lines 11804–11807, translation with the help of Yuri Tschinkel).

65. This is also the theme of Frenkel's (2013) book, entitled *Love and Math*. Neuroscientist Arjan Chatterjee is not convinced: in his analysis of the neurological basis of aesthetic response, he distinguishes "liking" from "desire" and is inclined to classify the response to mathematical beauty under the former heading (Chatterjee 2014, p. 62).

66. This may or may not disqualify an artificial theorem prover as a future subject of mathematical biography in the vein of Minds of Mathematics.

67. (Levinas 1987, p. 56).

CHAPTER β.5.

1. Taking their lead from (Hardy 2012), to which we return in chapter 10. For the use of the word *trick* in mathematics see chapter 8. The diagonalization trick is sketched in note 25 to chapter 7.

2. (Kontsevich and Zagier 2001).

3. (Turing 1936-1937). In the appendix, Turing shows that his computable numbers are equivalent to the "effectively calculable numbers" introduced by A. Church.

4. Rafael Bombelli, "the first mathematician to accept as valid the solutions of third-degree equations . . . containing imaginary numbers" called irrational numbers "impossible to name" in his *Algebra*, written in 1550 (La Nave 2012). The numbers I

am calling indescribable are much more radically impossible to name than π or $\sqrt{2}$. See, however, (Hamkins et al. 2013) for models of set theory in which every real number is definable in the sense of being "specified as the unique object with some first-order property." Hamkins explains in Math Overflow (where he is the reigning champion) that the Wikipedia article on definable numbers is mistaken in arguing that this is not the case.

5. See video.ias.edu/voevodsky-80th, around 33'30".

6. For the purposes of this dialogue I am considering only numbers that can be defined or described in the setting of a dinner party, where I think the subtle considerations of the article of Hamkins et al. do not apply.

7. (Weyl 1968, p. 177), quoted in (Schappacher 2010).

CHAPTER 7 THE HABIT OF CLINGING TO AN ULTIMATE GROUND

1. The Vimalakīrti Nirdésa Sūtra mentions three aspects of Dharma worship, among others: "attaining the tolerance of ultimate birthlessness and nonoccurrence of all things . . . realizing the ultimate absence of any fundamental consciousness; and overcoming the habit of clinging to an ultimate ground." See (Thurman 1976).

2. The reasons for this belief deserve to be explored. In what sense is such a belief rational?

3. For example, see (Pettie 2011). Erdős wasn't the first to define mathematicians in this way; the quotation is reportedly due to his fellow Hungarian and collaborator Alfréd Rényi: see http://en.wikipedia.org/wiki/Alfr%C3%A9d_R%C3%A9nyi. Erdős is, however, *famous* for having said it.

4. Fragment I, translated in (Kahn 1981, p. 29).

5. (Hofmann 1998). For Dr. Gonzo and his medicine cabinet, see Hunter S. Thompson, *Fear and Loathing in Las Vegas*, passim.

6. Quoted in (Jackson 2004, p. 1196). In his unpublished writings, Grothendieck frequently refers to his internal *rêveur*—the dreamer—which in one text he identifies with *Dieu*.

7. Godement's comments were reported by Leila Schneps. For the milk and bananas see (Jackson 2004, p. 1044). Godement told Schneps that, as a doctoral student in Nancy in the early 1950s, "Grothendieck would periodically suddenly decide that he was never going to eat anything again except for milk, cheese and bread. So if they invited him to lunch, he would come, but he would scold them all for eating too much, and demand to be served only milk, cheese and bread while everyone else ate whatever Godement's wife had prepared." The forty-five-day fast is recounted in (Scharlau 2010, pp. 250–251): "*Vielleicht wollte er . . . durch diesen Akt Gott zwingen, sich ihm zu offenbaren.*"

8. "Mirror Symmetry" in (Gowers et al. 2008, p. 528).

9. I. Grojnowski, "Representation Theory," in (Gowers et al. 2008, p. 421).

10. Compare this verbatim quotation: "I don't want to be too precise [about the definition] because it would take a quarter of an hour."

11. Translated in (Mazur 2004a), with Grothendieck's boldface restored. Here is the original text:

> *Contrairement à ce qui se passait en topologie ordinaire, on se trouve donc placé là devant une abondance déconcertante de théories cohomologiques différentes. On avait l'impression très nette qu'en un sens qui restait d'abord assez flou, toutes ces théories devaient "revenir au même", qu'elles "donnaient les mêmes résultats". C'est pour parvenir à exprimer cette intuition de "parenté" entre théories cohomologiques différentes, que j'ai dégagé la notion de "**motif**" associé à une variété algébrique. Par ce terme, j'entends suggérer qu'il s'agit du "motif commun" (ou de la "**raison** commune") sous-jacent à cette multitude d'invariants cohomologiques différents associés à la variété, à l'aide de la multitude des toutes les théories cohomologiques possibles à priori.*

12. For example, (Corry 2004, especially chapter 7; McLarty 2006).

13. (Drinfel'd 2012).

14. (Cook 2009, p. 156). Deligne, a Belgian, studied with Grothendieck in France and remained there for two decades before moving to the IAS. The Weil quotation is from an article entitled "*De la métaphysique aux mathématiques*," reproduced in volume II of his collected works; the English translation is from B. Mazur's (1977) article. Weil's title refers to the eighteenth-century mathematicians who used the word *metaphysics* to designate what Deligne is calling "philosophies" or "yogas." Grothendieck's allusion to "yoga" in the next paragraph is from (Colmez et al. 2003, p. 347).

15. (Krömer 2007, p. 9).

16. Since this chapter was originally part of a presentation at a philosophy conference, I was thinking of its use by philosophers like Brentano and Husserl. Compare also H. Weyl, "existence is only given and can only be given as intentional content of the conscious experience of a pure, sense-creating ego," quoted in (Schappacher 2010, p. 3274).

17. Worrying to what extent this is also the case in the natural sciences leads to another abyss that is not the topic of this paper.

18. Ian Hacking observes that Wittgenstein first used *übersichtlich* and its cognate *übersehbar* in connection with mathematics in his *Remarks on the Foundations of Mathematics* that dates from the 1930s. Translated "perspicuous" and "surveyable," respectively, these terms remain highly influential in contemporary philosophy of mathematics. Hacking points out that Wittgenstein's subsequent allusions to these words are comments on their first appearance in his work and advises philosophers to be more attentive to this context. See (Hacking 2014, chapter I, paragraph 24).

19. The dates and illustrations in English and French are taken from the *Oxford English Dictionary* and *Le Robert* or the *Dictionnaire de l'Académie*, respectively.

20. (Deligne 1971; Coleman et al. 1969). I thank IAS director Peter Goddard for drawing my attention to the latter article. The word had already appeared in 1929 in the very first recorded adventure of Deligne's most famous compatriot, in the Land of the Soviets: "*Le « Petit XXe », toujours désireux de satisfaire ses lecteurs et de les tenir au courant de ce qui se passe à l'étranger, vient d'envoyer en Russie Soviétique un de ses meilleurs reporters: Tintin! Ce sont ses multiples avatars que vous verrez défiler sous vos yeux chaque semaine*" (Hergé 1930).

21. This is indeed the spirit of the *Beilinson conjectures*, which strongly influenced Voevodsky's work.

22. The unifying principle was identified by the Milesians as water (Thales), air (Anaximenes), "the infinite" (*to apeiron*; Anaximander), or "being" (*to eon*; Parmenides). Plato frequently thematizes identity and difference, notably in the *Timaeus*, *Sophist*, and *Parmenides*; Aristotle treats the question in particular in the *Topics* and the *Metaphysics*. The Upanishads contain a monistic theory later summarized by Sankara's identification of atman with brahman; even in the Rig Veda one can read "The wise speak of what is One in many ways; they call it Agni, Yama, Matarishvan" (1.164.46); see (Doniger 2010). The logic of difference was analyzed by the Nyaya and Vaisesika schools as early as 2000 years ago: see (Ganeri 2009).

23. (Gowers 2000). Party planners are warned that Erdős's estimate is certainly far short of the real value of $R(k)$ [often written $R(k, k)$]; determining the exact values of $R(k)$ even for small k is one of the outstanding unsolved problems in combinatorics.

24. Ten years after the article's publication, the case no longer needs to be made. Thanks in large part to the successes of Gowers and a few of his colleagues and students, Erdős-type mathematics is now a reliable source of charisma at the highest levels: Erdős's collaborator Endre Szemerédi was awarded the Abel Prize in 2012 (and received the rarer honor of being compared to Leonardo, Einstein, and the Buddha in the pages of the *Guardian* by his compatriot, the novelist László Krasznahorkai) for his work in theoretical computer science as well as combinatorics; a July 2012 conference on additive combinatorics at the Institut Henri Poincaré held its official banquet at the Louvre.

25. Quotations from (Dauben 1979, pp. 290, 245–246, 144–146). The final quotation is from a letter to Cantor by Cardinal Johannes Franzelin.

The set denoted $P(S)$ is the *power set* of S, the set of all subsets of S. If $S = \{a, b\}$ has two elements, $P(S)$ has the four elements $\{a\}$, $\{b\}$, $\{a, b\}$, and the empty set. If S has n elements, then $P(S)$ has 2^n elements, more than S. The diagonalization trick shows that $P(S)$ is bigger than S even if S is infinite. Here is the trick in a few words. Suppose S and $P(S)$ had the same cardinality; in other words, for each s in S, there is a subset $f(s)$ in $P(S)$, and every member of $P(S)$ is labeled in this way. Now let T be the set of s such that s is not in $f(s)$. Suppose there were a t such that $T = f(t)$. Now if t were in T, then t would be in $f(t)$ and, hence, by definition, is not in T. Thus t must be in T; but then t fails the criterion to be in T. Hence there is no such t, and $P(S)$ cannot be indexed by the elements of S. The reader will note the resemblance to Russell's paradox.

The paradox described here is related to but not identical to the *Burali-Forti Paradox* regarding the endless sequence of ordinal numbers. See also (Garfield and Priest 2003, especially the discussion on pp. 17_19) for similar paradoxes in Nāgārjuna's philosophy (roughly AD 200).

26. Compare (Meillassoux 2006, pp. 115–118).

27. (Vallicella 2010).

28 *MMK*, chapters I, XXV, translations (Garfield 1995, pp. 3, 74; Yadav 1977, p. 463). See also (Nāgārjuna 2010). Compare the entry on Mādhyamika in the *Stanford Encyclopedia of Philosophy*: "There is ... broad agreement [among Mādhyamika thinkers] that language is limited to the everyday level of understanding and that the

truth of nirvana is beyond the reach of language and of the conceptualization that makes language possible."

29. (Stcherbatsky 1978, p. 215). These translations may be peculiar to Stcherbatsky.

30. "The Principle of Identity," in (Heidegger 1969). Heidegger adds on page 26 that "what is successful and fruitful about scientific knowledge is everywhere based on something useless," namely, the identity of an object with itself.

31. (Nāgārjuna 2010, IX, 16).

32. (Corry 2008).

33. (Mazur 2008b), the most readable account I've seen of how category theory treats equality.

34. http://ncatlab.org/nlab/show/identity+type. This is actually characteristic of *intensional type theories*, of which the homotopy type theory favored by Voevodsky is one example.

35. Quoted in (Garfield and Priest 2003, p. 15).

36. golem.ph.utexas.edu/category; ncatlab.org/nlab/show/nPOV.

37. (Langlands 2013).

38. (Grosholz 2005, p. 262).

39. (Azzouni 2010, p. 91). Azzouni writes that "the word 'exist' isn't amenable to an analysis that yields this, or any other, criterion"; that the word is "isolated and criterion-transcendent." Other philosophers naturally disagree. (Burgess and Rosen 2000) is a book-length analysis of the role of nominalism in philosophy of mathematics.

40. For Udayana, see (Yadav 1977). Udayana also appealed to the cosmological (or Uncaused Cause) argument, familiar from Aristotle and Aquinas. The Descartes and Kant arguments, in the Fifth Meditation and the *Critique of Pure Reason*, respectively, are summarized in plato.stanford.edu/entries/descartes-ontological/#1, where the connection is made to Russell's theory of descriptions.

41. Matters are not so simple in fluid mechanics. The Navier-Stokes equation describes the behavior—more precisely, the pressure and velocity—of an incompressible fluid (like the Mississippi River) in two- or three-dimensional space (within or sometimes overflowing the river's banks). "Existence" in this setting comes with a price tag: existence of smooth solutions to the three-dimensional Navier-Stokes equation is another Clay Millennium Problem. The two-dimensional case was settled long ago. See (Fefferman 2000).

42. Grothendieck, letter to D. Quillen, February 19, 1983; distributed in *Pursuing Stacks* (original in English). Over the next few days, he continued this letter, having found a "simple guiding principle" after all; but the general point remains valid.

43. (Hoffman 1998, p. 54).

44. Proving or investigating special cases or consequences of the Langlands conjectures is a full-time research career for (at least) hundreds of mathematicians and increasing numbers of physicists around the world. This could serve as a textbook example of the successful research program in mathematics; but neither philosophers nor sociologists nor historians have begun writing the textbook.

45. "Human spontaneous non-demonstrative inference is not, overall, a logical process. Hypothesis formation involves the use of deductive rules, but is not totally governed by them; hypothesis confirmation is a non-logical cognitive phenomenon: it is a

by-product of the way assumptions are processed, deductively or otherwise." (Sperber and Wilson 1995, p. 69).

46. Langlands drew profound conclusions from this particular avatar relation—which was also at the heart of Wiles's proof of Fermat's Last Theorem—in his "fairy tale" paper (Langlands 1979a). Drinfel'd, one of the great mathematical prodigies of the last century, announced his first results in this direction at the very Galois-appropriate age of 20; forty years later, he is not only still alive but fantastically creative.

47. A capsule version of the ontological conundrum at the heart of this chapter: did virtual reality exist before people like Lanier invented it? Since it's essentially a collection of computer programs, I believe a mathematical Platonist would have to answer that it did exist and that it was discovered rather than invented. I find that hard to accept.

Quotations from (Lanier 2010, pp. 185, 187). For VR vs. LSD, see (Rheingold 1992, 354–55).

48. *L'"évidence" de Grothendieck n'est pas liée à la proximité de deux termes dans une chaine déductive, mais à l'effet de "naturel" lié à l'abolition de l'espace entre le symbole qui capte et le geste qui est capté*" (Châtelet 1997, p. 16).

49. (Velleman 2008, pp. 421–422).

50. Mehrtens (1990, p. 8) continues: For what he calls the Moderns, *"Wahrheit und Sinn der Texte bestimmen sich in der Arbeit an ihnen, nicht mehr in der Repräsentation der gegebnene physischen Welt, auch nicht im Bezug auf eine transzendente Ordnung. Dagegen protestiert die Gegenmoderne und sucht den 'Ur-Grund,' in dem die Mathematik 'wurzelt' und aus dem sie ihre Wahrheit und ihre Ordnung bezieht."*

51. Grothendieck left Bourbaki after a few years because "he lacked humor and had difficulty accepting Bourbaki's criticism" according to Laurent Schwartz—but also because of his conflicts with Weil, especially over the latter's refusal to admit the categorical perspective into Bourbaki. See (Corry 2008). The term "programmatic manifesto," referring to *The Architecture of Mathematics* (written by Bourbaki founder Jean Dieudonné) is in Corry (2004, p. 303).

52. *ReS*, p. 48. Grothendieck's emphasis, my translation. Bourbaki was usually not a Platonist. The full quotation is

La structure d'une chose n'est nullement une chose que nous puissions "inventer". Nous pouvons seulement la mettre à jour patiemment, humblement en faire connaissance, la "découvrir". S'il y a inventivité dans [notre] travail, et s'il nous arrive de faire oeuvre de forgeron ou d'infatigable bâtisseur, ce n'est nullement pour "façonner", ou pour "bâtir", des "structures". Celles-ci ne nous ont nullement attendues pour être, et pour être exactement ce qu'elles sont! Mais c'est pour exprimer, le plus fidèlement que nous le pouvons, ces choses que nous sommes en train de découvrir et de sonder, et cette structure réticente à se livrer, que nous essayons à tâtons, et par un langage encore balbutiant peut-être, à cerner.

Compare Pynchon's *Vineland*: "I thought [theorems] sat around, like planets, and . . . well, every now and then somebody just, you know . . . discovered one."

53. Private communication with Leila Schneps, based on her conversations with Godement. Schneps is a founder of the Grothendieck Circle: http://www.Grothendieck circle.org/.

54. See (Friedlander and Suslin 2003) for an account of Voevodsky's construction.

Others are due to Marc Levine, M. Hanamura, and Madhav Nori. The first attempt to define mixed motives rigorously was due to Annette Huber, who followed Deligne's lead and worked with their avatars (called *realizations*, though they could just as well have been *incarnations*) since the ulterior items could not be defined.

55. *Phaedo*, 79c (translation from http://www.perseus.tufts.edu/hopper/text).

56. (McLarty 2007). See Serre's comments in chapter 8, note 80.

57. (Cartier 2014, p. 18). Grothendieck's elimination of dependence on choice should not be confused with working in the paradoxical "set of all sets." Mazur (2008b) also uses a maritime metaphor: "If we are of the make-up of Frege, who relentlessly strove to rid mathematical foundations of subjectivism, we look to universal quantification as a possible method of effacing the contingent—drowning it, one might say, in the sea of all contingencies. But this doesn't work [because of Russell's paradox]."

Of the mathematicians responding to Zarca's questionnaire, 38% defined themselves as atheists, 15%, as believers, 6%, as "having faith," with the rest offering a variety of answers (Zarca 2012, p. 252).

58. (Yadav 1977, p. 452). I have been unable to find an equivalent quotation in any translation of the *Prasannapada*. For infinite regress, see *MMK* VII, 3 and the commentary in the *Prasannapada*.

59. (Graves and De Morgan 1882).

60. The Abraham interview is on page 101 of (Kirn 1991); his photo is captioned "Prof Ralph Abraham believes in mixing math and acid." Mullis's drug use is well known; his Wikipedia page attributes the comment quoted here to "BBC Horizon—Psychedelic Science—DMT, LSD, Ibogaine—Part 5."

61. (Kaiser 2012, p. vii)

62. (Cook 2009, pp. 148, 138, 58, 32; Grothendieck 1988, p. 675). I use "hooked" to translate Grothendieck's "accrocher." The French word has the connotations of addiction and appears quite often in *ReS*; for example, he describes how in 1977 he got "strongly 'hooked' on a substance of exceptional richness" (p. 319).

63. (Scharlau 2010, p. 144). Scharlau writes, and I certainly agree, that "Grothendieck's life is so unusual and singular that it does not belong only to him." The Molinos quotation is from *Guía espiritual*, Capítulo XX.

64. (Hoffman 1998, pp. 21, 22, 243).

65. More precisely, an $(\infty, 1)$-category, for which the standard text is Lurie (2009); but it's only the beginning of the story, because an endless chain of (∞, n)-categories await their founding documents. Identity becomes problematic in ∞-categories, in the sense that the usual identity has many equivalent ways to be an identity; and the equivalences have many equivalent ways to be equivalences; and so on to ∞. There's no way to say this in a few words: the definitions are "too long," never mind the proofs; Lurie's book is 925 pages long.

66. Grothendieck (1988), letter to Quillen.

67. After writing this sentence, I discovered Manin's article "Foundations as Superstructure" and Reuben Hersh's "Wings, not Foundations!" which subject the foundations metaphor to a "deconstruction" (Hersh's word) similar to that carried out in the present chapter: (Manin 2012; Hersh 2005).

68. On page 39 of *Récoltes et Sémailles*, Grothendieck places his foundations at the base of a house, but it's the only instance of its kind in the book and it's clear he's just

using the conventional metaphor in order to talk about something else. He has built a house from the foundations up but he belongs in the open air . . . *Je me sens faire partie, quant à moi, de la lignée des mathématiciens dont la vocation spontanée et la joie est de construire sans cesse des maisons nouvelles. Chemin faisant, ils ne peuvent s'empêcher d'inventer aussi et de*◇*façonner au fur et à mesure tous les outils, ustensiles, meubles et instruments requis, tant pour construire la maison depuis les **fondations** jusqu'au faîte, que pour pourvoir en abondance les futures cuisines et les futurs ateliers, et installer la maison pour y vivre et y être à l'aise. Pourtant, une fois tout posé jusqu'au dernier chêneau et au dernier tabouret, c'est rare que l'ouvrier s'attarde longuement dans ces lieux, où chaque pierre et chaque chevron porte la trace de la main qui l'a travaillé et posé. Sa place n'est pas dans la quiétude des univers tout faits, si accueillants et si harmonieux soient-ils . . . . D'autres tâches déjà l'appelant sur de nouveaux chantiers, ... Sa place est au grand air* [my emphasis].

69. Quoted in Gray (2010, p. 240).

70. In the spirit of the theory of *actants* of sociologist Bruno Latour. We return to Krishna and Arjuna in the next chapter.

71. See Minhyong Kim's comment at (Mathoverflow 2012; Lurie, 2009, p. 50; Mazur 2008b).

72. (Carter 2004; Grosholz 2007). For Voevodsky's inaccessible cardinals, see (Shulman 2011).

73. (Kronecker 2001, p. 233, 228; Azzouni 2010, pp. 98–99).

74. For Bourbaki's manifesto, see note 51. Compare "The pathway of hypotheses sometimes snakes between, sometimes bridges, various domains of mathematical research" (Grosholz 2007, p. 60), reviewing Carlo Cellucci's theory of mathematical discovery.

75. Grattan-Guinness illustrates the problem by referring to one of André Weil's historical articles: " . . . the algebraic reading of Euclid has been discredited by specialist historians in recent decades. By contrast, it is still advocated by mathematicians, such as Weil . . . who even claimed that group theory is necessary to understand Book 5 . . . and Book 7 (introducing basic properties of positive integers)!" (Grattan-Guinness 2004).

76. (Tappenden 2008). See, specifically, Tappenden's discussion of quadratic reciprocity as a special case of class field theory. In algebra, prime numbers become *prime ideals*. Quotations from pages 269–70.

77. (Tappenden 2006). Nowadays the complementary approaches of Weierstrass and Riemann are taught side by side in every mathematics department in the world.

78. The homological mirror symmetry conjecture is an active research program, rooted in part in Grothendieck's work and overlapping with a geometric version of the Langlands program. Kontsevich, one of the most influential living mathematicians (1998 Fields Medal, 2008 Crafoord Prize, 2012 Shaw Prize, 2015 Breakthrough Prize in Mathematics. . . .), is currently a professor at IHES.

79. http://en.wikipedia.org/wiki/Talk%3ABarrington_Hall_%28Berkeley,_California%29.

80. (Garfield 1995, p. 91).

81. G. G. Joseph (2011, p. 345) hints cautiously at such a link in the chapter on ancient Indian mathematics; he suggests that "zero as a concept probably predated zero as

a number by hundreds of years" and speculates that the link between the counting and spiritual senses of *shunya* may be traced as far back as the *Rig Veda*. F. Gironi (2012) addresses the theme from the standpoint of continental philosophy, especially Derrida and Badiou. See also (Goppold 2012), especially section 3.

82. So much for Hans Reichenbach's distinction between "context of discovery" and "context of justification."

83. Example (for mathematically sophisticated readers): one categorifies a ring **A** by devising a category whose objects correspond to generators of **A** and morphisms to additive or multiplicative relations.

84. (Bellah 2011, p. 9).

85. Grothendieck's Wikipedia pages in most of the languages I can read describe him as stateless; however, Cartier (2014) writes that he "consented to apply for French citizenship" after 1980. As for Erdős, "There was even an interval when he was nearly stateless: after a trip abroad, U.S. officials refused to readmit him—on the grounds that he was a security risk! Israel came to the rescue with a passport" (Hayes 1998). Quotation from Hoffman (1998, p. 29).

86. (Pinker 1997, p. 21).

87. A *topos* is one of the notions Grothendieck introduced to look at space; in the hands of F. W. Lawvere and others, topos theory has developed into an alternative to set-theoretic logic.

88. http://mathoverflow.net/questions/59520/how-true-are-theorems-proved-by-coq.

89. (Stcherbatsky 1978, p. 44; Wittgenstein 1922). Juliette Kennedy (in press) quotes a private communication by the eminent logician S. Shelah expressing much the same sentiment in a much more technical context: "Considering classical model theory as a tower, the lower floors disappear—compactness, formulas, etc. . . . the higher floors do not have formulas or anything syntactical at all."

CHAPTER 8 THE SCIENCE OF TRICKS

1. This is from the the 1979 Penguin translation by Dim Cheuk Lau; however, I have replaced Lau's "plausible men" by "clever talkers," as in the 1938 Arthur Waley vintage translation (Confucius 1979, 1938). The term corresponds to the Chinese *ning ren*, elsewhere translated as "persuaders" or "specious talkers." "Wanton" is also translated "licentious," "obscène," etc.

2. Compare with Tao (2008b) in math.CA,tricks: " . . . Tim Gowers . . . has begun another mathematical initiative, namely a 'Tricks Wiki' to act as a repository for mathematical tricks and techniques. . . . today I'd like to start by extracting some material from an old post of mine on 'Amplification, arbitrage, and the tensor power trick' . . . " Note the seepage of financial metaphors into mainstream mathematics. I'm told the "Tricki" is currently more or less dormant.

3. See, for example, "The Mythological Trickster: A Study in Psychology and Character Theory" in (Manin 2007). For "mathematical wit," see (Hoffman 1998, p. 49).

4. A tensor product trick goes unmentioned in chapter 9 but was of crucial importance nonetheless.

5. Here is the incriminating sentence, from my article (Harris 2009): "The principal innovation is a tensor product trick that converts an odd-dimensional representation to an even-dimensional representation." Calling it "Harris' tensor product trick" was the idea of (Barnet-Lamb et al. 2014).

6. I'm still waiting.

7. For the trickster as creator and crosser of boundaries, figure at the crossroads, see (Hyde 1998, 2010, pp. 6, 119, 220–225; Gates 1988). Compare *"Kunstgriffe verblüffen, indem sie scheinbar verschiedene Aspekte miteinander verbinden und daraus Nutzen ziehen"* from (Rump 2011). The article mentions the famous "Kunstgriff von Rabinow-itsch" that appeared in van der Waerden's classic 1930 algebra text.

8. See (Dante 1981). I follow Barry Mazur's translation in (Mazur 2004a) but added my emphasis.

9. (Langlands 1979b). Grothendieck referred to himself as a *builder* (*bâtisseur*). The description in *Récoltes et Sémailles*, Prelude, section 2.5, deserves to be quoted in full. *"Il n'a que deux mains comme tout le monde, c'est sûr—mais deux mains qui à chaque moment devinent ce qu'elles ont à faire, qui ne répugnent ni aux plus grosses besognes, ni aux plus délicates, et qui jamais ne se lassent de faire et de refaire con-naissance de ces choses innombrables qui les appellent sans cesse à les connaître. Deux mains c'est peu, peut-être, car le Monde est infini. Jamais elles ne l'épuiseront! Et pourtant, deux mains, c'est beaucoup. . . ."*

10. Bas Edixhoven and Frans Oort, personal communications.

11. According to Wikipedia, the word is also used for acrobatic stunts and sec-ondarily for magic tricks, and a tryuk is typically "dangerous or unfeasible for the un-trained." http://ru.wikipedia.org/wiki/трюк. The primary word for magic tricks is *fokus*.

12. Compare *tour*, derived from "to turn," with the epithet *polutropos*, "of many turns," applied to Odysseus. Odysseus is also called *polumekhanos*, usually translated "of many devices" but sometimes "of many tricks."

13. " *Kunstgriffe gehen somit einer Theorie voraus, sie greifen ins noch Un-bekannte, verbinden das scheinbar Getrennte, damit dieses durch weiteres Nachdenken seinen natürlichen Platz in der allgemeinen Theorie erhält und damit bekannt wird*" (Rump 2011).

Michael Atiyah sees *trick* as the first way station, followed by *technique* and *method*, along the path "from problem to theory." "[I]n order to solve a problem you have got to have a clever idea, some kind of trick. If it is a sufficiently good trick and there are enough problems of a similar type you go on and develop this trick into a technique. If there is a large number of problems of this kind, you then have a method and finally if you have a very wide range you have a theory; this is the process of evolution from problem to theory" (Atiyah 1989, pp. 21116). In this connection, I note that Sanjoy Ma-hajan paraphrases George Pólya to the effect that "A tool . . . is a trick I use twice," in (Mahajan 2010). Tools and techniques are discussed later. See also the long discussion at http://mathoverflow.net/questions/48248/what-do-named-tricks-share.

14. (Hurston 1943, p. 452).

15. A complete translation is available online at http://ebooks.adelaide.edu.au/s /schopenhauer/arthur/controversy/chapter3.html. In Plato's *Cratylus*, 409 D, Socrates says "Now watch out for my special trick (*mekhane*) which I have for everything I can't solve!" The translation is from (Huizinga 1950, p. 150).

16. (Gheury 1909). Quotation marks are as in the original.

17. (O'Toole 1940). The Mathematical Association of America is the professional organization of mathematics teachers, whereas the American Mathematical Society is the professional association of mathematical researchers. Of course, there is a significant overlap.

18. (AMS 1962, p. 2).

19. In his second-most highly rated MathOverflow post, entitled "Proofs that require fundamentally new ways of thinking," Tim Gowers list six examples of what might be called tricks that have since been incorporated into mainstream mathematics; he finds them interesting because "they are the kinds of arguments where it is tempting to say that human intelligence was necessary for them to have been discovered." http://mathoverflow.net/questions/48771/proofs-that-require-fundamentally-new-ways-of-thinking.

20. (Restivo 1988; G. A. Miller 1908).

21. Unless the first computing machine was constructed by Wilhelm Schickard, as he described in his letter to Kepler dated September 20, 1623: "I constructed a machine [*machinam extruxi*], which includes eleven full and six partial pinion-wheels, which can calculate automatically, to add, subtract, multiply and divide." Text at http://history-computer.com/MechanicalCalculators/Pioneers/Schickard.html, in the original Latin and in English translation.

22. "Peter Sarnak, Professor in the School of Mathematics, compares the fundamental lemma [proved by B. C. Ngô; see chapter 2] to a screwdriver, functoriality to opening a screw, and the Langlands program to the big machine working to reveal the underlying structure of automorphic forms" (Kottwitz 2010).

23. Plutarch, *Table Talk*, VIII 2.1 [718e]; also *Life of Marcellus*, XIV, 5–6 (Bernadotte Perrin translation). Plato complained that the mathematicians in question had "descended to the things of sense" rather than "abstract thought" and moreover used "objects which required much mean and manual labour." This objection would in principle not apply to the canonical delooping machine or any other construction that is only metaphorically mechanical.

24. From a letter to Schumacher translated in (Ferreirós 2007, p. 216).

25. William Casselman, private communication, September 3, 2003. Casselman, a professor at the University of British Columbia, was one of the earliest collaborators on the Langlands program. The blurb for a planned semester program at the MSRI in Berkeley illustrates the continued appeal of the term *Langlands philosophy* (www.msri.org/web/msri/scientific/programs/show/-/event/Pm8951).

26. See Deligne's talk "*L'influence de la philosophie des motifs*," January 15, 2009, announced at http://www.ihes.fr/jsp/site/Portal.jsp?document_id=1661&portlet_id=999.

27. "By "yoga" he meant a unifying point of view, a lead in the search for concepts and proofs, a method one could use repeatedly." Pierre Cartier, February 25, 2011 (private communication).

28. (E. T. Bell 1944).

29. (Dumézil 1992, p. 101). In this passage, Dumézil also insists that tripartite organization is absent in the mythology of non-Indo-European cultures.

30. (Langlands 2007). Langlands cautions that "still unforeseeable difficulties . . . will make great demands on the inventive powers of analytic number theorists."

31. Or "of many tricks"—see note 12.

32. To save the Pandava brothers' army from slaughter at the hands of the invincible warrior Drona, Krishna devised a ruse to convince Drona that his beloved son Aśvatthāma had been killed. Bhima (the brother associated with brute force, a *kshatriya* in Dumézil's reading) killed an elephant named Aśvatthāma and shouted that Aśvatthāma was dead. Drona asked Yudishthira (the brother associated with the brahmin function) whether this was indeed true. Yudishthira, as Drona knew, was incapable of lying, but at Krishna's instigation, he replied "Aśvatthāma the elephant has been killed," pronouncing the word "elephant" inaudibly (or drowned out by Krishna's conch, in some versions). Brokenhearted at the news, Drona left the battlefield and went straight to heaven, as Krishna had anticipated.

Deceit is also a recurring feature in Ferdowsi's tenth-century epic of the Persian kings.

33. (Cook 2009, p. 138). Compare (Littlewood 1986, p. 195): a mathematician "has to be completely honest in his work, not from any superior morality, but because he simply cannot get away with a fake."

34. The word *Kunstgriff* appears on page 4 of *Die letzten Gründe der Hohern Analysis* by Joseph Nürnberger, 1815. It seems to be used in a slightly different sense in *Anfangsgründe der Mathematik: Anfangsgründe der Analysis . . .* , Volume 3, Issue 1, published by Abraham Gotthelf Kästner in 1767. . For the use of the word *trick* in English, see also Sadler (1773).

35. For *ta-mar*, see Friberg (2007). For Hellenistic mathematics, see Netz (2009, pp. 78, 67, 17ff, 54, 77–78). The *Lilavati* is (Brahmagupta and Colebrook 1817).

Netz identifies three characteristic "ludic" features of Hellenistic mathematics: "mosaic composition"—"the juxtaposition of apparently unrelated threads that, put together, delight with the surprise of a fruitful combination, or startle with the shock of incongruity" (p. 117)—"narrative surprise," and "generic experimentation" (including the so called carnival of calculation). We have seen that the first two (but not the third) of these features also help to identify contemporary mathematical tricks, but it seems to me there is an important difference: the latter don't call attention to themselves as tricks—as vehicles for entertainment or display—but primarily function as shortcuts on the way to a defined goal, possibly destined to be integrated in normal mathematics. The parallels and differences between Hellenistic and contemporary mathematics deserve further analysis.

36. From the verb root *ḥ-w-l*. Hans Wehr's dictionary lists the following meanings in Form VIII: "to employ artful means, resort to tricks, use stratagems. . . ." Also "to deceive, beguile, dupe, cheat, outwit, outsmart . . . or achieve by artful means, by tricks."

Jens Høyrup translates *'ilm al-ḥiyal* by "science of ingenuities" (following Gherard de Cremona's *scientia ingeniorum*—see below—and writes "the term *ingenio* can be read in its double Latin sense, as 'cleverness' and as 'instrument' (in agreement also with the Arabic text)—we might speak of the discipline as 'engineering' or as 'applied theoretical mathematics'" (Høyrup 1988). It's possible to write the tricks out of medieval Islamic mathematics. But why would one want to?

37. From the MacTutor History of Mathematics Archive, http://www-history.mcs .st-andrews.ac.uk/Biographies/Banu_Musa.html. Al-Khwarizmi's algebra also placed magnitudes and quantities on an equal footing: see chapter β.

38. Al-Fârâbî actually referred to *al-jabr wa al-muqabala*, from the name of al-Khwarizmi's ninth-century treatise. And it should be noted that *al-jabr* originally referred to the (mechanical?) operation of bonesetting. For Ibn Sina, see (Rashed 2011, p. 753). For *mekhane* see note 12. Most reproductions of the al-Khayyam quotation refer to the translation by Amir-Moez (1963); I have instead used the English translation by (Rashed and Vahabzadeh 2000, p. 171) but follow Amir-Moez in writing "trick" for *ḥila*. In their French translation, Rashed and Vahabzadeh write "artifice." This is, in any case, the *only* place the word *ḥila* appears in al-Khayyam's mathematical papers.

39. See http://plato.stanford.edu/entries/archytas/. Pierre Hadot, in his *Veil of Isis* (see chapter 11), points out that "[t]he idea of trickery—and ultimately, of violence—appears in the word 'mechanics,' since *mekhane* signifies 'trick.'" (Hadot 2006, p. 101)

40. G. de Cremona, *De Scientiis*, reprinted in (al-Fârâbî 2005; de Gandt 2000); for *Kunstgriff* as translation of *ingenium*, see (Glowatzki and Göttsche 1990, p. 23; Kausler 1840, p. 546).

Descartes defines *ingenium* in this passage from Rule XII: "In accordance with these diverse functions the same power is called now pure intellect, now imagination, now memory, now sense; and it is properly called mind (*ingenium*) when it is either forming new ideas in the phantasy or attending to those already formed." The roots of Descartes' use of the term are explored in (Robinet 1996) and (Sepper 1996). In contrast to de Gandt, neither Robinet nor Sepper mentions that *ingenium* is a translation of *mekhane*. The word *ingenium* has so many meanings in postcartesian philosophy that its association with *Kunstgriff* can hardly be considered primary.

The questions of where (or whether) to draw the line between art and technique or between the artist and the artisan dominate many of the aesthetic debates of the nineteenth and twentieth centuries, as reconstructed in (Rancière 2011).

41. Private communications of Kai-Wen Lan and T. Yoshida, September 2011.

42. *J. reine angew. Mathematik* **10** (1833): 154–166. The original text is "*Bei allen im Vorigen gezeigten Verwandlungen der Kettenbrüche in andere, wurde der Kunstgriff angewandt, dass man einen Theil des Kettenbruchs als summirt betrachtete, diese Summe durch einen Buchstaben ausdrückte, und durch die Verbindung dieses Buchstabens mit den übrigen Gliedern des Kettenbruchs neue Ausdrücke fand.*"

Apart from publishing the earliest *Kunstgriff* in *Crelle's Journal*, M. A. Stern has the distinction of being the first unconverted Jew to be named to a professorship at any German university, in any subject (Göttingen, in 1859). The reader may have noticed that a number of the tricks mentioned in this chapter are associated with Jewish authors, including the author of these lines. No doubt this is a coincidence.

43. One of the most apparently deceitful of mathematical tricks is called, depending on the context, either the *Eilenberg swindle* (after Samuel Eilenberg, one of the creators of category theory) or the *Mazur swindle* (after my thesis adviser Barry Mazur). Though highly counterintuitive, the trick is logically perfectly legitimate.

44. Oresme's comment is on page 188 of his translation of Aristotle. I thank Jean-Benoît Bost for directing my attention to Rey's dictionary and for providing these references, from 1967 and 1908 respectively: "*Astucieux est en train de perdre la nuance de malice et de duplicité qui y était attachée. Il signifie ingénieux, habile, perspicace*" [*Astucieux is losing the nuance of malice and duplicity with which it was associated. It means ingenious, clever, perspicacious.*] "*On dit: avec astuce pour « avec ingéni-*

osité," et aussi: dire une astuce, faire une astuce, où astuce désigne « quelque chose d'ingénieux," [*astuce* designates "something ingenious"] References online at http://www.cnrtl.fr/lexicographie/astucieux; http://www.cnrtl.fr/lexicographie/astuce.

45. The Greek word *kataskeue* for geometric construction, also used in connection with building, is not in Euclid but was used by later geometers, including Pappus.

46. On page 7 of (Lipschitz 1870), my translation. At the turn of the nineteenth century, the poet Novalis had already written "*Am Ende ist die ganze Mathemat[ik] gar keine besondre Wissenschaft—sondern nur ein allgem[ein] wissenschaftliches Werckzeug*" [in the end the whole of mathematics is not at all a special science—but rather a general scientific tool].

47. For *Technik*, see the 1835 book *Die darstellende Geometrie:(La géométrie déscriptive)* by S. Haindl (six results on Google books).

48. The *Jahrbuch* published abstracts and reviews of mathematical articles from 1868 through 1942; it was succeeded by *Zentralblatt für Mathematik und ihre Grenzgebiete* (1931—), which has made the *Jahrbuch*'s archives available on its online Web site, as well as *Mathematical Reviews*, published by the AMS since 1940, and the Russian *Referativnyi Zhurnal*, which began publication in the 1950s. Dieudonné's review is MR0179805 (31 #4047a), b.

49. (De Sigorgne, 1739–1741; Darboux 1866, the word *attaqué* is on p. 117; Dixon 1882). The word had already appeared in *The Analyst*—predecessor of the extremely distinguished *Annals of Mathematics*—in an 1876 article purporting to solve the equation of the fifth degree, which Abel and Galois had proved impossible.

50. (Bourbaki 1948, esp. pp. 40–41; 1950).

51. (Bourbaki 1950, p. 223).

52. (Gates 1988, p. 6). Analyzing Transylvanian trickster stories, L. Piasere sees the figure mediating a transition between potential and actual infinity, in Aristotle's sense: "*Ora, il trickster sembra proporsi come una figura dalle possibilità indefinitamente aperte . . . e quindi sembra essere più dal lato dell'infinito potenziale, dell'infinito aperto; ma il fatto di riunire su di sé, cioè nella propria identità, le infinite alterità concepibili, lo pone contemporaneamente dal lato dell'infinito attuale, dell'infinito chiuso*" (Piasere 2009). Dean A. Miller (2000, p. 243) sees the trickster-hero as "intermediary between [the supernatural] zone and the zone of heroic human action." Lewis Hyde (1998, 2010) points out that the trickster is not the Devil; but neither is Mephistopheles.

53. I've repeatedly drawn attention to clichés—you'll notice I just slipped one in myself—and this is probably the right place to explain that the abundant use of clichés in books about mathematics for the general public is not a response to contracts in which authors are paid by the word but rather express a judgment about what the audience can be expected to know. In a book about classical music pitched at the level of this one, you will not read that Bach had a lot of children or that Mozart was a prodigy or that Beethoven was deaf when he composed his Ninth Symphony, because that information belongs to general culture in a way that the irrationality of the square root of 2 or Cantor's diagonalization argument does not.

54. See "Aesthetics of Music" in W. Apel, *Harvard Dictionary of Music*, Harvard University Press (1972), p. 14. Just sticking to number theorists, Harvard's Noam Elk-

ies is a composer of (serious) classical music; Princeton's Manjul Bhargava (Fields Medal 2014) plays tabla at (close to) professional level.

The affinity mathematicians feel for music is not a myth. B. Zarca's sociological survey (cited in chapter 2) finds that, asked to choose from nine "worlds" those to which they felt close, mathematicians in France systematically placed the "world of music" at the top of their list; pure mathematicians were even more categorical. Scientists in allied fields felt closest to social sciences and "industry and finance." The "world of media" was at the bottom of everyone's list but was especially severely rejected by pure mathematicians (Zarca 2012, pp. 294–95). Honesty compels me to mention that the Zarca study did detect a significant preference for classical music over other genres among mathematicians in France, and this preference is more marked among mathematicians than among other scientists (p. 261).

55. (Christensen 1987).

56. "As the study of axioms eliminates the idea that axioms are something absolute, conceiving them instead as free propositions of the human mind, just so would this musical theory free us from the concept of major/minor tonality [. . .] as an irrevocable law of nature . . . Whereas geometric axioms are sufficiently justified if their combinations prove them to be both independent of and compatible with each other, the accuracy of musical axioms can be proved exclusively by their fitness for musical practical use" (Krenek 1939, pp. 206–7).

57. (Xenakis 1963).

58. David Lewin, Dmitri Tymoczko. http://dmitri.tymoczko.com/geometry-of -music.html, Guerino Mazzola. Since the year 2000, there has also been a regular seminar called MAMUPHI (Mathématiques, Musique, Philosophie) at the École Normale Supérieure.

59. (Dreyfus 1996, p. 9). C.P.E. Bach was famously quoted by his contemporary J. P. Kirnberger: "You can say out loud that my principles and those of my late father are anti-Rameau" (Kirnberger 1771).

60. (Berlioz 1836; Saint-Saëns 1900; Blackburn 1900). I thank Laurence Dreyfus for the Brahms reference.

61. (Schenker 1997).

62. (Richardson 2011; Hewitt 2011; Born 1995, p. 96).

63. (Lippman 1990, p. 92).

64. The book (Assayag et al. 2002) is also overwhelmingly concerned with classical music. I neglect jazz in this chapter because I have been unable to find any striking quotations linking jazz with mathematics. See also note 97.

65. Joan Wallach Scott notes that some of the founders of the American Association of University Professors considered "a faculty member's demeanor" important for establishing and maintaining legitimacy. She quotes the AAUP Committee on Academic Freedom and Academic Tenure, which argued in 1915 that academic responsibility presupposed "dignity, courtesy, and temperateness of language," "a peculiar obligation to avoid hasty or unverified or exaggerated statements and to refrain from intemperate or sensational modes of expression" (Scott 2009).

66. Pierre Bourdieu's study of *distinction* emphasizes the role of shared cultural values, including tastes in music and the arts, necessarily inclined to seriousness, in

qualifying *Homo Academicus* as the personification of legitimate authority: "nothing more clearly affirms one's 'class,' nothing more infallibly classifies, than tastes in music." See the long discussion of music's special role in defining the "aristocracy of culture" in *Distinction* (trans. Richard Nice), Harvard University Press (1984), pp. 18–19. Bourdieu also examines the relations between "cultural capital" and "academic capital" in *Homo Academicus* and *The State Nobility*.

67. The very word *culture* with its present associations only came into use in the English language in the mid–nineteenth century. Raymond Williams points out that what Coleridge called "cultivation, the harmonious development of those qualities and faculties that characterize our humanity" had previously been "an ideal of personality—a personal qualification for participation in polite society"; in the mid-nineteenth century, it had "to be redefined, as a condition on which society as a whole depended." For Matthew Arnold, culture was "the study of perfection" and the guardians of the standards of perfection, the condition of "general cultivation," were what Coleridge called the Clerisy, including "the sages and professors of . . . all the so-called liberal arts and sciences" (Williams 1958, pp. 61–64, 115). The Coleridge quotation is from *On the Constitution of Church and State,* V. The place of scientists and mathematicians in culture varied—Arnold wrote in his *Philistinism in England and America*, in connection with the mathematician Joseph Sylvester, that "for the majority of mankind a little of mathematics, even, goes a long way." Reprinted in volume X of *The Complete Prose Works of Matthew Arnold*, University of Michigan Press (1974).

68. Stephen Shapin argues that the credibility of research in early modern times presupposed the conformity of the researcher to the ideal of the *gentleman* whose word could be trusted. Thus, in England, Henry Peachum wrote in 1622 that "we ought to give credit to a noble or gentleman before any of the inferior sort." "Categories of people—women and the vulgar in particular—in whom [the higher intellectual] faculties were poorly developed might be . . . constitutionally prone to undisciplined and inaccurate perceptions." "The relevant maxim . . . in the civil conversations of early modern gentlemanly society, was 'believe those whose manner inspires confidence.'" Although the gentlemanly repertoire was alien to enlightenment France, manner mattered there as well: in the *éloges* composed by Fontenelle, perpetual secretary of the French Académie des Sciences, and by his successors, the "eighteenth century men of science were described as embodiments of Stoic fortitude and self-denial." The *Encyclopédie* characterized the ideal "philosopher" as "an *honnête homme*, polite, civil, and autonomous" who "serves Truth alone." (Shapin 1994, pp. 69, 77, 221; 2008, pp. 36–37).

69. ". . . a character who attracts our attention by his balanced pose and his ugliness." See http://www.diariodecultura.com.ar/web/news!get.action;jsessionid=B23 A495DF6437ED023AA643237DC57D9?news.id=7119. In the *catalogue raisonné* of Luca Giordano's works (Ferrari and Scavizzi 1992), this painting is called *Il filosofo*. For the painting, on display at the Museo Nacional de Bellas Artes in Buenos Aires, see http://www.mnba.gob.ar/coleccion/obra/2858. The commentary at the Buenos Aires Web site suggests that the inscription may be a representation of the "proto-Hebrew" believed to have been the universal language before Babel.

70. (Thorndike, 1923–58, Chapter LXI).

71. For Padova, (see Favaro 1888); for Galileo's astrology, see (Favaro 1881, p. 4).

Kepler's *De Fundamentis Astrologiae Certioribus* expresses some skepticism regarding astrology—"If [astrologers] at times do tell the truth, it ought to be attributed to luck, yet more frequently and commonly it is thought that this comes from some higher and occult instinct"—but like Brahe, he derived much of his income from horoscopes. The IBM timeline entries for Galileo and Kepler do not mention horoscopes.

72. From the Faustbuch 1580 (my emphasis). The English translation of this passage [online at (Wulfman 2001)] makes Faustus, incorrectly in my opinion, into a mathematician: ". . . being delighted . . . so well [in those diuelish arts] that he studied day and night therein: in so much that hee could not abide to be called Doctor of Diuinitie, but waxed a wordly man, and named himselfe an Astrologian, and a Mathematician . . ."

73. (Saiber 2003).

74. (Evenden and Freeman 2011, pp. 265–66). The Dee quotation, dated 1570, is taken from (Pimm and Sinclair 2006, chapter α). For his alchemical exploits, see the introduction to (Josten 1964, p. 90). John Dee is the hero of the semirock opera *Dr. Dee* and was also the alter ego of *Doctor Destiny*, a supervillain from the Justice League of America, known for creating the *Materioptikon*, "a device which allowed him to create reality from the fabric of dreams." (Source: Wikipedia entry on Doctor Destiny.)

75. (Josten 1964, p. 85). Huizinga, perhaps thinking of characters like Dee or Cardano, writes that "modern science, so long as it adheres to the strict demands of accuracy and veracity, is far less liable to fall into play as we have defined it, than was the case . . . right up to the Renaissance, when scientific thought and method showed unmistakable play-characteristics" (Huizinga 1950, p. 204).

76. (Weyl 1935, 1939; Hardy 2012, Chapter 11). "Very serious man" is from Peter Pesic's introduction to Weyl (2009), an edition of Weyl's writings on philosophy, mathematics and physics; the quotation on Newton's trick is on page 45 of the same book. See note 13 for similar comments by Pólya and Atiyah.

Høyrup, writing with Al-Fârâbî in mind, sees Jordanus de Nemore's thirteenth-century work as a stage in the legitimation of algebra, along very similar lines: "the *De numeris datis* transforms a mathematically dubious *ingenium* into a genuine piece of mathematical *theory*" (Høyrup 1988, unpublished, Høyrup's emphasis; see note 36).

77. (Hyde 1998, p. 100).

78. (Jackson 2004, p. 1203). Comparing Serre and Grothendieck, Leila Schneps finds that "Serre was the more open-minded of the two; any proof of a good theorem, whatever the style, was liable to enchant him, whereas obtaining even good results 'the wrong way'—using clever tricks to get around deep theoretical obstacles—could infuriate Grothendieck." [L. Schneps, "A biographical reading of the Grothendieck-Serre correspondence," p. 18; a short version was published in *Mathematical Intelligencer*, **29** (2007). See note 80].

Grothendieck banned tricks from his own work as well. Armand Borel mentioned in a lecture in Mumbai that Grothendieck refused to publish his proof of what is now known as the Grothendieck-Riemann-Roch theorem because (in positive characteristic) he made use of a trick; the proof was instead published by Borel and Serre. (I thank T. N. Venkataramana, who attended Borel's lecture, for this information.)

79. "Proof, in its best instances, increases understanding by revealing the heart of the matter" (Davis and Hersh 1981, p. 151).

80. (Langlands 2007). Serre briefly made a similar point in a 1985 letter to Grothendieck: "I know well that the very idea of "getting around a difficulty" [*contourner une difficulté*] is alien to you—and perhaps that's what you find the most shocking in Deligne's work ([an] example: in his proof of the Weil conjectures, he "gets around" the "standard conjectures"—that shocks you, but it delights me [*cela me ravit*])." (Colmez and Schneps 2003, letter of July 23, 1985).

Zarca found that "19% of mathematicians associate the idea of a trick [ruse] with the notion of an elegant proof (and none the opposite)." Zarca's findings on elegant versus natural proofs, analyzed on pages 267–72 of (Zarca 2012), deserve a longer discussion than we can provide here.

81. "*Was gute Kunst ist, läßt man sich nicht vom Kaiser und dem in ihm repräsentierten Volk sagen, auch wenn er die Ankäufe der großen Museen kontrolliert. Was gute Mathematik ist, läßt man sich nicht von Ingenieuren, Lehrern und Philosophen vorschreiben, die sich auf Prinzipien nach Art des Kaisertums berufen. Als Besonderung geht in die Gemeinsamkeit der Beziehung Profession-Klientel hier das "Schöpferische" ein, mit dem für die Mathematik in Anlehnung an die Kunst der Standort im kulturellen System markiert, der Wille zum Neuen bezeichnet und der Wert der eigenen Arbeit beschrieben wird.*" All quotations from (Mehrtens 1990, pp. 515, 543). In his best-selling novel *Die Vermessung der Welt* [*Measuring the World*[, Daniel Kehlmann has C. F. Gauss complaining (anachronistically?) 100 years before Liebmann about the obligation to respect court etiquette in order to obtain funding.

82. I thank Vladimir Tasic for this observation and for the reference to (Eagleton 1996, pp. 24–28), from which all quotations in this paragraph are drawn.

83. "*Hijo, esto de ser ladrón no es arte mecánica sino liberal.*" The Gauss character in Kehlmann (2005) has in common with Till Eulenspiegel a disdain for hypocrisy and a rejection of social decorum, sometimes taking extreme forms. Since Gauss and Eulenspiegel are the two most famous individuals associated with the city of Braunschweig (Brunswick), this may not be a coincidence. The Forsyth quotation is reproduced in (Heard 2004, p. 250).

On the other hand, one female colleague reminded me after reading this section that the freedom not to aspire to being taken seriously is a luxury not available in equal measure to all members of the profession.

For the transition from "rude and tumultuous mechanic persons" to the "mechanical philosophy" of the Royal Society, see (Hill 1975, pp. 255–60). The relation to tricks and magic, if any, is not a simple one.

84. Regarding finance, D. Steinsaltz, reviewing Patterson's *The Quants*, writes that an investment bank's "team of MIT Ph.D.s is a token of seriousness, like the Picasso in the lobby, the marble columns, and the expensive wristwatch" (Steinsaltz 2011, p. 703).

But I should make it clear that mathematics is hard work and that the value system of mathematics, if not that of Powerful Beings, rates this very highly. This must be why "work" occurs in G. H. Hardy's Apology as frequently as "beauty" and "beautiful" combined, though it's for the latter that his book is known. Tricks thus conflict with seriousness as conceived by Powerful Beings as well as by the value system of mathematics. But finding the right trick is also hard work!

85. In fact I did. From http://www.math.utah.edu/~ptrapa/finalreport/linernotes.html:

[I]n 1998 . . . several bored IAS members concocted a radical plan to break the grip of academic tedium. What the staid Institute needed, they believed, was a red-lining, full-blown, hard-driving Rock band . . . The band conspirators thought that . . . "music in the popular idiom" would complement [musician-in-residence Robert] Taub's classical menu . . . So the Institute's first (and, no doubt, last) band was born. It took the name "Do Not Erase". . . .

At least three of the band members were mathematicians—representation theorists, in fact. See also note 90.

The lyrics are from *I Ain't No Joke* by Eric B. and Rakim.

86. Space limitations prevent me from exploring mathematical music in other languages—for example, the ranchera *Las Matemáticas* by *Los Alteños de la Sierra*, the tango *El Algebrista, Matemáticas* by the Spanish MC ToteKing, *Per ogni matematico* by the Italian folk singer Angelo Branduardi, and so on.

87. *La Tribu*, "the tribe," was and perhaps still is the name of the Bourbaki association's internal bulletin. See (Corry 2004, p. 293).

88. http://genealogy.math.ndsu.nodak.edu/.

89. The two kinship systems are not quite independent. There is not exactly an incest prohibition: it is perfectly acceptable to collaborate with your offspring, since the normal result is a published article rather than another descendent, and indeed the shortest path between me and Toby Gee includes an article he wrote with his "grandfather," Richard Taylor. But it is considered poor form for an adviser to cosign a student's first published article, since that might create an impression of dependence, harming the student's career prospects.

90. http://www.spinner.com/2011/06/06/battles-gloss-drop/. *Atlas* is online at http://www.youtube.com/watch?v=IpGp-22t0lU. Another example: When at 24 the "cool guy," Peter Scholze, became the youngest professor in Germany with a *Lehrstuhl* (named chair) at the University of Bonn—in part for his work on the Langlands program—his high school teacher reminisced in the press: "Peter is not a typical mathematician . . . He is popular, involved, and played bass guitar in a band" (Harmsen 2012).

91. The Patti and Fred Smith lyrics are from *It Takes Time* on the soundtrack album from *Until the End of the World* by Wim Wenders. The Grebenshchikov quotation is from http://www.aquarium.ru/misc/aerostat/aerostat137.html, downloaded February 17, 2012, and translated (from a different site) on September 15, 2011. For Lynch, see (Nochimson 1997). Other rock numbers expressing a skeptical attitude to mathematics include *Algebra* by Soul Hooligan, *Algernon's Awfully Good at Algebra* by Malcolm McLaren, and *In Real Life There Is No Algebra* by Beardfish. More recently, Lynch and Smith claimed an abiding interest in mathematics when they teamed up with geometer Misha Gromov for the exhibition *Mathematics: A Beautiful Elsewhere* at the Fondation Cartier in Paris.

92. And another track entitled "Who taught you Math?" under his own name. Hood once reminded "the younger generation" to "remember it's all about the rhythm . . . and not to get so serious and so caught up in the hype of minimalism" (2Sheep4Coke 2009).

93. "Because when I grew up in the Bronx in the Eighties . . . there was a part of rap music where guys would battle each other on the streets, and sometimes the battling . . .

would be talking about science, for example, and they even called it *droppin' science*, like Eric B and Rakim" (Stephon Alexander 2008).

The term "dropping mathematics" was also common, according to Alexander (private communication). See (Enders 2009, p. 84): "to this day, a lot of us old Gees still be dropping mathematics." No doubt this is due in part to the influence of the *Nation of Gods and Earths*, also known as the *Five Percenters*, an offshoot of the Nation of Islam. The Five Percent Nation's cosmological system centers on *Supreme Mathematics* and the *Supreme Alphabet*. Rakim was only one of many hip-hop leaders to be a Five Percenter. For an early report on Five Percenter influence on hip-hop, see (Ahearn 1991); see also (Gibbs 2003). This is not intended as an explanation of the phenomenon under discussion: it should be obvious that a creative community predisposed to see mathematics as antithetical to their art would have rejected the message of Supreme Mathematics.

94. Lyrics copied from http://rapgenius.com/Mos-def-mathematics-lyrics, where you can also read that "Rap—like physics—is fundamentally a mathematical discipline."

95. Also known as Ronald Maurice Bean. Allah Mathematics, like other members of the Wu-Tang Clan, has been associated with the Five Percent Nation. The quotation is online at http://wu-international.com/Killabeez/wuelements_pro.htm.

96. Mos Def's *Mathematics* lyrics mentioned in the previous note are mainly devoted to quantifying this marginalization. A longer quotation from the same song opens the article (Terry 2011).

Mathieu Guillien, whose book (Guillien 2014) focuses on the minimal techno of Detroit and specifically the music of Robert Hood, agrees (private communication): "Rock is the music of a white middle class that can permit itself to criticize an education to which it had access. The politics of rap have identified education as one of the things the white establishment has denied the Afro-American population, and it seems to me that for that reason mathematics is still perceived with respect." It's probably no accident that a *Washington Post* staff writer, wondering why Hollywood has yet to give the civil rights movement the treatment it deserves, proposed Mos Def for the role of Robert Moses, legendary organizer of the 1964 Mississippi Freedom Summer. Moses used his 1982 MacArthur Fellowship to found the Algebra Project, whose motto is "Math literacy is the key to 21st century citizenship" (Hornaday 2007).

97. Quotations respectively from (Gates 2010) and from Run-DMC, "It's Tricky." Readers who suspect that the point of this chapter is to suggest that improvisation, rather than rule following, is the essence of mathematics, may be surprised that *jazz* has been omitted from the discussion. Here is an inadequate answer. On the one hand, no one these days (Theodor Adorno notwithstanding) denies that jazz is serious music, and the exclusion of jazz from the classical canon looks increasingly arbitrary and untenable. On the other hand, the little I have read suggests that jazz musicians and classical musicians feel much the same way about mathematics. There are exceptions, however: mathematical physicist and jazz saxophonist Stephon Alexander draws astonishing parallels between John Coltrane's chord progressions in *Giant Steps* and the principles of galaxy formation (Stephon Alexander 2012).

98. (Williams 1958, p. 306).

99. Boulez, in (Born 1995, p. 139); Leavis, quoted in (Williams 1958, p. 253).

100. Arthur Waley explains in the notes to his Vintage translation that "ordinary as

opposed to serious-minded people had the same feelings" to the music Confucius approved "as they have towards our own classical music today. Waley cites the following passage from the *Liji* (*Book of Rites*): "'How is it,' [asked] the Prince of Wei . . . 'that when I sit listening to old music . . . I am all the time in terror of dropping off asleep; whereas when I listen to the tunes of Cheng and Wei, I never feel the least tired?'" (Confucius 1938, p. 250).

101. (Ferreirós 2007, p. 220); Russell, *Mysticism and Logic*; (Hardy 2012).

102. See *Meno*, 86b and note 55 to the previous chapter.

103. I once heard Serre claim at the beginning of a lecture that nothing is easier than to make an auditorium full of mathematicians laugh; within two minutes he had proved his claim. For Serre, see (Huet 2003); for Villani, see (Delesalle 2011) and http://math.univ-lyon1.fr/~villani/cv.html.

Chapter 28 of (Villani 2012), with nearly 10 pages detailing Villani's eclectic listening habits, should put to rest any naive assumption that mathematicians are naturally inclined to prefer classical to popular music, or vice versa. French pop singer Catherine Ribeiro is a special favorite of Villani.

104. (Dreyfus 1996, p. 31).

105. F. Schiller, *On the Aesthetic Education of Man* (Gopnik 2009, p. 14; Pinker 1997, p. 525). Now, cognitivists use the word *spandrel* instead of *cheesecake*, referring to an architectural detail that appears not by design but through the interaction of two or more structural features and, by extension (following a well-known article by S. J. Gould and R. Lewontin), an indirect result of natural selection that is not itself selected by the environment.

To say that mathematics is the structure of music is only a way of saying that mathematics is never far when we talk about structure. If mathematical structure is more visible in music than in other fine arts it's because there is no other distracting source of structure; this is what it means to say that music and mathematics are both abstract.

106. (Oort 2014). Oort is translating Serre's "*Cela me ravit*" (see note 80).

107. (Levi-Strauss 1969, pp. 15–16; Elias 1993, p. 131).

108. Bruce Bethke, who invented the word *cyberpunk*, traces it to "cross-pollination going on between the synthesizer circuit hackers (that is, we totally cool musician types) and the homebrew computer hackers (a.k.a., those terminally unhip nerds)" (Bethke 1997).

109. (Rucker 1982 (one of the first cyberpunk novels), chapter 4). The speaker in this case is an artificial intelligence. Rucker himself is a professional mathematician and computer scientist. For mathematics and anarchism in Pynchon, see (Engelhardt 2013).

110. In prehacker science fiction, the logic of the confrontation was different: "Time and again when confronted with the superior strength and knowledge of some robot enemy, Master Mechanics defeat . . . the machine's ideology by forcing the logic-bound robot to attempt to process information which is fundamentally irreducible to true/false resolutions. Thus the ultimate superiority of human intelligence is asserted, and that superiority is fugured as based on our emotions and their theorized ability to accept contradictory assertions . . ." (Allen 2009). Still, a human hero without an understanding of the machine's ideology had no hope of prevailing.

111. A decisive (?) confrontation of the Counterforce and the Firm (one avatar of They) in *Gravity's Rainbow* takes the form of a banquet scene in a bad taste that is

extraordinary, even for a novel that contains many scenes in bad taste, in which chance triumphs, provisionally, over determinism.

CHAPTER γ HOW TO EXPLAIN NUMBER THEORY (CONGRUENCES)

1. Equations (E1) and (E5) are examples of *elliptic curves*; there are graphic representations in chapter δ. At the launching in 2007 of the Fondation des Sciences Mathématiques de Paris, a grouping of mathematics research laboratories including the one where I work, one could read that "The most abstract mathematical theories are intimately linked to technologies omnipresent in our daily life. This is true of elliptic curves" (FSMP 2007). This was an allusion to the applications of the theory of approximate solutions to equations like (E1) and (E5) to modern cryptography and data security; some less precise versions of the same claim are quoted at the beginning of chapter 10. The story starts (in one version) with the patenting in 1983 by Ron Rivest, Adi Shamir, and Leonard Adelman of a public-key cryptosystem, the *RSA algorithm*, based on Fermat's little theorem. Elliptic curves came into use a few years later: see chapter 14 of Koblitz (2007), written by one of the inventors of elliptic curve cryptography, quoted in chapter 10.

The National Security Agency has approved elliptic curve cryptography in its "Suite B" for "the secure sharing of information among Department of Defense, coalition forces, and first responders" (NSA 2009).

CHAPTER 9 A MATHEMATICAL DREAM AND ITS INTERPRETATION

1. The idea recently resurfaced in the work of Wei Zhang, presently at Columbia. Both of the ideas needed to carry out this program—presented, respectively, by Rallis and Rapoport in 1992—have matured in the intervening years. Zhang has a thorough understanding of all the main questions and has been making rapid progress, in part in collaboration with Rapoport—but I'm getting ahead of my story.

2. I will need to refer to the letter and have reproduced the mathematical argument in its entirety in this note, though I don't think it will be of much interest to most readers.

> We start with a representation ρ of D^* (D the division algebra with invariant $1/n$) and use it to construct a local system $L(\rho)$ on the Drinfeld upper half space Ω. We assume ρ is associated to a supercuspidal representation of $GL(n)$ (I'm thinking in terms of mixed characteristic, by the way) and we make the following two hypotheses:
>
> > (a) $H^*(\Omega,L(\rho))$ is concentrated in degree $n-1$;
> >
> > (b) $H^*(\Omega,L(\rho))$ is discrete series.
>
> Verifying these hypotheses may turn out to be the hard part. In any case, (b) guarantees that the representation of $GL(n)$ on $H^*(\Omega,L(\rho))$ is determined by the values of its character on the elliptic regular set. Now any elliptic regular element g in $GL(n)$ has exactly n fixed points in Ω. The following hypotheses involve the development of technology but the technology shouldn't depend on the specific situation.

(c) The character of g on $H^*(\Omega, L(\rho))$, i.e. the value of the function representing the distribution character at g, is given by the Lefschetz fixed point formula. (This is actually a series of hypotheses, including something about the summability of the character and something about l-adic cohomology – or whatever cohomology – of rigid analytic spaces not of finite type.)

Assuming (c), one can calculate the character of g as the sum over the fixed points x of $Tr(g;L(\rho)_x)$. Assuming (a), we get

$$(1)^{(n-1)}\ Tr(g;\ H^{n-1}(\Omega, L(\rho))) = \sum_x (Tr(g;L(\rho)_x))$$

Note that the $(-1)n-1$ (I lost the $-$ sign somehow) is exactly the sign relating the characters of ρ and the corresponding representation of GL(n). Sorry about the missing parentheses. We need another hypothesis:

(d) At each fixed point x of g, $Tr(g;L(\rho)_x) = Tr(\rho(g'))$, where g' is an element of $D^\times$ corresponding to g.

Assuming all four hypotheses, we then conclude that the representation of GL(n) on $H^{n-1}(\Omega, L(\rho))$ is isomorphic to n copies of the representation corresponding to ρ; alternatively, is isomorphic to (representation corresponding to ρ)$\otimes\sigma$ where σ is some n-dimensional representation of the Weil group . . .

Hypothesis (d) would follow from the identification of the action of g on the fiber of the Drinfeld cover of Ω with the action of g', a sort of reciprocity law at the fixed points, which ought to extend to the Weil group as well. I suggested that the fixed points ought to be obtained as follows: let $K = $ field generated by g (or by its eigenvalues), and let LT be the Lubin-Tate formal group with CM by K. For some embedding of K in D, we can construct

$$D \otimes_K LT$$

and this should be a special formal O_D-module (the tensor should involve rings of integers). This should also provide an alternative proof of Genestier's theorem on the irreducibility of the Drinfeld cover (by letting the elliptic torus vary). Genestier has been working on this and he seems to have verified what is needed (it isn't clear whether the indicated method would also imply geometric irreducibility; Genestier thinks it doesn't). Of course the association of g' to g depends on some choices, but it seems that these choices are already implicit in the moduli problem, just as in the classical theory of elliptic curves, the relation to the upper half-plane depends on the choice of a basis for the homology.

I have some vague ideas about (a) and (b) involving Poincare duality and a cohomological interpretation of the Jacquet module, but I realize that I can't go anywhere until I actually learn the theory. The article of Boutot-Carayol is clear but it isn't easy. Drinfeld's original article is not so helpful. But of course if I want to think about intersection theory I will have to read everything. Laumon has made a number of suggestions regarding (c); he doesn't like the shape of the cohomology (inverse limits vs. direct limits) but he thinks it may be possible to rewrite everything in terms of vanishing cycles on the special fiber, where the cohomology theory really exists. Problems (a) and (b) are closer to things I have thought about in the past, but as Laumon points out, they are both false when the coefficients are trivial (Schneider-

Stuhler) so they can't be easy to prove in the supercuspidal case. On the other hand, any proof would have to find a characterization of the supercuspidals in terms of the representations ρ of D^*.

3. They still are. Although I eventually went on to write several articles on the local Langlands correspondence for GL(n) and, with Richard Taylor, wrote a book containing the first proof of this correspondence, it is significant that our work used only the most formal properties of representations of GL(n). The representation theory of the group J appears only in a minor role in a counting argument due to Henniart and quoted in a crucial way, until 2010, in every proof of a local Langlands correspondence, including the one found by Henniart soon after my work with Taylor. Henniart's proof is certainly the most natural one, but it also uses very little of the detailed representation theory developed by Bushnell, Henniart, and Kutzko. In 2010, Rapoport's student Peter Scholze found a new proof of the local Langlands correspondence that does not make use of Henniart's counting argument. The relation between the local Langlands correspondence and representation theory, in the sense of Bushnell, Henniart, and Kutzko, remains an important open question, on which Jared Weinstein has recently made progress.

4. A counting argument of the kind developed by Henniart had first been used in this context by Jerry Tunnell to give a nearly complete proof of the local Langlands correspondence for GL(2); the first complete proof in this case was discovered by Kutzko. Tunnell studied with me at Harvard and will reappear near the end of this story.

5. (Harris 2012).

6. (Max Weber 1922).

7. Most of my time with Rapoport in Wuppertal was spent trying to understand the Gross-Zagier formula in the framework of his book with Zink. On this we made no progress whatsoever, but Rapoport continued to think about the question. A few months later, he began a collaboration with Steve Kudla, who had been developing a much more detailed and flexible framework for understanding the Gross-Zagier formula in a very general setting. This collaboration continues to this day. More recently Rapoport has written a paper with Wei Zhang (see note 1) and Ulrich Terstiege that revisits some of the ideas we discussed in Wuppertal.

8. More precisely, a version of (b) at the level of group characters, especially that the character of the cohomology of coverings of Ω on the Euler characteristic was concentrated on elliptic elements.

9. The letter continues with the following details: " . . . since the spectral sequence on page 30 of your article degenerates for supercuspidals. Moreover, by using a (global) argument of Taylor, one obtains a Galois representation of the right dimention in the cohomology of the Shimura variety (but independent of the global realization; more precisely, its restriction to the decomposition group at p depends only on the original supercuspidal representation)."

10. In reality, the local and global approaches were not so different. The global approach was based on the Arthur-Selberg trace formula rather than the Lefschetz trace formula, but it is well known that the two are closely related and in many cases are different ways of saying the same thing. As I had already hinted in my December message to Rapoport, one of the principal sources of my intuition in this unfamiliar subject was Wilfried Schmid's work, some twenty years earlier, on the representation theory of real

Lie groups. During the period described in this essay I was quite conscious that Schmid had made use of global as well as local trace formulas, perfectly analogous to the ones to which I allude here, in his articles of the 1970s. More recently, he had discovered a new geometric approach to the local trace formula in joint work with my Brandeis colleague Kari Vilonen that I cited explicitly in the first paper written on the basis of my dream. More about Schmid shortly.

11. This refers to Henniart's counting argument, mentioned in note 3.

12. The dream gave rise directly to the publication of (Harris 1997). The local Langlands conjecture was proved in (Harris and Taylor 2001). Not only our own ideas were involved: an observation made by Pascal Boyer in his thesis was crucial, as was the generous support of Vladimir Berkovich, already mentioned previously.

13. (Schmid 1971).

14. I can be even more specific. My five-year plan was to find a p-adic analogue of Schmid's calculation of the distribution characters of discrete series of real Lie groups as "Lefschetz numbers of certain complexes." This is in an obvious sense absurd, because Schmid's calculation is apparently dependent on the use of Hilbert space methods, and there is nothing of the sort available in the setting of the Berkovich cohomology theory. But my letter to Rapoport already invokes "Zuckerman's algebraic version of Schmid"—meaning Schmid's PhD thesis as well as the article (Schmid 1971)—and I had strong hopes that I could define an algebraic version of the Berkovich cohomology of the Drinfel'd coverings to which Schmid's methods could be applied. I still suspect this is possible, and in a more general setting. But I have no idea how to carry this out.

CHAPTER 10 NO APOLOGIES

1. " . . . or relativity," the sentence continues. All Hardy quotations not otherwise attributed are from (Hardy 2012). On page 120, Hardy was quoted defining utility as accentuation of inequalities or destruction of human life. In the original quotation from 1915, he continues, "The theory of prime numbers satisfies no such criteria."

2. (Wolfe 2009: Singh 2000). Singh adds, perhaps to spite Hardy's pacifism, that "military and government encryption is also based on prime number theory." Also: "Hardy's claim that real mathematics is almost wholly useless has been over-played and, to my mind, is now very dated, given the importance of cryptography and other pieces of algebra and number theory devolving from very pure study." J. Borwein, in (Pimm and Sinclair 2006, p. 24).

3. Tom Waits, quoted in (Adams 2011). Also: "Everyone's afraid of Amazon," said Richard Curtis, a longtime agent who is also an e-book publisher. "If you're a bookstore, Amazon has been in competition with you for some time. If you're a publisher, one day you wake up and Amazon is competing with you too. And if you're an agent, Amazon may be stealing your lunch because it is offering authors the opportunity to publish directly and cut you out " (Streitfeld 2011).

4. This thesis was tested empirically and found to be erroneous: see (Stern 1978).

5. (Wallace 2003, p. 25).

6. This is taken from Jacobi's letter to Legendre, responding to criticism by the late Fourier. The full quotation is *"Il est vrai que M. Fourier avait l'opinion que le but principal des mathématiques était l'utilité publique et l'explication des phénomènes na-*

turels; mais un philosophe comme lui aurait dû savoir que le but unique de la science, c'est l'honneur de l'esprit humain, et que sous ce titre, une question des nombres vaut autant qu'une question du système du monde."

7. (Gauss 1808). Gauss spoke on taking over the direction of the Göttingen Astronomical Observatory and astronomy was the *Wissenschaft* he felt obliged to justify, but he made it clear that he also had in mind "the more beautiful parts" of mathematics, including what is now called number theory—a field largely shaped by Gauss's *Disquisitiones Arithmeticae*. See the extended discussion in (Ferreirós 2007), from which much of the translation is taken. I thank Ian Hacking for bringing this reference to my attention.

8. (Pieper 2007, pp. 211–12).

9. *"zu einem wichtigen Wettbewerbsfaktor für die Wirtschaft geworden"* in the original (MFO undated; (Greuel et al. 2008, p. ix).

10. Neunzert's list is on p. 111 of (Greuel et al. 2008). SIAM's careers page is http://www.siam.org/careers/sinews.phpf.

11. (Deloitte 2013). In view of these conclusions, one would expect the coalition government to be throwing money at mathematical research, but there is no sign of this happening.

12. Quotations here and following are from (Roy and Bonami 2010) and from the texts and videos available online at the Web site (Maths à Venir 2009). The quotations from the *décideurs* on the closing panel are transcribed from the video at http://www.maths-a-venir.org/2009/les-math%C3%A9matiques-ressource-strat%C3%A9gique-pour-lavenir, at approximately 1:00:00, 1:06:00, and 1:20:10, respectively.

13. (Zinoman 2006), a review of *A First Class Man*.

14. Hardy's aphorism serves as epigraph to the preface of a book by Manfred Schroeder entitled *Number Theory in Science and Communication*, whose (treacherous) purpose is to show just how useful number theory is after all.

15. (Mehrtens 1990, p. 538). The history is far from exhausted. A few months after he received his Fields Medal, Cédric Villani, already well on his way to becoming the national celebrity he is today, was invited (along with fellow Fields Medalist Ngô Bảo Châu and Gauss Prize winner Yves Meyer) to speak about mathematics before the *Office parlementaire d'évaluation des choix scientifiques et technologiques*. Asked to explain his comparison between mathematics and "certain aspects of art," Villani began by remarking that the comparison "seems so natural to mathematicians, because the artistic aspect of our discipline is so obvious, even more than for other sciences"—note that for Villani a science can also be an art. The rest of his answer covered familiar ground (*"ce qui fait généralement avancer un mathématicien, c'est le désir de produire quelque chose de beau"* [what generally makes a mathematician progress is the desire to produce something beautiful]), but, after claiming that, "sociologically speaking," mathematicians and artists work in much the same way, he justified his claim by pointing out that "in mathematics there are tendencies and tastes," concluding that "[t]he world of mathematics like that of art" is *"parcouru par des modes et des groupes"* [crisscrossed by fashions and groups]. Villani was definitely **not** recycling the mathematical romanticism, symbolized by Galois, which we encountered in chapter 6. See also (Hasse 1952).

16. For example, (Lockhart 2009), which is subtitled *How School Cheats Us out*

of Our Most Fascinating and Imaginative Art Form, includes this statement on page 34: "Mathematics should be taught as art for art's sake. [The] mundane 'useful' aspects would follow naturally as a trivial by-product. Beethoven could easily write an advertising jingle, but his motivation for learning music was to create something beautiful."

17. (Pater 1873, preface and conclusion).

18. (Ruskin 2009), W. S. Jevons, *Methods of Social Reform*, London: Macmillan (1883), quoted in (Maas 1999). See also (Mosselmans and Mathijs, 1999).

For the educational benefits of the arts in France, see (Rancière 2011, chapter 8, especially pp. 173–75). The aesthetic theorists Rancière treats in this chapter, which covers a period stretching from Ruskin through the Paris Exposition Universelle of 1900 to the Bauhaus, have in common a vision of art as "the power to order the forms of individual life and those in which the community expresses itself as community in a single spiritual unity" (p. 178). This ethic of art is much more familiar than the model on which Hardy draws and it is hard to imagine its application to mathematics in any way.

19. (Heard 2004). The extended quotation, reproduced on page 238 of Heard's thesis, is from (Salmon 1883). Regarding Pater, Heard argues that "in Pater's philosophy the value of a beautiful thing lay entirely in its ability to give pleasure to the individual, rather than in some moral truth that it conveyed, and from which the individual and society would benefit. And in the realisation of this philosophy by the aesthetes, the individuals who gained pleasure were their own elite group."

20. (Williams 1976, p. 42). Williams's book is about the use of words in English, but the timing was not so different in other European languages. Jacques Rancière claims that art or Art, as we now understand it, is a notion that dates back at most to the eighteenth century and, specifically, to the publication in 1764 of Johann Joachim Winckelmann's *Geschichte der Kunst des Alterthums* [*The History of Ancient Art*]. With Winckelmann, art is "no longer the competence of makers of paintings, statues, or poems, but as a sensible medium of coexistence of their works . . . Art [with a capital A] becomes an autonomous reality with the idea of history as a relation between a medium, a form of individual life, and the possibility of individual invention." (Rancière 2011, pp. 31, 33). Pater (1873, preface), on the other hand, writing a century and a half before Rancière, sees Winckelmann as belonging "in spirit to another age . . . the last fruit of the Renaissance." Undoubtedly both are right.

21. This sort of talk was not uncommon in those days: in the 1850s, for example, William Stanley Jevons had entitled an unpublished manuscript *On the Science and Art of Music*. The explanation of Kronecker's defense is taken from André Weil's (1976) review of Eisenstein's collected works. According to Borel, an oral communication of Kronecker, reported by Paul Du Bois-Reymond, is the only source for the claim about Eisenstein (Borel 1983; Du Bois-Reymond 1910). No less tragic a figure than Abel and Galois, Eisenstein died at age 29 and was largely forgotten, according to Weil, for over a century.

22. *Nicomachean Ethics*, III.3, VI.3, VI.4, respectively.

23. Marwan Rashed, private communication. The grounds for the claim that a science applies to its specific ontological domain can be found in Aristotle's *Posterior Analytics*, section 7 (from which this quote is taken) and section 12.

24. (Al-Khwarizmi 1831, pp. 5, 87). Contemporary algebra replaces "thing" by the variable *x* and rewrites the first sentence as the equation

$$3 \ 1/2 + 2/5 \ x = x$$

and the second sentence as the instruction to subtract $2/5 \ x$ from both sides. Al-Khwarizmi devotes several long chapters to applications of his new methods to financial questions.

25. (Høyrup 1988). Høyrup writes that the *De numeris datis* of the medieval European mathematician Jordanus de Nemore "transforms a mathematically dubious *ingenium*"—algebra —"into a genuine piece of mathematical *theory*."

Just for completeness, here is how the late antique pseudo-Aristotelian text *Problemata mechanica* distinguishes tricks from art (or from the other "parts" of art):

When we have to produce an effect contrary to nature, we are at a loss, because of the difficulty, and require skill [*techne*, art]. Therefore we call that part of *techne* which assists such difficulties, a device [*mekhane*].

26. The interpretation of medieval Islamic Aristotelianism is based on conversations with Marwan Rashed.

27. The first quotation is from (Fried and Unguru 2001, p. 26). The al-Khayyām quotations are from *Al-Khayyam's Treatise on Algebra* (1070) in (Rashed and Vahabzadeh 2000, pp. 111–13). In the translations I have twice replaced "mathematical" by "scientific" to represent the Arabic word *at-t'alimiya* in order to emphasize the derivation from *'ilm* (science, *epistêmê*). This is the only appearance of the term "scientific art" (*sina'a 'ilmiya*) in Khayyām's extant work. The reader should not assume, however, that the boundaries of the domains identified as *sina'a* and *'ilm* were less fluid than those of *art* and *science* today.

One century after al-Khayyām, al-Jazari combined *'ilm, sina'a*, and *ḥiyal* in the title of his best-known work: *Al-Jami' bayn al-'ilm wa 'l-'amal al-nafi' fi sina'at al-ḥiyal*, translated *A Compendium on the Theory* [Science?] *and Useful Practice of the Mechanical Arts*. Al-Jazari would now be called a mechanical engineer.

The dispute about the nature of algebra was still lively in nineteenth century, when William Rowan Hamilton contrasted three schools of thought, "the Practical, the Philological, or the Theoretical, according as Algebra itself is accounted an Instrument, or a Language, or a Contemplation." "Instrument" comes closest to art, "Contemplation" to science. See (Fisch 1999)

28. (Cardano 1968, Chapter XXXVII, p. 220). Cardano was specifically alluding to his use of the square roots of the negative number -15, what we would now call imaginary numbers, but to which he could assign no meaning.

29. (Cottingham 1976, pp 47–48). I thank Olivia Chevalier-Chandeigne for this reference.

30. Among philosophers, Hegel, Heidegger, and Collingwood, among others, have drawn very different conclusions from the fact that the Greek word *tekhne* refers to arts as well as crafts. More recently, we have also to contend with the fashionable notion of *technoscience*.

31. (Max Weber 1922, p. 7).

32. (Sinclair and Pimm 2010).

33. (Lavin and Lavin 2012).

34. Quotations are from a pamphlet entitled "Our Debt to Mathematics," prepared by Veblen and Julian Coolidge because they were "pessimistic . . . about the chances of getting a substantial endowment grant from the Rockefeller Foundation" and therefore "focused on potential industrial patrons." The pamphlet also "explained how mathematics develops ('by the devoted labors of highly trained scientists who pursue their studies with indefatigable patience, unnoticed by the world, undeterred by the supercilious pity of those who are unable to appreciate their work') [and] asserted that abstract mathematical principles often evolve into practical results" (Feffer 1998).

35. The quotations, in order: (Max Weber 1922); quoted in (Mulkay 1976, p. 457), but Mulkay adds on p. 459 that "in general it appears that there is a pronounced bias in Britain in favour of fields with high scientific status but low economic relevance." (Galison et al. 1992, pp. 54–55): Webster worried that "physicists would occupy their time producing knowledge for litigation rather than for innovation); (Shapin 2008, p. 19): from a speech by Robert C. Dynes, chancellor of UCSD at the time (and later president of the UC system), in his State of the Campus address, March 22, 2001; from Whyte's *The Organization Man*, 226–27, quoted on page 176 of (Shapin 2008; Conference des presidents d'université 2011).

36. (Shapin 2008, pp. 125, 104).

37. (Gelfand 2009; SIAM 2012). Manin stressed that his remarks apply only to mathematicians; the SIAM report is concerned with mathematical research in the corporate setting.

38. (Mehrtens 1990, p. 57).

39. Blair's government listed the thirteen industies in 1998: film, TV and radio, publishing, music, performing arts, arts and antiques, crafts, video and computer games, architecture, design, fashion, software and computer services, and advertising. See (Ross 2009, pp. 28, 25).

40. Hardy's function theory is applied throughout mathematics—in particular, to differential equations like the Black-Scholes equation—for example, in (Arendt and De Pagter 2002). The art that contributes to that 7.3% of the British economy naturally differs from the kind of art Hardy had in mind, whose economy is examined in (Muntadas 2011) and (Stallabrass 2006), especially chapter 4 of the latter; the point is rather that pure mathematics and what we might call "pure art" engender similar Golden Geese.

41. (Smith 1759).

42. Or so claimed the minister. Those who spoke up at the roundtable I attended were either business executives or representatives of various levels of government; the only researchers visible were those who had created either partnerships with industry or their own startups.

43. The scope of the Loan, as initially planned, was not limited to Madame Pécresse's ministry (10 billion euros for higher education, 3.5 billion for "valorisation" of public research, of a total of 35 billion); and it's hard to tell to what extent the Loan's initial priorities survived the realities of the economic crisis and the transition to Hollande's socialist government. Some information is available at the Web site (progfrance 2009–2013). The contribution to the operating budget of my university, at least, seems to be much less than anticipated.

44. My handwriting here is blurry and I can hardly believe anyone really said this, but Google finds the expression "quête de rente" in 101,000 Web pages, so who knows?

45. No enlightenment on this score seemed likely at the other three roundtables, on health/well-being/food (biotechnology), *l'urgence environmentale* (ecotechnology, especially energy and transport), or information/communication (nanotechnology and telecommunications).

46. This is an old idea. John Dewey already wrote in 1902 that "Teaching . . . is something of a protected industry; it is sheltered. . . . There is always the danger of a teacher's losing something of the virility that comes from having to face and wrestle with economic and political problems on equal terms with competitors." Dewey adds that "Specialization . . . leads the individual . . . into bypaths still further off from the highway where men, struggling together, develop strength." Reprinted in (Dewey 1976).

47. Also, for the record: (3) There is concern about the condition of the researcher, especially in the creation of spinoffs; and how to manage the return of the researcher to the university if the spinoff is sold; (4) Universities have to agree to finance their offices of transfer/innovation on their overhead budgets; "then they will be viable"; (5) Mao's "hundred flowers" image is still alive among French *décideurs*.

48. Respectively, 20%, 15%, and 15% (AFP 2006).

49. (Koblitz 2007). In a private communication, Koblitz clarified his position: "My own personal feeling is that it is unseemly for academic mathematicians to go lusting after money. We receive decent salaries as full-time professors and have near-perfect job security. That's enough, it seems to me."

50. (Biagioli 2003, p. 255); European Commission 2008; European Commission 2009, Section 9, "Communication on University Business Dialogue"). The three keywords in 2008 were "governance, commercialization, entrepreneurial mindset." Commitment to these values crosses party lines: Geneviève Fioraso, the socialist party Minister of Higher Education and Research, called for instilling the entrepreneurial mindset as early as preschool—and for *décloisonnement* of research and business (Talmon 2014).

51. (SIAM 2012). Neunzert is hearing those same voices: *"Auf einer breiten und soliden Basis allgemeiner Mathematik **muss**, zumindest für die Mehrheit der Studenten, die später ihr Auskommen in eben dieser Praxis finden **müssen**, ein vertieftes Wissen in mathematischer Modellierung und in numerischen Methoden gesetzt werden. Das geht nicht ohne Kenntnisse in reiner Mathematik, aber es geht auch nicht ohne Kenntnisse in angewandter Mathematik, in Modellierung und Numerik. Außerdem **sollte** in den jungen Mathematikern das Interesse für die Anwendungswissenschaften und die Bereitschaft der Zu- sammenarbeit mit deren Vertretern geweckt werden"* (Greuel et al. 2008, p. 119 [emphasis added]).

And in France, the socialist government has drafted a bill to include "transfer of results [of scientific research] to the socio-economic world" among the official "missions" of higher education and research, alongside the six "missions" defined in the law of 2007 (listed in the Code de l'éducation—Article L123-3, at http://www.legifrance.gouv.fr). See http://www.enseignementsup-recherche.gouv.fr/cid70881/une-nouvelle-ambition-pour-la-recherche-mesure-15.html.

52. It is quoted in *The Beginning of the End* (2004) by Peter Hershey, p. 109.

53. Neunzert in (Greuel et al 2008, p. 118; ADEC 2011).

54. It is understood—and understandable—that social scientists would want to avoid "going native" at all costs and are subject to a professional requirement not to accept the beliefs of their informants uncritically. But reflexivity works both ways: mathematicians are (mostly) no fools and know what it means to act as informants. There's no simple way out of this conundrum.

The Deligne, Sally, and MacPherson quotations are from (Simons Foundation, undated). In (Cook 2009, p. 156). Deligne said much the same thing: "It was a wonderful surprise to learn one could at the same time play and earn one's living." And one finds a similar statement in the biography that accompanied the announcement, posted by the Norwegian Academy of Science and Letters, that Deligne had been awarded the 2013 Abel Prize.

55. (Gopnik 2009).

56. *Philebus*, 66a–c. First on the list is "measure, moderation, fitness." We have seen that Aristotle had little patience with those who identified pleasure with "the good," but he acknowledged that "each activity has a proper pleasure" (*oikeia hedone*; *Nicomachean Ethics* X, 5)—for example, the "proper pleasure" of tragedy, which "is the pleasure that comes from pity and fear by imitation," (*Poetics*, XIV). In his *Metaphysics*, on the other hand, we read that 'The mathematical sciences particularly exhibit order, symmetry, and limitation; and these are the greatest forms of the beautiful" (*Metaphysics* XIII, 3.107b).

57. Quotations from (Cook 2009, pp. 68, 160, 188). This is probably the place to mention that Belgian mathematicians can aspire to honors not available elsewhere: Deligne is a Vicomte and Daubechies is a Baroness.

58. Transcribed from BBC Horizon/PBS Nova video entitled *The Pleasure of Finding Things Out*, at around 22:30–23:30. Some of this was published in a book by the same name, but the word *relaxed* was omitted.

59. (Hume 1739, Part III, Book II, Section X).

60. (Greuel et al. 2008, p. 115; SIAM 2012, p. 37). Do these quotations contradict Hardy's claim that only the "dull and elementary parts of applied mathematics" are useful?

61. (Edgeworth 1881, p. 72; Cohen, undated). The second postulate is perhaps more familiar: "that making social utility as great as possible is the ultimate moral imperative."

62. (European Commission 2009, Section 9). Isabelle Bruno, a specialist in EU research and innovation policy, wrote an 832-page thesis on "benchmarking" in EU policymaking; the word *pleasure* [*plaisir*] appears only in the acknowledgments and inconspicuously in a footnote, whereas the word *happiness* [*bonheur*] is only used in connection with her historical review of utilitarianism.

Compare these thoughts of Harvard President Drew Gilpin Faust (Faust 2009), reacting to the economic crisis that started in 2008: "As the world indulged in a bubble of false prosperity and excessive materialism, should universities—in their research, teaching and writing—have made greater efforts to expose the patterns of risk and denial? Should universities have presented a firmer counterweight to economic irresponsibility? Have universities become captive to the immediate and worldly purposes they serve? Has the market model become the fundamental and defining identity of higher education?"

63. (Burke 1757; Hume 1739, Part II, section V; Kant 1790).

64. (O'Doherty et al. 2003; Ishizu and Zeki, 2011). The author of a recent survey (Chatterjee 2014) writes that he is "not aware of any studies that have examined the neuroaesthetics of math." Here and in what follows, the mOFC stands in for any "ensemble of neural subsystems" that might some day be found responsible for the human aesthetic response.

65. It is! At least that's what Zeki and his collaborators—including no less an authority than Fields Medalist Sir Michael Atiyah—announced in February 2014, after this chapter was written. More precisely, "the experience of mathematical beauty correlates parametrically with activity in the same part of the emotional brain, namely, field A1 of the medial orbito-frontal cortex (mOFC), as the experience of beauty derived from other sources." See (Zeki et al. 2014).

I have to say I'm a little skeptical of the research protocol, which consisted in asking subjects to rate equations as beautiful, neutral, or ugly (before and after MRI scans), given that Euler's identity $1 + e^{i\pi} = 0$ won the beauty contest, while an formula of Ramanujan—much less elaborate than the Frenkel et al. Love Equation—came in dead last.

66. Quotations from *An Essay in Aesthetics*, in (Fry 1920).

67. This chapter suggests that Hardy's *Apology* is aligned with Bloomsbury values; but these can be read in more than one way. Regarding the relation of Bloomsbury ethics to utilitarianism, Craufurd Goodwin writes "The Bloomsburys were convinced that the imaginative life could not be understood simply as responses to utilitarian stimuli, coordinated through the market mechanism. . . . On the supply side of the labour market in the imaginative life, artists, writers, and scientists did not typically perceive their activities as onerous and a source of disutility that had to be compensated by utility derived from expenditures of income. Quite the contrary, they were driven to action by an irrepressible urge to communicate . . . " (Goodwin 2001). On the other hand, according to Raymond Williams, Bloomsbury "was against cant, superstition, hypocrisy, pretension and public show. It was also against ignorance, poverty, sexual and racial discrimination, militarism and imperialism. But . . . [w]hat it appealed to, against all these evils, was not any alternative idea of a whole society. Instead it appealed to the supreme value of the civilized *individual* . . . The profoundly representative character of this perspective and commitment . . . is today the central definition of bourgeois ideology" (Williams 1980).

68. (Borel 1989).

69. (Grothendieck 1988, p. 202, text and footnote), my (rough) translation.

70. (Langlands 2013). The "division of elliptic integrals" to which Langlands refers is an early stage in the study of cubic equations (see chapters γ and δ) and thus represents part of the historical background to the Birch-Swinnerton-Dyer conjecture.

71. Psychologically, this is very different from Aristotle's list: order, symmetry, and limitation, "the greatest forms of the beautiful" (*Metaphysics*, XIII, 3.107b); or, in another vein, the triplet "coldness, tedium and irrelevance" of "cultural perceptions of

mathematics" cited by Pimm and Sinclair (*The Many and the Few*). Michael Atiyah, in his video interview at [Simons Foundation undated], opts for "elegance, simplicity, structure, and form . . . all sorts of things," which sounds more aristotelian than Hardy's list. Alain Badiou suggested a somewhat different three-term list—economy, rational totalization, and fruitfulness—in a public lecture in Paris on June 16, 2011. He did not cite Hardy.

72. Mathematical tricks, as we characterized them in chapter 8, qualify as beautiful on all three of Hardy's counts: economy is represented by what we called a "short cut" while unexpectedness and inevitability come together in a trick to create an Aha! experience. Nevertheless, Hardy himself used the word *trick* infrequently and pragmatically rather than aesthetically. Google Scholar returns 457 articles authored by G. H. Hardy—there are many repetitions—and only two of them include the word *trick*. In "Prolegomena to a chapter on inequalities" [*J. London Math. Soc.* (1929): 61–78], he stressed the need for those working in function theory to master both the "main results and the tricks of the trade"; in "Note on the theory of series (XIX)" with Littlewood [*Quarterly J. Math.* (1935) 304–15] it is said that "some special trick" is needed to prove a certain statement.

73. Much of the following discussion inevitably overlaps with the exhaustive work on mathematical aesthetics by Nathalie Sinclair and her collaborators. See, in particular, (Sinclair 2011).

74. (Fry 1956, p. 8).

75. For example, see (Fry 1920, p. 54).

76. This reading of the *Apology* is from (Kanigel 1991). The quotation is from (Moore 1903, section 113). In his biography of Paul Dirac, Graham Farmelo suggests that "Moore's common-sense approach to beauty probably influenced his scientific colleagues at Trinity, including Rutherford and . . . G. H. Hardy" as well as Dirac: (Farmelo 2010, p. 74).

77. All Bell quotations from (Clive Bell 1914).

CHAPTER δ HOW TO EXPLAIN NUMBER THEORY (ORDER AND RANDOMNESS)

1. The actual condition is that the equation define a *curve of genus 0*. Equations in two variables of degree higher than 2 can define curves of genus 0, but this is the exception rather than the rule. Wiles's description of the Birch-Swinnerton-Dyer conjecture on the Clay Millenium Problem Web site (Wiles 2000) gives (Hilbert and Hurwitz 1890) as the source for the complete characterization of curves of genus 0 with infinitely many rational points.

2. Tunnell's insight was to reinterpret the congruent number problem in terms of the Birch-Swinnerton-Dyer conjecture and to use the latter to provide a criterion that can easily be tested (but remains, for the moment, purely conjectural).

For this and for much more about elliptic curves and the BSD conjecture, at a level of mathematical sophistication that may not be appropriate for all dinner parties, see *Elliptic Tales* (Ash and Gross 2011).

3. Actually, it plots the scaled difference $(p - S(p))/\sqrt{p}$, so that the result always fits between -2 and 2. The histogram was produced by William Stein and was published in (Mazur 2008a).

4. Quotations here and following are from (Gray 2006).

5. P. A. The *Bhagavad-Gita* is unambiguous:

> When your intelligence has passed out of the dense forest of delusion, you shall become indifferent to all that has been heard and *all that is to be heard.* (P.A.'s emphasis)

N. T. Is that what the *Gita* says? In that case, the author should have been more careful and not have trusted Weil blindly . . .

P. A. Weil is using the *Gita* to say very clearly that the indifference applies to all knowledge, past or future.

N. T. But a few lines later he says the opposite:

> *Heureusement pour les chercheurs, à mesure que les brouillards se dissipent, sur un point, c'est pour se reformer sur un autre.* [Fortunately for researchers, insofar as the fog dissipates around one point, it is only to reappear around another.]

(Translation of verse 52, chapter 2 of the *Bhagavad-Gita* from a lecture of Barry Mazur in February 2014, available on his home page at http://www.math.harvard .edu/~mazur/.)

6. (Plassard 1992, p. 35; Rancière 2011, p. 207).

AFTERWORD: THE VEIL OF MAYA

1. (Nietzsche 1974), Preface to the second edtion, paragraph 3; translation slightly modified. Stephen Strogatz's book *The Joy of x*, published in 2012 may not be about the same joy, but it must be about the same x.

2. The discussion of the last two paragraphs is taken from chapter 21 of (Hadot 2006).

3. (Hausdorff 2011).

4. Hausdorff made major contributions to set theory, probability, functional analysis, and several other branches of mathematics as well as mathematical physics. His notion of continuous (or fractional) dimension is the starting point of Benoît Mandelbrot's theory of fractals. But he is best known among mathematicians for his development of topology.

5. The comedy, entitled *The Surgeon of his Honor* [*Der Arzt seiner Ehre*], ridiculed the Prussian code of honor that came to an end with the war.

6. Quotations from (Purkert 2008; Monist 1900); see also (Epple 2006). Mongré was also concerned to disprove Nietzsche's doctrine of eternal return, and to this end he appealed to Cantor's infinities.

The author was not surprised to learn that philosophers have revived the term "transcendental nihilism," specifically in connection with the book *Nihil Unbound* by Ray Brassier. There are parallels with the thoughts of Mongré's Truth-seeker, but neither Mongré nor Hausdorff is cited as a precursor.

7. Translation from (Purkert 2008, p. 40).

8. (Stegmaier 2004).

9. See (Epple 2007): "According to Hausdorff, experience limits the range of possible mathematical conceptions of notions like time, space, motion, temperature, dimension, . . . but it is the task of mathematics to explore the full *Spielraum* of possibilities within these limits." Hausdorff's considered empiricism in some ways anticipated the positions of the Vienna Circle, specifically the early work of Moritz Schlick; see (Epple 2006: Stegmaier 2004).

10. Hausdorff, quoted in (Stegmaier 2004, note 22). For "very large indeed," see (Epple 2006, p. 282).

11. In contrast, the word *freedom* is nowhere to be seen in Hardy's *Apology*, nor in Russell's *Principles of Mathematics*. When it appears in Jacques Hadamard's *Psychology of Mathematical Invention*, it is associated with the arts, or with dreams; Poincaré uses the word *liberté* in *Science et Hypothèse* only to talk about the "*limite imposée à la liberté*" [the limits of freedom] and to remind us that "*la liberté n'est pas arbitraire*" [freedom is not arbitrary].

12. Hausdorff could not resist inserting a "glimmer of [dark] humor" in his suicide note. Mongré had published poetry; before advising the friend to whom the letter was addressed how to dispose of his property, Hausdorff alluded to the likelihood of deportation from Endenich, where he and his wife were about to be interned, in a two-line verse:

> *auch Endenich*
> *ist noch vielleicht das Ende nich!*
> [Endenich may not be the End either!]

13. (Purkert 2008), from the introduction by David E. Rowe, p. 36 and p. 37. The expression "Dionysian mathematician" was due to Hausdorff's contemporary, the writer Paul Lauterbach.

bibliography

Abattouy, Mohammed. "Nutaf Min Al-Hiyal: A partial Arabic version of pseudo-Aristotle's 'Problemata Mechanica.' " *Early Science and Medicine* **6**, no. 2 (2001): 96–122 .

Aczel, Amir. D. *The Mystery of the Aleph: Mathematics, the Kabbalah, and the Search for Infinity*. New York: Simon and Schuster, 2001.

Adams, Tim. "Tom Waits: 'I look like hell but I'm going to see where it gets me.' " *The Observer*, October 23, 2011.

ADEC. "In a survey from ADEC: The teachers' professional satisfaction rate in Abu Dhabi Schools is as high as 78.3%." Abu Dhabi Education Council (ADEC). August 15, 2011. www.adec.ac.ae/en/MediaCenter/News/Pages/Pressitem418.aspx (accessed June 6, 2013).

AFP. "Le premier but de la recherche est d'améliorer le bien-être (sondage)." *Le Monde*, May 31, 2006.

Ahearn, Charlie. "The five percent solution" *Spin Magazine* (February 1991): 54ff.

Alexander, Amir. *Duel at Dawn: Heroes, Martyrs, and the Rise of Modern Mathematics*. Cambridge: Harvard University Press, 2011.

Alexander, Stephon. "Physics and jazz." http://bigthink.com/videos/physics-and-jazz, June 20, 2008 (accessed December 12, 2013).

———. "The physics of jazz." June 26, 2012. http://www.youtube.com/watch?v=90bjRafoP0Q (accessed December 15, 2013).

Allen, G. S. *Master Mechanics and Wicked Wizards, Images of the American Scientist as Hero and Villain from Colonial Times to the Present*. Amherst, MA: University of Massachusetts Press, 2009.

Alvey, J. E. "An introduction to economics as a moral science." *International Journal of Social Economics* **27**, no. 12 (2000): 1231–52.

Amir-Moez, A. R. "A paper of Omar Khayyám." *Scripta Mathematica* **26** (1963): 323–37.

AMS. "Manual for authors of mathematical papers." *Bulletin of the AMS* **68** (1962): 429–44.

Arendt, Wolfgang, and Ben De Pagter. "Spectrum and asymptotics of the Black-Scholes partial differential equation in (L^1, L^∞)-interpolation spaces." *Pacific Journal of Mathematics* **202**, no. 1 (2002): 1–36.

Ash, Avner, and Rob Gross. *Elliptic Tales*. Princeton: Princeton University Press, 2011.

Assayag, G., H. G. Feichtinger, and J. F. Rodrigues. *Mathematics and Music: A Diderot Mathematical Forum*. Berlin: Springer-Verlag, 2002.

Atiyah, Sir Michael. In *Collected Works*, Vol. 1: *Early Papers, General Papers*. Oxford: Oxford University Press, 1989, 297–307.

Audin, Michèle. *Remembering Sofya Kovalevskaya*. Berlin: Springer-Verlag, 2008.

Azzouni, Jody. "Ontology and the word 'exist': Uneasy relations." *Philosophia Mathematica* **18** (2010): 74–101.

Bachem, Anna. *Theorien des Spiels: Beispiele, Unterschiede und Parallelen und die Bedeutung für kindliche Bildungsprozesse*. Munich: GRIN Verlag, 2012.

Baillet, Adrien. *La Vie de Monsieur Descartes*. Paris: D. Horthemels, 1691. http://gallica
.bnf.fr/ark:/12148/bpt6k75560v/f563.image..

Barany, Michael J., and Donald MacKenzie. "Chalk: Materials and concepts in mathematical research." In *Representation in Scientific Practice Revisited*, ed. C. Coopmans et al. Cambridge, MA: MIT Press, 2014, 107–29.

Barnet-Lamb, Thomas, Toby Gee, David Geraghty, and Richard Taylor. "Potential automorphy and change of weight." *Annals of Mathematics*, **179.2** (2014), 501–609.

Beebee, Helen. "Philosophy, impact, and the research excellence framework." Letter dated July 12, 2010. http://www.bpa.ac.uk/uploads/2010/09/bpa-willetts-12710.pdf
?location=resources/uploads/2010/09/bpa-willetts-12710.pdf (accessed May 18, 2013).

Bell, Clive. *Art*. London: Chatto and Windus, 1914.

Bell, Eric Temple. "Gauss and the early development of algebraic numbers." *National Mathematics Magazine* **18**, no. 5 (February 1944): 188–204.

———. *Men of Mathematics*. New York: Touchstone, 1986.

Bellah, Robert N. *Religion in Human Evolution*. Cambridge, MA: Belknap Press, 2011.

Benacerraf, Paul. "Mathematical truth." *The Journal of Philosophy* **70**, no. 19 (1973): 661–79.

Berlioz, Hector. "Antoine Reicha." *Journal des débats* (3 juillet 1836).

Besson, Sylvain. "Michel Rocard: 'Le déclin de l'empire romain a commencé comme ça.' " *Le Temps*. 22 octobre 2008. www.LeTemps.ch.

Bethke, Bruce. "The Etymology of 'Cyberpunk.' " *Bruce Bethke, Writer*. 1997, 2000, 2010. http://www.brucebethke.com/articles/re_cp.html (accessed July 20, 2012).

Biagioli, Mario. "Rights or rewards? Changing frameworks of scientific authorship." In *Scientific Authorship: Credit and Intellectual Property in Science*, ed. Mario Biagioli and Peter Galison. New York: Routledge, 2003, 253–79.

Biruni, Abu al-Rayhan Muhammad b. Ahmad al-. *The Exhaustive Treatise on Shadows*. Trans. E. S. Kennedy. Aleppo: Institute for the History of Arabic Science, University of Aleppo, 1976.

BIS (Bank for International Settlements). "Statistical release: OTC derivatives statistics at end-December 2012." Basel, May 2013.

Blackburn, Vernon. *Pall Mall Gazette*, London, February 28, 1900.

Bloor, David. *Knowledge and Social Imagery*. Chicago: University of Chicago Press, 1991.

Borel, Armand. "Mathematics: Art and science." *The Mathematical Intelligencer* **5**, no. 4 (1983): 9–17. (Trans. from "Mathematik: Kunst und Wissenschaft." Carl Friedrich von Siemens Stiftung, 1984.)

———. "The school of mathematics at the Institute for Advanced Study." In *A Century of Mathematics in America*, Part III, ed. Peter Duren et al. Providence, RI: American Mathematical Society, 1989, 119–48.

Born, Georgina. *Rationalizing Culture: IRCAM, Boulez, and the Institutionalization of the Musical Avant-Garde*. Berkeley: University of California Press, 1995.

Boudine, Jean-Pierre. *Homo mathematicus : Les mathématiques et nous*. Paris: Vuibert, 2000.

Boudine, Jean-Peirre, and Olivier Courcelle. "Clairaut, 'Problème physico-mathematique.' " *Quadrature* **36** (avril-juin 1999): 31–37.

Boulding, Kenneth E. "Economics as a moral science." *The American Economic Review* (1969): 1–12.

Boulton, G. "University rankings: Diversity, excellence, and the European initiative." League of European Research Universities Advice Paper, No. 3. June 2010.

Bourbaki, Nicolas. "L'architecture des mathématiques." In *Les grands courants de la pensée mathématique*, by François Le Lionnais. Paris: Hermann, coll. Histoire de la pensée, 1948, 35–47.

———. "The architecture of mathematics." *American Mathematical Monthly* **57**, no. 4 (April 1950): 221–32.

Bourdieu, Pierre. *Distinction*. Trans. Richard Nice. Cambridge: Harvard University Press, 1984.

Bourdieu, Pierre, and J.-C. Passeron. *Les Héritiers*. Paris: Minuit, 1964, 1985.

Bourdieu, Pierre, and Monique de Saint Martin. *Academic Discourse: Linguistic Misunderstanding and Professorial Power*. Redwood City: CA: Stanford University Press, 1996.

Bowles, Samuel, and Herbert Gintis. *A Cooperative Species*. Princeton: Princeton University Press, 2011.

Brahmagupta, Bhaskara, and Henry Thomas Colebrook. *Algebra, with Arithmetic and Mensuration*. London: John Murray, 1817. http://www.archive.org/stream/algebra witharith00brahuoft#page/n5/mode/2u.

Braudel, Fernand. *The Structures of Everyday Life: The Limits of the Possible*. Berkeley: University of California Press, 1981.

Broad, William J. "Billionaires with big ideas are privatizing American science." *New York Times*, March 15, 2014.

Browder, Felix. "The Stone Age of Mathematics on the Midway." In *A Century of Mathematics in America*, Part II, ed. P. L. Duren et al. Providence, RI: American Mathematical Society, 1989, 191–94.

Brown, Frederic. "A dilettante of consequence." *New York Times Book Review*, May 4, 1997.

Bruno, Isabelle. "Déchiffrer l' 'Europe compétitive,' tome I." *FASOPO* (12 décembre 2006). http://www.fasopo.org/reasopo/jr/these_tome1_bruno.pdf (accessed July 13, 2012).

Burgess, J. P., and G. G. Rosen. *A Subject with No Object, Strategies for Nominalistic Interpretation of Mathematics* . Oxford: Oxford University Press, 2000.

Burghardt, Gordon M. *The Genesis of Animal Play: Testing the Limits*. Cambridge, MA: The MIT Press, 2005.

Burke, Edmund. *A Philosophical Enquiry into the Origin of Our Ideas of the Sublime and Beautiful*. London: J. Dodsley, 1757.

Burris, Val. "The academic caste system: Prestige hierarchies in PhD exchange networks," *American Sociological Review* **69**, no. 2 (2004): 239–64.

Capgemini and RBC Wealth Management, World Wealth Report 2013, downloaded August 23, 2013 from http://www.worldwealthreport.com/ 2013.

Caporaso, Lucia, and Roberto Natalini. "Ricominciamo da Tre" MaddMaths! editorial, July 10, 2013. http://maddmaths.simai.eu/rubriche/l-editoriale/ricominciamo-da-tre/ (accessed October 26, 2013).

Cardano, Girolamo. *Ars Magna (The Great Art) or the Rules of Algebra*. Trans. T. Richard Witmer. New York: Dover, 1968.

Carlyle, Thomas. "Signs of the times." *Edinburgh Review*, June 1829.

Carter, J. "Ontology and mathematical practice." *Philosophia Mathematica* (2004): 244–67.

Cartier, Pierre. "A country of which nothing is known but the name: Grothendieck and 'motives,' " in (Schneps 2014), 269–98.

Casazza, Peter G. "Pete Casazza's home page." *A Mathematician's Survival Guide*. http://www.math.missouri.edu/~pete/ (accessed September 21, 2011).

Châtelet, Gilles. "De la victoire de Platon et d'un certain techno-populisme hostile aux mathématiques." *Gazette des Mathématiques* **74** (octobre 1997).

Chance, Don M., and Robert Brooks. *Introduction to Derivatives and Risk Management*. 9th ed., Mason OH: South-Western, Cengage Learning, 2012.

Chang, Alexandra. "New IBM app presents nearly 1,000 years of math history." *Wired* (April 6, 2012). http://www.wired.com/gadgetlab/2012/04/new-ibm-app-presents-nearly -1000-years-of-math-history/ (accessed April 6, 2012).

Chatterjee, Anjan. *The Aesthetic Brain*. Oxford University Press, 2014.

Chemillier-Gendreau, Denis, and Elyès Jouini. "Haro sur la finance !" *Le Monde* (4 novembre 2008).

Christensen, T. "Eighteenth-century science and the 'Corps Sonore:' The scientific background to Rameau's 'Principle of Harmony.' " *J. Music Theory* **31**, no. 1 (1987): 23–50.

Chung, S.-Y., and C. Berenstein. "Omega-harmonic functions and inverse conductivity problems on networks." *SIAM J. Appl. Math.* **65** (2005): 1220–26.

Clairaut, Alexis Claude. "Problème physico-mathematique." In *Recueil de ces Messieurs*. Amsterdam: Chez les frères Westein, 1745, 355–58.

Clarke, M. "Why Hasn't Scientific Publishing been Disrupted Already?" *The Scholarly Kitchen* (January 4, 2010). http://scholarlykitchen.sspnet.org/2010/01/04/why-hasnt -scientific-publishing-been-disrupted-already (accessed February 3, 2012).

Clay Mathematics Institute. "PROMYS." http://www.claymath.org/programs/outreach /PROMYS/ (accessed December 1, 2013).

Clerval, Anne. *Paris sans le peuple*. Paris: La Découverte, 2013.

Cohen, Gilles. "Charles Torossian, l'inventeur." *Tangente éducation* (janvier 2012): 6–9.

Cohen, Shiri. "Economics and Ethics meet under the same umbrella: Edgeworth's 'exact utilitarianism' (1877–1881)." Center for the History of Political Economy at Duke University. undated. hope.econ.duke.edu/sites/default/files/Edgeworth%20 for%20CHOPE.pdf (accessed June 6, 2013).

Cole, J., and S. Cole. *Social Stratification in Science*. Chicago: University of Chicago Press, 1973.

Cole, S. "Professional Standing and the Reception of Scientific Discoveries." *Am. J. Sociology* **76** (1970): 286–306.

Cole, S., and J. R. Cole. "Scientific output and recognition: A study in the operation of the reward system in science." *Am. Sociological Review* 32 (1967): 377–90.

———. "Visibility and the structural basis of awareness of scientific research." *Am. Sociological Review* **33** (1968): 397–413.

Coleman, Sidney, David Gross, and Roman Jackiw. "Fermion avatars of the Sugawara model." *Physical Review* **180** (1969): 1359–66.

Collini, Stefan. "From Robbins to McKinsey." *London Review of Books* (August 25, 2011).

———. "Sold Out." *London Review of Books* (October 24, 2013).

Colmez, Pierre, Jean-Pierre Serre, and Leila Schneps, eds. *Grothendieck-Serre Correspondance*. Trans. Catriona Maclean. Bilingual edition. Providence, RI: American Mathematical Society, 2003.

Compétences. "La taxe Tobin inquiète les experts." *Les Echos.fr*. (10 février 2005). http://www.lesechos.fr/10/02/2005/LesEchos/19348-066-ECH_la-taxe-tobin -inquiete-les-experts.htm (accessed April 18, 2013).

Comptes rendus de l'Office parlementaire d'évaluation des choix scientifiques et technologiques. "Débat sur les mathématiques en France et dans les sciences d'aujourd'hui." Sénat. November 17, 2010. http://www.senat.fr/compte-rendu-commissions /20101115/office.html (accessed November 26, 2010).

Conférence des présidents d'université. "Sommet Mondial des Universités 2011: Besançon les 28, 29, et 30 avril 2011; Dijon les 5, 6, et 7 mai 2011, 07/04/2011." Conférence des présidents d'université (7 avril 2011). http://www.cpu.fr/Actualites.240.0 .html?&no_cache=1&actu_id=304 (accessed April 16, 2013).

Confucius. *Analects*. Trans. Arthur Waley. New York: Vintage, 1938.

———. *Analects*. Trans. Dim Cheuk Lau. London: Penguin, 1979.

Cook, Mariana. *Mathematicians: An Outer View of the Inner World*. Princeton: Princeton University Press, 2009.

Corfield, David. "Narrative and the Rationality of Mathematical Practice." In (Doxiadis and Mazur 2012, 244–80).

Corry, Leo. *Modern Algebra and the Rise of Mathematical Structures*. Basel: Birkhäuser, 2004.

———. "Writing the ultimate mathematical textbook: Nicolas Bourbaki's Éléments de mathématique,." In *The Oxford Handbook of the History of Mathematics*, edited by E. Robinson and J. Stedall. Oxford: Oxford University Press, 2008, 565–88.

"The cost of knowledge." http://www.thecostofknowledge.com/.

Cottingham, John. *Descartes' Conversation with Burman*. Oxford: Clarendon Press, 1976.

Crouhy, Michel. "Risk management failures during the financial crisis." *Math Phi*. (January 26, 2009). http://stage-finance.mathphi.com/dossier-crise-financiere/Fed -Chicago-012609.pdf (accessed March 31, 2012).

Cziszar, Alex. "Stylizing rigor; or, why mathematicians write so well." *Configurations* **11**, no. 2 (2003): 239–68.

Dalsgaard, Inger H. "Science and technology," In *The Cambridge Companion to Thomas Pynchon*, ed. I. Dalsgaard, L. Herman, B. McHale. Cambridge: Cambridge University Press, 2012, 156–67.

Dante, Waldemar. "No soy dado a las abstracciones," an interview in *Vogue* from 1981. Sololiteratura. 1981. http://sololiteratura.com/ggm/marquezvogue.html (accessed September 7, 2011).

Darboux, Gaston. "Sur les surfaces orthogonales." *Annales scientifiques de l'École Normale Supérieure*, Sér. 1, 3 (1866): 97–141.

Daston, Lorraine, and Peter Galison. *Objectivity*. New York: Zone Books, 2007.

Dauben, J. W. *Georg Cantor: His Mathematics and Philosophy of the Infinite*. Princeton: Princeton University Press, 1979.

Davis, Philip, and Reuben Hersh. *The Mathematical Experience*. Boston: Birkhäuser , 1981.

de Gandt, F. "Technology." In *The Greek Pursuit of Knowledge*, ed. J. Brunschwig and G.E.R. Lloyd. Cambridge, MA: Belknap Press, 2000, 340.

De Sigorgne, M. "A physico-mathematical demonstration of the impossibility and in-sufficiency of vortices." *Philosophical Transactions of the Royal Society* (1683–1775) **41** (1739–1741): 409–35.

Delesalle, Nicolas. "Cédric Villani, 'la Lady Gaga des maths.' " *Télérama*, October 22, 2011.

Deligne, Pierre. "Travaux de Shimura." In *Séminaire Bourbaki*, 23ème année (1970/71), *Lecture Notes in Math*, Vol. 244. Berlin: Springer, 1971, 123–165.

Deloitte. "Measuring the economic benefits of mathematical science research in the UK." *ESPRC* (June 2013). http://www.cms.ac.uk/files/Submissions/article _EconomicBenefits.pdf.

Derman, Emanuel, and Paul Wilmott. "Financial modelers' manifesto." Wilmott.com. January 8, 2009. http://www.wilmott.com/blogs/paul/index.cfm/2009/1/8/Financial -Modelers-Manifesto.

Descartes, René. *Principles of Philosophy*. Trans. Valentine Rodger Miller and Reese Miller. Dordrecht: Kluwer Academic Publishers, 1991.

Dewey, John. *Logic—The Theory of Inquiry*. New York: Henry Holt and Company, 1938.

———. "Academic freedom." In *John Dewey: The Middle Works*: 1899–1924, ed. Jo Ann Boydston. Carbondale and Edwardsville: Southern Illinois University Press, 1976, 53–56

———. *Art as Experience*. Vol. 10, in *John Dewey: The Late Works*, 1925–1933. Carbondale: Southern Illinois University Press, 1987.

Diderot, Denis, and F. M. Grimm. "Notice sur Clairaut." June 1, 1765. http://www .clairaut.com/n1juin1765po1pf.html (accessed June 12, 2014).

Dijkstra, Edsger W. "Real mathematicians don't prove." January 24, 1988. http://www .cs.utexas.edu/~EWD/ (accessed April 5, 2013).

Dixon, T.S.E. "A general algebraic method for the solution of equations." *The Analyst* **9** (January 1882): 1–8.

Doniger, Wendy. *The Uses and Misuses of Polytheism and Monotheism in Hinduism*. Divinity School at the University of Chicago|Publications. January 2010. http:// divinity.uchicago.edu/sites/default/files/imce/pdfs/webforum/012010 /monotheism%20for%20religion%20and%20culture-titlecorr.pdf (accessed March 2012).

Donoghue, Frank. *The Last Professors: The Corporate University and the Fate of the Humanities*. New York: Fordham University Press, 2008.

Douroux, Philippe. "Alexandre Grothendieck." *Images des Mathématiques*, CNRS. 8 février 2012. http://images.math.cnrs.fr/Alexandre-Grothendieck.html (accessed June 11, 2014).

Doxiadis, Apostolos. *Uncle Petros and Goldbach's Conjecture*. New York: Bloomsbury USA, 1992.

Doxiadis, Apostolos, and Barry Mazur. *Circles Disturbed*. Princeton: Princeton University Press, 2012.

Dreyfus, Laurence. *Bach and the Patterns of Invention*. Cambridge, MA: Harvard University Press, (1996).

———. *Wagner and the Erotic Impulse*. Cambridge: Harvard University Press, 2010.

Drinfel'd, Vladimir. "On a conjecture of Deligne." *Moscow Math. J.* **12** (2012): 515–42.

Du Bois-Reymond, Paul. "Was will die Mathematik und was will der Mathematiker?" *Jahresbericht der Deutschen Mathematiker-Vereinigung* **19** (1910): 190–98.

Dumézil, Georges. *Mythes et dieux des indo-européens*. Paris: Flammarion, 1992.

Eagleton, Terry. *Literary Theory: An Introduction*. Malden, MA: Wiley-Blackwell, 1996.

Edgeworth, Francis Ysidro. *Mathematical Psychics. An Essay on the Application of Mathematics to the Moral Sciences*. London: C. Kegan Paul and Co., 1881.

Ehrhardt, Caroline. "Évariste Galois and the social time of mathematics." *Revue d'histoire des mathématiques* **17**, no. 2 (2011): 175–96.

Einstein, Albert. "Why socialism?" *Monthly Review* (1949).

Ekeland, Ivar. "Response to Rogalski." *The Mathematical Intelligencer* (Summer 2010): 9–10.

El Karoui, Nicole, and Gilles Pagès. "Quel parcours scientifique en amont du Master 2 pour devenir 'Quant'?" *Maths-Fi*. http://www.maths-fi.com/devenirquant.html (accessed November 24, 2009).

Elias, Norbert. *Time: An Essay*. Trans. Edmund Jephcott. Oxford: Blackwell, 1993.

Ellerson, Lindsey. "Obama to Bankers: I'm standing 'between you and the pitchforks.' " ABC News, April 3, 2009. http://abcnews.go.com/blogs/politics/2009/04/obama-to-banker/ (accessed March 3, 2013).

Ellmann, Richard. *James Joyce*. Oxford: Oxford University Press, 1983.

Emmer, Michele. "The mathematics of enigma." In *Mathematics, Art, Technology, and Cinema*, ed. Michele Emmer and Mirella Manaresi. Berlin: Springer-Verlag, 2003, 145–51.

Enders, Jerome. *Yonkers the Lost City of Hip Hop*. Bloomington, IN: AuthorHouse, 2009.

Engelhardt, Nina. "Mathematics, reality and fiction in Pynchon's against the day." In *Thomas Pynchon and the (De)vices of Global (Post)modernity*, ed. Zofia Kolbuszewska. Lublin: Wydawnictwo KUL [John Paul II Catholic UP], 2013a.

———. "Thomas Pynchon: Against the day," chapter from forthcoming book, 2013b.

Epple, Moritz. "Felix Hausdorff's considered empiricism." In (Ferreirós and Gray 2006, pp. 263–89).

———. "Beyond metaphysics and intuition: Felix Hausdorff's views on geometry," lecture at Luminy 2007. www.cirm.univ-mrs.fr/videos/2007/exposes/11b/Epple.pdf (accessed September 2, 2013).

Esposito, Elena. *Die Fiktion der wahrscheinlichen Realität*. Berlin: Suhrkamp Verlag, 2007.

European Commission. "First European forum on cooperation between higher education and the business community." ec-European Commission. February 28–29, 2008. http://ec.europa.eu/education/higher-education/doc/business/forum08/report_en.pdf (accessed June 20, 2013; site now discontinued).

———. "Commission legislative and work programme 2009—Priority initiatives." The European Commission at Work. 2009. http://ec.europa.eu/atwork/pdf /clwp2009_roadmap_priority_initiatives_en.pdf (accessed November 13, 2009).

Evenden, E., and T. S. Freeman. *Religion and the Book in Early Modern England: The Making of John Foxe's 'Book of Martyrs.'* Cambridge University Press, 2011.

Fârâbî, al-. *Über die Wissenschaften/De Scientiis, Philosophische Bibliothek Band* 568. Hamburg: Felix Meiner Verlag, 2005.

Farmelo, Graham. *The Strangest Man.* London: Faber and Faber, 2010.

Faust, Drew Gilpin. "The university's crisis of purpose." *New York Times*, September 1, 2009.

Favaro, A. "Galileo astrologo secondo i documenti editi e inediti." *Mente a Cuore*, (1881): 1–10.

———. *Galileo Galilei e lo studio de Padova.* Florence, 1888.

FCIC (Financial Crisis Inquiry Commission). *The Financial Crisis Inquiry Report.* Washington, DC: U.S. Government Printing Office, January 2011.

Feffer, L. Butler. "Oswald Veblen and the capitalization of American mathematics: Raising money for research, 1923–1928 ." *Isis* **89**, no. 3 (September 1998): 474–97.

Fefferman, Charles. "Existence and smoothness of the Navier-Stokes equation." Clay Mathematics Institute. 2000. http://www.claymath.org/sites/default/files/navierstokes .pdf (accessed June 12, 2014).

Ferguson, Charles. *Predator Nation: Corporate Criminals, Political Corruption, and the Hijacking of America.* New York: Crown Business, 2012.

Ferrari, Oreste, and Giuseppe Scavizzi. *Luca Giordano: L'opera completa.* Napoli: Electa,1992.

Ferreirós, José. " 'O Theos' Arithmetizei: The rise of pure mathematics as arithmetic with Gauss." In *The Shaping of Arithmetic after C. F. Gauss's Disquisitiones Arithmeticae*, ed. Catherine Goldstein, Norbert Schappacher, and Joachim Schwermer. Berlin, Heidelberg: Springer-Verlag, 2007, 235–68.

Ferreirós, José, and Jeremy Gray, eds. *The Architecture of Modern Mathematics.* Oxford: Oxford University Press, 2006.

Fisch, Menachem. "The making of Peacock's treatise on algebra: A case of creative indecision ." *Arch. Hist. Exact Sci.* **54** (1999): 137–79 .

Fischer, K. "Fehlfunktionen der Wissenschaft." *Erwägen Wissen Ethik* **18** (2007): 3–16.

Fisher, C. S. "Some social characteristics of mathematicians and their work." *American J. Sociology* **78** (1973): 1094–1118.

Fisher, D., with Steven Bertoni and Devon Pendleton. "Survivor's guide for the affluent." *Forbes* (May 11, 2009).

Flexner, Abraham. "The usefulness of useless knowledge." *Harper's Magazine*, (1939): 544–52.

Florensky, Pavel. *Iconostasis.* Trans. D. Sheehan and O. Andrejev. Yonkers, NY: St. Vladimir's Seminary Press, 1996.

————. *The Pillar and Ground of Truth* [*Столп и утверждение истины*]. Trans. Boris Jakim. Princeton: Princeton University Press, 1997.

Foucault, Michel. "What is an Author?" In *The Foucault Reader*, ed. Paul Rabinow, 101–20. 1984.

Frege, Gottlob. *Begriffsschrift*. Halle: Verlag von Louis Nebert, 1879.

————. *Grundlagen der Arithmetik*. Breslau: W. Koebner, 1884.

Frenkel, E. *Love and Math*. New York: Basic Books (2013).

Frenkel, E., A. Losev, and N. Nekrasov. "Instantons beyond topological theory. I." *Journal of the Institute of Mathematics of Jussieu* **10**, no. 3 (2011): 463–565.

Freud, Sigmund. "The Relation of the Poet to Daydreaming." In *On Creativity and the Unconscious*. New York: HarperCollins, 1958, 1986, 44–54.

Friberg, Jöran. *A Remarkable Collection of Babylonian Mathematical Texts*. New York: Springer-Verlag, 2007.

Fried, Michael N., and Šabbetay Unguru. *Apollonius of Perga's Conica: Text, Context, Subtext*, Vol. 222. Leiden: Brill Academic Pub, 2001.

Friedlander, Eric, and Andrei Suslin. "The work of Vladimir Voevodsky." *Notices of the AMS*, (February 2003): 214–16.

Fry, Roger. *Vision and Design*. London: Chatto and Windus, 1920.

————. *Transformations*. New York: Doubleday Anchor Paperback, 1956.

Frye, Northrop. *The Anatomy of Criticism*. Princeton: Princeton University Press, 1957.

FSMP. "De la conjecture de Langlands à la carte à puce." FSMP (Fondation Sciences Mathématiques de Paris). 28 septembre 2007. http://old.sciencesmaths-paris.org /docs/280907_dp_web/6c-NEKOVAR-FOUQUET.pdf (accessed May 25, 2013).

Galison, P. L., B. Hevly, and R. Lowen. "Controlling the monster: Stanford and the growth of physics research, 1935–1962." In *Big Science: The Growth of Large-scale Research*, ed. Peter Louis Galison and Bruce Hevly. Stanford: Stanford University Press, 1992, 46–77.

Gamelin, Theodore W. "In memoriam: Raymond Redheffer" (2005). http://www .universityofcalifornia.edu/senate/inmemoriam/raymondredheffer.htm (accessed April 7, 2012).

Ganeri, Jonardon. "Analytic philosophy in early modern India." In *The Stanford Encyclopedia of Philosophy* (Winter 2012 Edition), ed. Edward N. Zalta. March 10, 2009. http://plato.stanford.edu/archives/win2012/entries/early-modern-india/ (accessed March 2012).

Garfield, Jay L. *The Fundamental Wisdom of the Middle Way*. Oxford: Oxford University Press, 1995.

————. "Taking conventional truth seriously: Authority regarding deceptive reality." *Philosophy East and West* **60** (July 2010): 341–54.

Garfield, Jay L., and Graham Priest. "Nāgārjuna and the limits of thought." *Philosophy East and West* 53 (2003): 1–21.

Gates, Henry Louis, Jr. *The Signifying Monkey*. Oxford: Oxford University Press, 1988.

————. "An Anthology of Rap Lyrics." *Financial Times* (November 5, 2010).

Gauss, Carl Friedrich. "Prolegomena zur Astronomie: Studium, Einteilung, Wert, 1808." In *Göttinger Universitätsreden aus zwei Jahrhunderten*: 1737–1934, ed. Wilhelm Ebel. Göttingen: Vandenhoeck & Ruprecht, 1978, 203–19.

Gelfand, Mikhail. "We do not choose mathematics as our profession, it chooses us: Interview with Yuri Manin." *Notices of the AMS* **56**, no. 10 (2009): 1268–74.

Gessen, Masha. *Perfect Rigor: A Genius and the Mathematical Breakthrough of the Century*. New York: Houghton Mifflin Harcourt, 2009.

Gheury, M.E.Y. "Item 303." *The Mathematical Gazette* (1909): 163–65.

Gibbs, Melvin. "Thug gods: Spiritual darkness and hip-hop," In *Everything But the Burden: What White People are Taking from Black Culture*, ed. G. Tate. New York: Broadway Books, 2003.

Gillies, D., ed. *Revolutions in Mathematics*. Oxford: Oxford University Press, 1995.

Gironi, Fabio. "Sunyata and the zeroing of being: A reworking of empty concepts." *Journal of Indian Philosophy and Religion* **15** (2012).

Glowatzki, E., and H. Göttsche. *Die Tafeln von Regiomontanus*. Institut für Geschichte der Naturwissenschaften, 1990.

Godechot, Olivier. "Finance and the rise of inequalities in France," Paris School of Economics, Working Paper No 2011–13, April 11, 2011. http://halshs.archives -ouvertes.fr/docs/00/58/48/81/PDF/wp201113.pdf (accessed August 23, 2013).

Goldstein, R. *Incompleteness: the Proof and Paradox of Kurt Gödel*. New York: W. W. Norton and Co., 2005.

Goodwin, Craufurd. "The value of things in the imaginative life: Microeconomics in the Bloomsbury Group." *History of Economics Review* **30** (2001): 56–73.

Gopnik, Alison. *The Philosophical Baby: What Children's Minds Tell Us about Truth, Love & the Meaning of Life*. New York: Farrar, Straus and Giroux, 2009.

Goppold, A. "Die Logik der Lehre von der Leere: Die Shunyata des Nāgārjuna." August 29, 2012. http://www.noologie.de/shunya01.htm.

Gould, Roger V. "The origins of status hierarchies: A formal theory and empirical test." *American Journal of Sociology* **107** (March 2002) 1143–78.

Gowers, W. Timothy. "The two cultures of mathematics." In *Mathematics: Frontiers and Perspectives*, ed. V. I. Arnold, M. Atiyah, P. Lax, and B. Mazur. Providence, RI: American Mathematical Society, 2000, 65–78.

———. "Finding moonshine: A mathematician's journey through symmetry." *Times Higher Education Supplement*, February 21, 2008.

———. "Elsevier—my part in its downfall." Gowers's Weblog. January 21, 2012. http://gowers.wordpress.com/2012/01/21/elsevier-my-part-in-its-downfall/ (accessed January 21, 2012).

———. "The Work of Pierre Deligne." The Abel Prize 2013. 2013. http://www .abelprize.no/c57681/binfil/download.php?tid=57753 (accessed March 20, 2013).

Gowers, W. Timothy, June Barrow-Green, and Imre Leader, eds. *The Princeton Companion to Mathematics*. Princeton: Princeton University Press, 2008.

Graham, Loren, and Jean-Michel Kantor. *Naming Infinity*. Cambridge: Harvard University Press, 2009.

Grattan-Guinness, Ivor. "The mathematics of the past: Distinguishing its history from our heritage." *Historia Mathematica* **31**, no. 2 (May 2004): 163–85.

Graves, Robert Perceval, and Augustus De Morgan. *Life of Sir William Rowan Hamilton, Andrews Professor of Astronomy in the University of Dublin, and Royal Astronomer of Ireland, Including Selections from His Poems, Correspondence, and Miscel-*

laneous Writings. Dublin: Hodges, Figgis, 1882. http://archive.org/details/lifeof sirwilliam01gravuoft.

Gray, Jeremy J. "Anxiety and abstraction in nineteenth-century mathematics." *Science in Context* **17**, no. 2 (2004) 23–47.

———. "A history of prizes in mathematics." In *The Millennium Prize Problems*, ed. J. Carlos, A. Jaffe, and A. Wiles. Providence, RI: American Mathematical Society (for the Clay Mathematics Institute), 2006, 3–27.

———. *Plato's Ghost*. Princeton: Princeton University Press, 2010.

Greenspan, Alan. *The Age of Turbulence: Adventures in a New World*. New York: The Penguin Press, 2007.

Greiffenhagen, Christian. "Video analysis of mathematical practice? Different attempts to 'open up' mathematics for sociological investigation." *FQS* **9**, no. 3 (September 2008): Art. 32.

———. "The materiality of mathematics: presenting mathematics at the blackboard." *British Journal of Sociology* (published online March 12, 2014).

Greiffenhagen, Christian, and W. Sharrock. "Mathematical equations as Durkheimian social facts." In *Sociological Objects*, ed. A. King, R. Rettie, and G. Cooper, 119–35. London: Ashgate, 2009a.

———. "Two concepts of attachment to rules." *J. Classical Sociology* **9** (2009b): 403-425.

———. "Does mathematics look certain in the front, but fallible in the back?" *Social Studies of Science* **41** (2011): 839–66.

Greuel, Gert-Martin, Reinhold Remmert, and Gerhard Rupprecht, eds. *Mathematik— Motor der Wirtschaft*. Berlin, Heidelberg: Springer-Verlag, 2008.

Grosholz, Emily. "Jules Vuillemin's La Philosophie de l'algèbre: the philosophical uses of mathematics." In *Philosophie des mathématiques et théorie de la connaissance. L'Œuvre de Jules Vuillemin*, ed. R Rashed and P. Pellegrin. Paris: Albert Blanchard, 2005, 253–69.

———. *Representation and Productive Ambiguity in Mathematics and the Sciences*. Oxford: Oxford University Press, 2007.

Grothendieck, Alexander. "Récoltes et Sémailles." Unpublished manuscript, 1988.

Guedj, Denis. "Ces mathématiques vendues aux financiers." *Libération*, 10 décembre 2008.

Guerzoni, Guido. "Liberalitas, magnificentia, splendor: The classic origins of Italian Renaissance lifestyles." In *Economic Engagements with Art*, ed. N. N. De Marchi and C.D.W. Goodwin. Durham, NC: Duke University Press, 1999, 332–78.

Guillien, Mathieu. *La techno minimale*. Paris: Éditions Aedam Musicae, 2014.

Gustin, B. H. "Charisma, recognition, and the motivation of scientists." *Am. J. Sociology* (1973): 1119–34.

Hacking, Ian. *Why is there Philosophy of Mathematics at All?* Cambridge: Cambridge University Press, 2014.

Hadot, Pierre. *The Veil of Isis*. Trans. Michael Chase. Cambridge: Harvard University Press, 2006.

Hagstrom, W. O. "Anomy in scientific communities." *Social Problems* **12**, no. 2 (Autumn 1964): 186–95.

Hagstrom, W. O., and L. Hargens. "Scientific consensus and academic status attainment patterns." *Sociology of Education* **55** (1982): 183–96.

Halimi, Serge. "L'audace ou l'enlisement." *Le Monde Diplomatique* (avril 2012).

Hamkins, J. D., D. Linetsky, and J. Reitz. "Pointwise definable models of set theory." *J. Symbolic Logic*, **78** (2013): 139–56.

Hardy, G. H. "Mathematical proof." *Mind*, New Series **38**, no. 149 (1929): 1–25.

———. *A Mathematician's Apology*. Cambridge: Cambridge University Press, 2012.

Harford, Tim. "Black-Scholes: The maths formula linked to the financial crash." BBC News, April 27, 2012. http://www.bbc.co.uk/news/magazine-17866646 (accessed June 22, 2013).

Hargens, L. L., N. C. Mullins, and P. K. Hecht. "Research areas and stratification processes in science." *Social Studies of Science* **10** (1980): 55–74.

Harkinson, Josh. "Meet the financial wizards working with Occupy Wall Street." *Mother Jones* (December 13, 2011).

Harmsen, Torsten. "Mathematikgenie aus Berlin." *Berliner Zeitung* (October 15, 2012).

Harris, Michael. Supercuspidal representations in the cohomology of Drinfel'd upper half spaces; elaboration of Carayol's program, *Inventiones Math.* **129** (1997): 75–119.

———. Potential automorphy of odd-dimensional symmetric powers of elliptic curves, and applications, Vol. 270 in *Algebra, Arithmetic, and Geometry*, Volume II: In Honor of Yu. I. Manin, Ed. Yuri Tschinkel and Yuri Zarhin, Boston: Birkhäuser, Progress in Mathematics, 2009, 1–21.

———. "Do androids prove theorems in their sleep?" In (Doxiadis and Mazur 2012, 130–82).

Harris, Michael, and Richard Taylor. "The geometry and cohomology of some simple Shimura varieties," *Annals of Mathematics Studies* **151** (2001).

Harris, Michael, "Pynchon's postcoloniality." In *Thomas Pynchon, Reading from the Margins*, ed. N. Abbas. Danvers: Rosemont Publishing & Printing Corp., 2003, 199–214.

Hasse, Helmut. *Mathematik als Wissenschaft, Kunst, und Macht*. Wiesbaden: Verlag für Angewandte Wissenschaften, 1952.

Hausdorff, Felix. *Gesammelte Werke. Band VII: Philosophisches Werk*. Ed. Werner Stegmaier. Berlin: Springer-Verlag, 2004.

———. *Gesammelte Werke. Band VIII: Literarisches Werk*. Eds. F. Vollhardt and U. Roth. Berlin: Springer-Verlag, 2011.

Hayes, Brian. "Odd numbers." *The Sciences* **38** (September/October 1998).

Heard, J. M. "The Evolution of the Pure Mathematician in England, 1850–1920." EThOS. 2004. http://ethos.bl.uk/OrderDetails.do;jsessionid=A9FC981F210E2E4F7 643F25A9878207D?did=1&uin=uk.bl.ethos.411630 (accessed April 6, 2012).

Heidegger, Martin. *Being and Time*. Trans. John Macquarrie and Edward Robinson. New York: Harper and Row, 1962.

———. *Identity and Difference*. Trans. Joan Stambaugh. New York: Harper Torchbooks, 1969.

Heintz, Bettina. *Die Innenwelt der Mathematik: Zur Kultur und Praxis einer beweisenden Disziplin*. Vienna: Springer-Verlag, 2000.

———. "When is a proof a proof?" *Social Studies of Science* **33** (2003): 929–43.

Hergé. *Tintin au pays des soviets*. Brussels: Casterman, 1930.

Hermanowicz, J. C. "Argument and outline for the sociology of scientific (and other) careers." *Social Studies of Science* **37** (2007): 625–46.

Hersh, Reuben. "Mathematics has a front and a back." *Eureka* **48** (1988): 27–31.

———. "Mathematics has a front and a back." *Synthèse* **88** (1991): 127–33.

———. *What is Mathematics, Really?* Oxford: Oxford University Press 1997.

———. "Wings, not foundations!" In Giandomenico Sica, ed.. *Essays on the Foundations of Mathematics and Logic*, Vol. 1, Monza: Polimetrica International Scientific Publisher, 2005, 155–64.

Hewitt, Ivan. "Why music as maths just doesn't add up." *The Telegraph*, June 24, 2011.

Higgins, Robert C. *Analysis for Financial Management*. 6th ed. New York: McGraw Hill, 2001.

Hilbert, David. "Mathematische Probleme, Vortrag, gehalten auf dem internationalen Mathematiker-Kongreß zu Paris 1900." *Nachrichten von der Königl. Gesellschaft der Wissenschaften zu Göttingen. Mathematisch-Physikalische Klasse*, 1900, 253–97.

Hilbert, David, and A. Hurwitz. "Über die diophantischen Gleichungen von Geschlecht Null." *Acta Mathematica* **14** (1890): 217–24.

Hill, Christopher. *Change and Continuity in Seventeenth-Century England*. Cambridge: Harvard University Press, 1975.

Hirsch, J. E. "An index to quantify an individual's scientific research output." *PNAS* **102**, no. 46 (2005): 16569–72.

Hoffman, Paul. *The Man Who Loved Only Numbers*. Boston: Hyperion, 1998.

Hofmann, K. H. "Commutative diagrams in the fine arts." *Notices of the AMS* **49**, no. 6 (June–July 2002). 663–68.

Hood, Robert. "TH3 M4TH3M4T1C 4554551N5 - C4LCUL4T0R." YouTube. November 19, 2008. http://www.youtube.com/watch?v=iDdsS0BSbw8 (accessed December 21, 2011).

Hooker, C., et al. "The group as mentor." In *Group Creativity*, eds. P. B. Paulus and B. A. Nijstad. Oxford University Press, 2003, 225–44.

Hornaday, Ann. "Waiting for 'action'!" *Washington Post*, July 10, 2007.

Høyrup, Jens. "Jordanus de Nemore: A case study on 13th century mathematical innovation and failure in cultural context." *Philosophica* **42** (1988): 43–77.

Hubey, H. M. "Catastrophe theory and the human sexual response." In Third International Symposium on Systems Research, Informatics, & Cybernetics, Baden-Baden, Germany, August, 12–18, 1991.

Huet, Sylvestre. "Boss des maths." *Libération*, May 28, 2003.

Huizinga, Johan. *Homo Ludens*. Boston: Beacon Press, 1950.

Hume, David. *A Treatise of Human Nature*. London: John Noon, 1739.

Hurston, Zora Neale. "High John de Conquer." *The American Mercury* (October 1943): 450–58.

Hyde, Lewis. *Trickster Makes This World*. New York: Farrar, Straus and Giroux, 1998, 2010.

Ikeda, Ryoji, and Frédéric Bonnet. "Le nombre premier représente la beauté." *Journal des Arts* (Octobre 2008).

Ishizu T and S. Zeki. "Toward a brain-based theory of beauty." *PLoS ONE* **6**, no. 7, e21852. doi:10.1371/journal.pone.0021852 (July 6, 2011).

O'Doherty, J., et al. "Beauty in a smile: the role of medial orbitofrontal cortex in facial attractiveness." *Neuropsychologia* **41**, no. 2 (2003): 147–55.

Jackson, Allyn. "Million-dollar mathematics prizes announced." *Notices of the AMS* (September 2000): 877–79.

———. "Comme Appelé du Néant: As if summoned from the void." *Notices of the AMS* **51**, no. 9, 10 (October, November 2004): 1038–56, 1196–1212.

Joseph, George Gheverghese. *The Crest of the Peacock*, 3rd ed. Princeton: Princeton University Press, 2011.

Josten, C. H. "A translation of John Dee's 'Monas Hieroglyphica' (Antwerp, 1564), with an introduction and annotations." *Ambix* **xii**, nos. 2, 3 (June and October 1964): 83–222.

Jouve, François. "Vive les mathématiques financières!" *Images des Mathématiques* (18 octobre 2011). http://images.math.cnrs.fr/Vive-les-mathematiques-financieres.html (accessed November 11, 2011).

Joyal, André. "The theory of quasi-categories and its applications." Notes of a course given in 2008 online at http://mat.uab.cat/~kock/crm/hocat/advanced-course /Quadern45-2.pdf (accessed April 29, 2011).

Jung, Carl Gustav. *The Psychology of the Transference*. Vol. 16. Princeton: Princeton University Press, 1969.

Justens, Daniel. "Dieu, le sexe et la mathématique." Haute-Ecole Francisco Ferrer Papiers de Recherche. undated. http://www.he-ferrer.eu/uer-mathematique/papiers-de -recherche (accessed June 6, 2010).

Kästner, Abraham Gotthelf. *Geschichte der Mathematik Erster Band*, Göttingen: Rosenbusch, 1799.

Kahane, J.-P., D. D. Talay, and M. Yor. "Finance, politique et mathématiques, quels liens?" *Images des Mathématiques* (2 mai 2009). http://images.math.cnrs.fr/Finance -politique-et-mathematiques.html (accessed November 4, 2012).

Kahn, Charles H. *The Art and Thought of Heraclitus: An Edition of the Fragments with Translation and Commentary*. Cambridge: Cambridge University Press, 1981.

Kaiser, D. *How the Hippies Saved Physics*. New York: W. W. Norton, 2012.

Kanigel, Robert. *The Man Who Knew Infinity*. New York: Washington Square Press, 1991.

Kant, Immanuel. *The Critique of Judgment*. Trans. James Creed Meredith. 1790.

Kausler, E. *Reimchronik von Flanderen*. Tübingen: Ludwig Friedrich Fues, 1840.

Kehlmann, Daniel. *Die Vermessung der Welt*. Reinbek bei Hamburg: Rowohlt, 2005.

Kennedy, Juliette. "On formalism freeness: implementing Gödel's 1946 Princeton bicentennial lecture." *Bulletin of Symbolic Logic*, in press.

Khwarizmi Al-. *The Algebra of Mohammed ben Musa*. Trans. Frederic Rosen. London: The Oriental Translation Fund, 1831.

Kim, Walter. "Valley of the nerds." *GQ* (July 1991).

Kirnberger, Johann Philipp. *Die Kunst des reinen Satzes in der Musik* **2 Teil**, Abteilung 3 Berlin und Königsberg: Decker & Hartung, 1771.

Kitcher, Philip. *The Nature of Mathematical Knowledge*. Oxford: Oxford University Press, 1984.

Koblitz, Ann Hibner. *A Convergence of Lives: Sofia Kovalevskaia—Scientist, Writer, Revolutionary*. New Brunswick: Rutgers University Press, 1983.

Koblitz, Neal. *Random Curves: Journeys of a Mathematician*. New York: Springer-Verlag, 2007.

Kontsevich, Maxim, and Don Zagier. "Periods." In *Mathematics Unlimited, Year 2001 and Beyond*, ed. B. Engquist and W. Schmid. New York: Springer-Verlag, 2001.

Konzelmann Ziv, A. "Naturalized rationality—a glance at Bolzano's philosophy of mind." *The Baltic International Yearbook of Cognition, Logic and Communication* **4** (2009): 1–20.

Kottwitz, Robert. "The fundamental lemma: From minor irritant to central problem." Institute for Advanced Study (2010). http://www.ias.edu/about/publications/ias-letter/articles/2010-summer/fundamental-lemma (accessed April 5, 2013).

Krömer, R. *Tool and Object, A History and Philosophy of Category Theory*. Basel: Birkhäuser, 2007.

Krantz, Steven G. *Mathematical Apocrypha*. Washington, DC: The Mathematical Association of America, 2002.

Kreisel, Georg. "Informal rigour and completeness proofs." *Problems in the Philosophy of Mathematics* (1967): 138–86.

Krenek, Ernst. *Music Here and Now*. New York: W. W. Norton, 1939.

Kronecker, Leopold. "Über den Begriff der Zahl in der Mathematik." *Revue d'histoire des mathématiques* **7** (2001): 207–75.

Krugman, Paul. "Rewarding bad actors." *New York Times*, August 2, 2009.

Krugman, Paul, and Robin Wells. "The busts keep getting bigger: Why?" *New York Review of Books*, July 14, 2011.

Kuhn, Thomas. *The Structure of Scientific Revolutions*. Chicago: University of Chicago Press, 1996.

Lafforgue, Laurent. "Notre vrai problème est de retrouver le réel." *Le Figaro* (30 octobre 2008).

Lakatos, Imre. *The Methodology of Scientific Research Programmes: Philosophical Papers* Vol. 1. Cambridge: Cambridge University Press, 1978.

Lambert, A., and L. Mazliak. "E la nave va?" *Gazette des mathématiciens* **120** (avril 2009): 103–105.

La Nave, Federica. "Deductive narrative and the epistemological function of belief in mathematics: On Bombelli and imginary numbers." In (Doxiadis and Mazur 2012, pp. 79–104).

Lang, Serge. "Mordell's review, Siegel's letter to Mordell, diophantine geometry, and 20th century mathematics." *Notices of the AMS* **42**, no. 3 (March 1995).

Langlands, R. P., Jr. "Automorphic representations, Shimura varieties, and motives. Ein Märchen." In *Proc. Symp. Pure Math.*, **33**, II , eds A. Borel and W. Casselman, Providence: American Mathematical Society, 1979a, 205–446.

———. "On the zeta-functions of some simple Shimura varieties." *Canadian J. Math.* **31** (1979b): 1121–16.

———. "Reflexions on receiving the Shaw prize." (2007). http://publications.ias.edu/rpl/.

———. "Is there beauty in mathematical theories?" In Vittorio Hösle, *The Many Faces of Beauty*. Notre Dame, IN: University of Notre Dame Press, 2013.

Lanier, Jaron. *You Are Not a Gadget*. New York: Alfred A. Knopf, 2010.

Lave, Jean. *Cognition in Practice: Mind, Mathematics, and Culture in Everyday Life*. Cambridge: Cambridge University Press, 1988.

Lavin, Irving, and Marilyn Aronberg Lavin. "Truth and beauty at the Institute for Advanced Study." Institute for Advanced Study. 2012. www.ias.edu/files/pdfs/publications/ias-seal.pdf (accessed October 27, 2012).

Lear, Jonathan. "Katharsis." In *Essays on Aristotle's Poetics*. Princeton: Princeton University Press, 1992, 315–40.

Legge, James. "Top economist Jeffrey Sachs says Wall Street is full of 'crooks' and hasn't changed since the financial crash." *The Independent*, April 29, 2013.

Levinas, Emmanuel. "Philosophy and the idea of infinity." In *Collected Philosophical Papers*, trans. A. Lingis. Dordrecht: Martinus Nijhoff, 1987.

Lévi-Strauss, Claude. *The Raw and the Cooked*. Trans. John and Doreen Weightman. New York: Harper and Row, 1969.

———. *Tristes Tropiques*. Trans.J. and D. Weightman. London: Penguin, 1992.

Lindsay, Richard R., and Barry Schachter. *How I Became a Quant*. Hoboken, NJ: John Wiley and Sons, 2007.

Lippman, Edward A. "Objectivity of music." In *Musical Aesthetics: A Historical Reader*. Vol III, *The Twentieth Century*. Stuyvesant, NY: Pendragon Press, 1990.

Lipschitz, R. "Fortgesetzte Untersuchungen in Betreff der ganzen homogenen Functionen von *n* Differentialen." *J. reine angew. Mathematik* **72** (1870): 1–56.

Littlewood, John E. *Littlewood's Miscellany*, ed. B. Bollobás. Cambridge: Cambridge University Press, 1986.

Lockhart, Paul. *A Mathematician's Lament*. New York: Bellevue Literary Press, 2009.

Lorentz Center. *Mathematics: Algorithms and Proofs*. 2011. http://www.lorentzcenter.nl/lc/web/2011/467/description.php3?wsid=467 (accessed June 6, 2013).

Lurie, Jacob. *Higher Topos Theory*. Princeton: Princeton University Press, 2009.

Maas, H. "Pacifying the workman: Ruskin and Jevons on labor and popular culture." In *Economic Engagements with Art*, ed. N. De Marchi and C.D.W. Goodwin. Durham, NC: Duke University Press, 1999, 85–120.

MacCannell, Juliet Flower. "Alain Badiou: Philosophical outlaw." In Alain Badiou: *Philosophy and Its Conditions*, ed. Gabriel Riera. Albany, NY: SUNY Press, 2005, 137–85.

MacIntyre, Alasdair. *After Virtue*. Notre Dame, IN: Notre Dame University Press, 1981.

MacKenzie, Donald. "Slaying the Kraken: The sociohistory of a mathematical proof." *Social Studies of Science* **29** (1999): 7–60.

———. *Mechanizing Proof: Computing, Risk, and Trust*. Cambridge, MA: The MIT Press, 2001.

———. "An equation and its worlds: Bricolage, exemplars, disunity and performativity in financial economics." *Social Studies of Science* **33** (2003): 831–68.

Mahajan, Sanjoy. *Street-Fighting Mathematics: The Art of Educated Guessing and Opportunistic Problem Solving*. Cambridge, MA: MIT Press, 2010.

Mancosu, Paolo, ed. *From Brouwer to Hilbert. The Debate on the Foundations of Mathematics in the 1920s*, Oxford: Oxford University Press, 1998.

———. "Between Russell and Hilbert: Behmann on the foundations of mathematics." *The Bulletin of Symbolic Logic*, 5 (1999) 303–30.

————, ed. *Philosophy of Mathematical Practice*. New York: Oxford University Press, 2008.

Manin, Yuri Ivanovich. "Georg Cantor and his heritage." *Proc. Steklov Inst. Math.* 246 (2004): 195–203.

————. *Mathematics as Metaphor: Selected Essays of Yuri I. Manin*. Providence, RI: American Mathematical Society, 2007.

————. "Foundations as superstructure (reflections of a practicing mathematician)." arXiv.org. May 28, 2012. http://arxiv.org/pdf/1205.6044v1.pdf (accessed April 28, 2013).

Martin, John Levi. *Social Structures*. Princeton: Princeton University Press, 2009.

Martin, Ursula. "Computers, reasoning and mathematical practice " In *Computational Logic, Proceedings of the NATO Advanced Study Institute on Computational Logic, held in Marktoberdorf, Germany, July 29–August 10, 1997*, ed. U. Berger, H. Schwichtenberg. Berlin: Springer-Verlag, 1998. 301–46.

Marx, Karl. *Grundrisse*, 1858.

mathbabe (Catherine O'Neil). "Is mathbabe a terrorist or a lazy hippy? (#OWS)", *mathbabe* (January 1, 2013). http://mathbabe.org/2013/01/01/is-mathbabe-a-terrorist-or-a-lazy-hippy-ows/ (accessed June 11, 2014).

MathOverflow. "Philosophy behind Mochizuki's work on the ABC conjecture." *MathOverflow* (September 9, 2012). http://mathoverflow.net/questions/106560/philosophy-behind-mochizukis-work-on-the-abc-conjecture (accessed September 9, 2012).

Maths à Venir. *Maths à Venir* (2009). http://www.maths-a-venir.org/ (accessed May 2013).

Mazur, Barry. "Conjecture." *Synthèse* **111** (1997): 197–210.

————. *Imagining Numbers (particularly the square root of minus fifteen)*. New York: Picador, 2004a.

————. "What is . . . a Motive?" *Notices of the AMS* **51**, no. 10 (November 2004b): 1214–16.

————. "Finding meaning in error terms." *Bulletin of the AMS* (April 2008a): 185–228.

————. "When is one thing equal to some other thing." In *Proof and Other Dilemmas: Mathematics and Philosophy*, ed. B. Gold and R. Simons. Washington, DC: Mathematical Association of America Spectrum Series, 2008b, 221–42.

Mazzucato, Luca. "La formula dell'amore: matematici al cinema." *Oggi Scienza* (April 2010).

McCain, K. W. "The view from Garfield's shoulders: Tri-citation mapping of Eugene Garfield's citation image over three successive decades." *Ann. Library Information Studies* **57** (2010): 261–70.

McLarty, Colin. "Emmy Noether's 'set theoretic' topology: From Dedekind to the rise of functors." In (Ferreiros and Gray 2006, pp. 187–208).

————. "The Rising Sea: Grothendieck on Simplicity and Generality." Articles |Philosophy|Case Western Reserve University. 2007. www.cwru.edu/artsci/phil/RisingSea.pdf (accessed May 1, 2013).

Mehrtens, Herbert. *Moderne Sprache, Mathematik*. Frankfurt: Suhrkamp, 1990.

Meillassoux, Quentin. *Après la Finitude*. Paris: Éditions du Seuil, 2006.

Menand, Louis. *The Metaphysical Club*. New York: Farrar, Strauss, and Giroux, 2001.

————. "Do the math." *The New Yorker*, November 27, 2006.

Menkes, Suzy. "Designers outline a new geometry." *New York Times*, March 5, 2010.

Merton, Robert K. "The Matthew Effect in science." *Science* **159** (1968): 56–63.

Metz, Christian. *Le signifiant imaginaire—Psychanalyse et cinéma*. Paris: Union Générale d'Editions, Coll, October 18, 1977.

MFO. "Further Publications—MFO." *Mathematisches Forschunginstitut Oberwolfach.* undated. http://www.mfo.de/scientific-programme/publications/further-publications (accessed March 24, 2013).

Miller, Dean A., *The Epic Hero*, Baltimore: Johns Hopkins University Press, 2000.

Miller, G. A. "Definitions of the term 'Mathematics.' " *The American Mathematical Monthly* **15**, no. 11 (November 1908): 197–99.

Mirowski, Philip. *The Effortless Economy of Science?* Durham, NC: Duke University Press, 2004.

Moore, G. E. *Principia Ethica*. Cambridge: Cambridge University Press, 1903.

Mosselmans, B., and E. Mathijs. "Jevons's music manuscript and the political economy of music," In *Economic Engagements with Art*, ed. N. De Marchi and C.D.W. Goodwin. Durham, NC: Duke University Press, 1999, 121–56.

μ. "Review of Das Chaos in kosmischer Auslese," *The Monist* **11**. (1900) 141–42.

Mulkay, Michael. "The mediating role of the scientific elite." *Social Studies of Science* **6** (1976): 445–70.

Muntadas. *Between the Frames*. Barcelona: Museu d'Art Contemporani de Barcelona, 2011.

My[confined]Space. "Jump you fuckers! |My[confined]Space." *My[confined]Space*. September 29, 2008. www.myconfinedspace.com/2008/09/29/jump-you-fuckers/ (accessed March 3, 2013)

Nāgārjuna. "Mūlamadhyamikakārikā." *Bibliotheca Polyglotta*. https://www2.hf.uio.no/polyglotta/index.php?page=volume&vid=27, input by F. Liland, 2010 (accessed December 23, 2011).

Natalini, Roberto. "David Foster Wallace and the mathematics of infinity." In M. Boswell and S. J. Burn, eds., *A Companion to David Foster Wallace Studies*. New York: Palgrave Macmillan, 2013, 43–58.

Netz, Reviel. *Ludic Proof*. Cambridge: Cambridge University Press, 2009.

Newfield, Christopher. *Unmaking the Public University: The Forty-Year Assault on the Middle Class*. Cambridge, MA: Harvard University Press, 2008.

Ngô, Bảo Chau. "Le lemme fondamental pour les algèbres de Lie." *Publications mathématiques de l'IHES* **111** (2010): 1–169.

Nietzsche, Friedrich. *The Gay Science*. Trans. Walter Kaufmann. New York: Vintage Books, 1974.

Nightingale, Andrea Wilson. *Spectacles of Truth in Ancient Greek Philosophy*. Cambridge: Cambridge University Press, 2009.

Nocera, Joe. "Poking holes in a theory of markets." *New York Times*, June 5, 2009.

Nochimson, M. P. *The Passion of David Lynch: Wild at Heart in Hollywood*. Austin: University of Texas Press, 1997.

NSA. "Suite B Cryptography / Cryptographic Interoperability." National Security Agency. January 15, 2009. http://www.nsa.gov/ia/programs/suiteb_cryptography/index.shtml (accessed June 16, 2013).

Nürnberger, Joseph Emil. *Die letzten Gründe der höheren Analysis*. Halle: Renger, 1815.

OMI. "Going from strength to strength. OMI in its fifth year." Oxford-Man Institute of Quantitative Finance. August 2012. http://www.oxford-man.ox.ac.uk/news/annual _report_2011_2012 (accessed July 2, 2013).

O'Neil, Catherine. "A view on the transition from academia to finance." *Notices of the AMS* **55**, no. 6 (June/July 2008): 700–2.

Oort, Frans. "Did earlier thoughts inspire Grothendieck?" in (Schneps 2014a, 231–68).

O'Toole, A. L. "Insights or trick methods?" *National Mathematics Magazine* **15**, no. 1 (October 1940): 35–38.

Overbye, Dennis. "They tried to outsmart Wall Street." *New York Times*, March 9, 2009.

Özdural, A. "A mathematical sonata for architecture: Omar Khayyam and the Friday mosque of Isfahan." *Technology and Culture* **39**, no. 4 (1998): 699–715.

Parliamentary Commission on Banking Standards. "Changing banking for good." June 12, 2013. http://www.publications.parliament.uk/pa/jt201314/jtselect/jtpcbs/27 /27.pdf (accessed July 2, 2013).

Parshin, Alexey Nikolaevich. *Путь. Математика и другие миры* [*The Way. Mathematics and Other Worlds*]. Moscow: Dobrosvet, 2002.

———. *Зеркала и отражения* [*Mirrors and Reflections*] pdf file at http://iph.ras.ru /uplfile/histph/yearbook/2005/5.pdf, pp. 9-12.

———. *Лестница отражений* [*Staircase of Reflections*], in *Антропологические матрицы XX века. Л.С. Выготский – П.А. Флоренский: несостоявшийся диалог— приглашение к диалогу* [*Anthropological matrices of the twentieth century. L.S. Vygotsky—P.A. Florensky*], A. A. Andryshkov et al., eds. Moscow: Progress-Traditsii, 2007, 205–219.

———. *Зеркало и двойник у Гоголя и Достоевского* [*Mirror and Double in Gogol and Dostoevski*]. In *А.Л.Бем и гуманитарные проекты русского зарубежья* (Conference proceedings) Moscow: Russkii Put', 2008, 136–50.

Pater, Walter. *Studies in the History of the Renaissance*. 1873.

Patterson, Scott. *The Quants*. New York: Crown Business, 2010.

Pettie, Andrew. "Marcus du Sautoy interview: 'There's a geek chic about maths now.' " *The Telegraph*, July 26, 2011.

Piasere, L. Il trickster e l'infinito: alcuni riflessioni a partire da esempi Rom. *I Quaderni del Ramo d'Oro*, online n. 2 (2009).

Pieper, Herbert. "Network of scientific philanthropy: Humboldt's relations with number theorists." In *The Shaping of Arithmetic, Number Theory after Carl Friedrich Gauss's "Disquisitiones Arithmeticae,"* ed. C. Goldstein, N. Schappacher, and J. Schwermer. Berlin, Heidelberg: Springer-Verlag, 2007, 210–33.

Pimm, David, and Nathalie Sinclair. *Mathematics and the Aesthetic*. Berlin: Springer-Verlag, 2006.

Pinker, Steven. *How the Mind Works*. New York: W. W. Norton and Co., 1997.

progfrance. 2009–2013. http://www.grandemprunt.net/ (accessed May 8, 2013).

Plassard, D. *L'acteur en effigie*, Paris: Editions L'Âge d'homme, 1992.

Pólya, G. *Mathematical Discovery: On Understanding, Learning, and Teaching Problem Solving*, Vol. II. New York: John Wiley and Sons, 1962.

Purkert, Walter. "The double life of Felix Hausdorff/Paul Mongré." *Mathematical Intelligencer* **30** (2008) 36–50.

Pyman, Avril. *Pavel Florensky: A Quiet Genius*. New York: Continuum, 2010.

Pynchon, Thomas. *Gravity's Rainbow*. New York: Viking Press, 1973.

———. "Is it OK to be a Luddite?" *New York Times Book Review*, October 28, 1984.

———. *Mason & Dixon*. New York: Henry Holt and Co., 1997.

———. *Against the Day*. New York: Penguin Press, 2006.

Quinn, Frank. "A revolution in mathematics? What really happened a century ago and why it matters today." *Notices of the AMS*, January 2012: 31–37.

Rancière, Jacques. *Aisthesis*. Paris: Editions Galilée, 2011.

Rashed, Roshdi. "La modernité mathématique: Descartes et Fermat." In *Philosophie des mathématiques et théorie de la connaissance, L'oeuvre de Jules Vuillemin*, by R. Rashed and P. Pellegrin. Paris: Eds. Albert Blanchard, 2005, 238–52.

———. *D'al-Khwarizmi à Descartes*. Paris: Hermann, 2011.

Rashed, Roshdi, and B. Vahabzadeh. *Al-Khayyam, mathématicien*. Paris: Eds. Albert Blanchard, 1999.

———. *Omar Khayyam, the Mathematician*. Persian Heritage Foundation, 2000.

Restivo, Sal. "The social life of mathematics." *Philosophica* **42** (1988): 5–20.

———. *Mathematics in Society and History: Sociological Inquiries*, Vol. 20. Springer London, 2001.

Rheingold, Howard. *Virtual Reality*. New York: Simon & Schuster, 1992.

Richardson, Nick. "Set on being singular, review of *Arnold Schoenberg* by B. Bujic." *London Review of Books* (October 20, 2011).

Robinet, André. *Aux sources de l'esprit cartésien*. Paris: J. Vrin , 1996.

Rocard, Michael. "La crise sonne le glas de l'ultralibéralisme." *Le Monde* (1 novembre 2008).

Rogalski, Marc. "Mathematics and finance: An ethical malaise." *Mathematical Intelligencer*, 32 (2010): 6–8.

Rosental, Claude. *La trame de l'évidence: sociologie de la démonstration en logique*. Paris: Presses Universitaires de France-PUF, 2003.

Ross, Andrew. *Nice Work if You Can Get It*. New York: NYU Press, 2009.

Rothman, Tony. "Genius and biographers: The fictionalization of Evariste Galois." *American Mathematical Monthly* **89** (February 1982): 84–106.

Rotman, Brian. *Signifying Nothing*. Stanford: Stanford University Press, 1987.

Roy, Marie-Françoise, and Aline Bonami, eds. "Maths à Venir 2009." *Gazette des Mathématiciens* **124** (April 2010): 49–64.

Rucker, Rudy. *Software (The Ware Tetralogy, Book* 1). New York: Ace Books, 1982.

Ruelle, David. *Chance and Chaos*. Princeton: Princeton University Press , 1991.

Rump, Wolfgang. "Mathematik ohne Kunstgriffe." http://www.iaz.uni-stuttgart.de/LstAGeoAlg/Rump/kunst.html (accessed May 26, 2011).

Ruskin, John. "The project Gutenberg eBook of modern painters, Vol. I." Project Gutenberg. September 4, 2009. http://www.gutenberg.org/files/29907/29907-h/29907-h.htm (accessed June 12, 2013).

Sadler, Thomas. "A complete system of practical arithmetic (both vulgar & decimal); on an entire new plan to which is added, a large collection of new questions, with only the answers thereto." Online at http://archive.org/stream/acompletesystem

00sadlgoog#page/n7/mode/2up. Shrewsbury: printed for the author by W. Williams, 1773.

Saiber, Arielle. "Ornamental flourishes in Giordano Bruno's geometry." *The Sixteenth Century Journal* **34**, no. 3 (2003): 729–45.

Saint-Saëns, Camille. "Portraits et Souvenirs." *Société d'édition artistique* (1900).

Salmon, G. "Arthur Cayley." *Nature* **XXVIII**, no. 725 (September 1883): 481–85.

Sante, Luc. "Inside the time machine." *The New York Review of Books* **54**, no.1, (January 11, 2007).

Schappacher, Norbert. "Rewriting points." In *Proceedings of the International Congress of Mathematicians, Hyderabad, India 2010*, 3258–91. New Delhi: Hindustan Book Agency, 2010.

Scharlau, Winfried. *Wer ist Alexander Grothendieck, Anarchie, Mathematik, Spiritualität, Einsamkeit,* Teil 3: *Spiritualität.* Norderstadt: Books On Demand, 2010.

Schenker, Heinrich. "Rameau or Beethoven? Paralysis or spiritual life in music?" translated and reprinted in *The Masterwork in Music*, Volume III, Cambridge: Cambridge University Press. 1997.

Schiller, Friedrich. *On the Aesthetic Education of Man.* Trans. Reginald Snell. Mineola, NY: Dover Publications, 2004.

Schmid, Wilfried. "On a conjecture of Langlands." *Annals of Mathematics* **93** (1971), 1–42.

Schneps, Leila, ed. *Alexandre Grothendieck: A Mathematical Portrait.* Somerville, MA: International Press, 2014a.

———. "The Serre-Grothendieck correspondence." (2014b) In (Schneps 2014a, 193–230).

Schreiber, Urs. "What is homotopy type theory good for?" *n-Category Café.* (May 10, 2012).

Schwartz, Laurent. *A Mathematician Grappling with His Century.* Boston: Birkhäuser, 2001.

Scott, Joan Wallach. "Knowledge, power, and academic freedom." *Social Research* **76** (Summer 2009): 451–80.

Segal, Sanford L. *Mathematicians under the Nazis.* Princeton: Princeton University Press, 2003.

Sepper, D. L. *Descartes's Imagination: Proportion, Images, and the Activity of Thinking.* Berkeley, Los Angeles, Oxford: University of California Press, 1996.

Shapin, Steven. *A Social History of Truth: Civility and Science in Seventeenth-century England.* Chicago: University of Chicago Press, 1994.

———. *Scientific Life: A Moral History of a Late Modern Vocation.* Chicago: University of Chicago Press, 2008.

———. "An example of the good life." *London Review of Books*, December 15, 2011.

Shils, E. "Charisma." In *International Encyclopedia of the Social Sciences.* New York: Macmillan, 1968, 386.

Shreve, Steven. "Don't blame the quants." *Forbes,* October 10, 2008.

Shulman, Mike. "Homotopy type theory IV." *n-Category Cafe* (April 5, 2011). http://golem.ph.utexas.edu/category/2011/04/homotopy_type_theory_iv.html (accessed November 22, 2012).

SIAM. "Mathematics in industry." SIAM Society for Industrial and Applied Mathematics. (2012). http://www.siam.org/reports/mii/2012/report.pdf (accessed May 4, 2013).

Simmel, Georg. *The Philosophy of Money*, ed. David Frisby. Trans.Tom Bottomore and David Frisby. London: Routledge and Kegan Paul, 1978.

Simons Foundation. "Science lives." n.d. https://www.simonsfoundation.org/category /features/science-lives/alphabetical-listing/ (accessed 2013).

Simons, James. "Mathematics, common sense, and good luck: My life and careers," December 9, 2010, video lecture at http://video.mit.edu/watch/mathematics-common -sense-and-good-luck-my-life-and-careers-9644/ (accessed July 24, 2013).

Sinclair, Nathalie. "Aesthetics as a liberating force in mathematics education?" *ZDM Mathematics Education* (2009): 45–60.

———. "Aesthetic considerations in mathematics." *J. Humanistic Mathematics* **1**, no. 1 (2011): Article 3.

Sinclair, Nathalie, and David Pimm. "The many and the few: Mathematics, democracy and the aesthetic." *Educational Insights* **13**, no. 1 (2010).

Singh, Simon. "The beauty of numbers." *New Statesman* (January 2000).

Skovsmose, Ole. "Symbolic power and mathematics." *Proceedings of the International Congress of Mathematicians, Hyderabad, India* 2010. New Delhi: Hindustan Book Agency, 2010, 690–705.

Slade, J. W. "Communication, group theory, and perception in *Vineland*." *Critique: Studies in Contemporary Fiction* **32**, no. 2 (1990): 126–44.

Smith, Adam. *The Theory of Moral Sentiments*. London: A. Millar, 1759.

———. *The Wealth of Nations*. London: W. Strahan and T. Cadell, 1776.

Sperber, D., and D. Wilson. *Relevance*, 2nd ed. Oxford: Blackwell, 1995.

Spice, Nicolas. "Is Wagner bad for us?" *London Review of Books* (April 11, 2013).

Spinney, Laura. "Erotic equations: Love meets mathematics on film." *New Scientist* (April 2010).

Stallabrass, Julian. *Contemporary Art: A Very Short Introduction*. Oxford: Oxford University Press, 2006.

Stark, Harold M. *An Introduction to Number Theory*, 5th ed. Cambridge: MIT Press, 1987.

Stcherbatsky, T. *The Conception of Buddhist Nirvana*. Delhi: Motilal Banarsidass, 1978.

Stegmaier, Werner. "Einleitung der Herausgebers." In (Hausdorff 2004).

Steinsaltz, David. "The value of nothing: A review of the quants." *Notices of the AMS* (May 2011): 699–704.

Stephenson, Neil. *Cryptonomicon*. New York: Avon Books, 1999.

Stern, N. "Age and achievement in mathematics: A case-study in the sociology of science." *Social Studies of Science* **8** (1978): 127–40.

Streitfeld, David. "Amazon signs up authors, writing publishers out of deal." *New York Time*, October 16, 2011.

Tacopino, Joe. "Battles carry on as a trio, condemn 'ridiculous, boring' math-rock label." *Spinner* (June 6, 2011). http://www.spinner.com/2011/06/06/battles-gloss -drop/ (accessed November 23, 2011).

Talmon, Valérie. "G. Fioraso: Il faut enseigner la culture de l'entreprise dès la maternelle," *Les Echos* (February 5, 2014).

Tao, Terence. "The Black-Scholes equation." *What's New* (July 1, 2008a). http://terrytao
.wordpress.com/2008/07/01/the-black-scholes-equation/(accessed February 13, 2013).

———. "Tricks Wiki article: The tensor power trick." *What's New* (August 25, 2008b). http://terrytao.wordpress.com/2008/08/25/tricks-wiki-article-the-tensor-product -trick/ (accessed August 15, 2011).

Tao, Terrence, and Henry Cohn. "Math 2.0—what's the point of journals?" *Math* 2.0. (February 2012). http://publishing.mathforge.org/discussion/7/whats-the-point-of -journals/#Item_0 (accessed February 10, 2012).

Tappenden, Jamie. "The Riemannian Background to Frege's Philosophy." In (Ferreirós and Gray 2006) 97–132.

———. "Mathematical Concepts: Fruitfulness and Naturalness." In (Mancuso 2008 276–301).

Taylor, Charles. *A Secular Age*. Cambridge, MA: Harvard University Press, 2007.

Terry, Clarence L., Sr. "Mathematical counterstory and African American male students: Urban mathematics education from a critical race theory perspective." *Journal of Urban Mathematics Education* 4 (2011): 23–49.

Thorndike, Lynn. *A History of Magic and Experimental Science*, Chapter LXI. 8 vols. New York: Macmillan, 1923–1958.

Thorp, Edward O. *Beat the Dealer*. New York: Vintage, 1966.

Thurman, Robert A. F., trans. "Vimalakīrti Nirdeśa Sūtra." Kenyon College—Department of Religious Studies. 1976. http://www2.kenyon.edu/Depts/Religion/Fac/Adler /Reln260/Vimalakirti.htm (accessed June 9, 2013).

Turing, Alan. "On computable numbers, with an application to the Entscheidungsproblem." *Proc. London Math. Soc.* (1936–1937): 230–65.

2Sheep4Coke. "Robert Hood, le preacher du rythme." *Jekyll & Hyde* (December 21, 2009). http://jekyllethyde.fr/2009/12/robert-hood-interview-video (accessed April 7, 2012).

Tymoczko, Thomas. "Value judgments in mathematics: Can we treat mathematics as an art?" In *Essays in Humanistic Mathematics*, ed. A. White. Washington, DC: The Mathematical Association of America, 1993, 62–77.

Vallicella, William F. "Divine simplicity." *The Stanford Encyclopedia of Philosophy*, ed. Edward N. Zalta (Fall 2010 edition). http://plato.stanford.edu/archives/fall2010 /entries/divine-simplicity/.

Velleman, D. "Bodies, selves." *American Imago* 65 (2008): 405–26.

Villani, C. *Théorème Vivant*. Paris: Grasset, 2012.

Voevodsky, Vladimir. "What if current foundations of mathematics are inconsistent?" Institute for Advanced Study. September 15, 2010. http://video.ias.edu/voevodsky -80th (accessed June 14, 2013).

Wagner, G. "Die Soziale Ordnung der Wissenschaft." *Erwägen Wissen Ethik* 18 (2007): 60–62.

Walker, Tim. "Oscars of science: Breakthrough awards hands out $21m to transform physicists into rockstars." *The Independent*, December 13, 2013.

Wallace, David Foster. "Rhetoric and the math melodrama." *Science* **290**, no. 5500 (December 2000): 2263–67.

———. *Everything and More: A Compact History of Infinity*. New York: W. W. Norton, 2003.

Weber, A.-G., and A. A. Albrecht. "Évariste Galois ou le roman du mathématicien." *Revue d'histoire des mathématiques* 17 (2011): 401–33.

Weber, Max. "Science as a vocation." In *Gesammelte Aufsätze zur Wissenschaftslehre*. Tübingen: J.C.B. Mohr, 1922, 524–55.

———. *Economy and Society*. Berkeley: University of California Press, 1978.

Weil, André. "Review of Gotthold Eisenstein's Mathematische Werke." *Bulletin of the AMS* **82** (September 1976): 658–63.

———. "History of mathematics: Why and how." *Proceedings of the International Congress of Mathematicians Helsinki 1978*. Helsinki: Academia Scientiarum Fennica, 1980.

Weyl, Hermann. "Geodesic fields in the calculus of variation for multiple integrals." *Annals of Mathematics* **36** (July 1935): 607–29.

———. "Mean motion, II." *American Journal of Mathematics* **61** (January 1939): 143–48.

———. *Gesammelte Abhandlungen* (in four volumes). Ed. K. Chandrasekharan. Berlin: Springer-Verlag, 1968.

———. *Mind and Nature*. Ed. Peter Pesic. Princeton: Princeton University Press, 2009.

Wikipedia. "High-net-worth individual." http://en.wikipedia.org/wiki/High-net-worth _individual#Ultra_high_net_worth_individuals (accessed April 21, 2013).

Wiles, Andrew. "Birch and Swinnerton-Dyer conjecture." Clay Mathematics Institute. 2000. http://www.claymath.org/sites/default/files/birchswin.pdf (accessed June 11, 2014).

Williams, Raymond. *Culture and Society*. London: Chatto & Windus, 1958.

———. *Keywords*. Oxford: Oxford University Press, 1976.

———. "The Bloomsbury fraction." In *Problems in Materialism and Culture: Selected Essays*, by Raymond Williams. London: Verso, 1980, 165.

Wilmott, Paul. "The use, misuse and abuse of mathematics in finance." *Philosophical Transactions: Mathematical, Physical and Engineering Sciences* **358**, no. 1765 (January 2000): 63–73.

Wittgenstein, Ludwig. *Tractatus Logico-Philosophicus*. New York: Harcourt, Brace and Co., 1922.

Wolfe, Josh. January 23, 2009. http://blogs.forbes.com/joshwolfe (accessed December 12, 2010).

Wright, Crispin. "Wright's philosophical ramblings." *The Northern Institute of Philosophy* (2011). http://www.abdn.ac.uk/philosophy/nip/page?id=39 (accessed January 9, 2013).

Wulfman, Clifford E. *Faust Book, Of his Parentage and Birth*. Chapter 1. The Perseus Garner (Perseus Digital Library). April 2001. www.perseus.tufts.edu/hopper/text?do c=Perseus:text:1999.03.0001&lang=original (accessed June 21, 2013).

Xenakis, Iannis. "Musiques formelles." *La Revue Musicale Double numéro spécial* **253–254** (1963): 16–17.

Yadav, B. S. "Nirvana and nonsense." *J. American Academy of Religion* **45** (1977): 451–71.

"Yasiin Bey—mathematics lyrics." *rapgenius*. http://rapgenius.com/Mos-def-mathematics-lyrics (accessed January 6, 2013).

Zamyatin, Yevgenii. *We*. Trans. Gregory Zilboorg. New York: E. P. Dutton, 1924.

Zarca, Bernard. *L'univers des mathématiciens*. Rennes: Presses Universitaires de Rennes, 2012.

Zeki, Semir, John Paul Romaya, Dionigi M. T. Benincasa, and Michael F. Atiyah. "The experience of mathematical beauty and its neural correlates." *Frontiers in Human Neuroscience* (February 13, 2014).

Zinoman, Jason. "Longing, friendship and calculations between mathematicians." *New York Times*, October 21, 2006.

———. "There's nothing to fear from infinity, review of *A Disappearing Number*." *New York Times*, July 4, 2010.

index of mathematicians

Note: The individuals listed below are all cited in the text and are nearly all clearly identified as mathematicians. There are a few borderline cases on this list; a few names listed in the subject index could plausibly be described as mathematicians. No deep or systematic principles were applied in deciding what to do about such cases, and the author apologizes to anyone who feels he or she was placed in the wrong list.

subject index

246, 254, 324, 345n62, 375n56, 395n11;
  "medium of free becoming," 179
Freeman, David, 286
French Libertines, 143
Freud, Sigmund, 71, 135, 248, 254
Fry, Roger, 304–5, 307–10; *Art and Science*,
  305, 308
fuck-you money, 102
Fujiwara, Dai, 159
Fukaya category, 216
fundamental lemma, 31, 85, 266

Galois group, 19–20, 25, 124–26, 189, 197,
  220, 270, 353–54n14
Galois representations, 34, 200, 202, 220,
  384n9
Galois theory, 19, 20, 25, 32, 110–11, 126,
  147, 189, 266,
Garcia, Jerry, 203
García Márquez, Gabriel, 225
Garfield, Jay, 343n45
Gekko, Gordon, 100
genealogical, 29–30, 34
geometric Langlands program, 202, 221
geometry, 7, 14, 15, 25, 32, 44, 62, 86, 104,
  120, 130, 133, 149, 159–60, 189, 202–3,
  215–16, 229, 232, 238, 246, 267, 273, 283,
  291, 315, 322–23, 337n63, 342n31
Germany, 38, 55, 105, 282, 324
Gherard of Cremona, 233
Giants, 9, 11, 17, 24, 39, 76, 280, 300, 324,
  328n4, 331nn 23, 27; Supergiants, 9, 39, 76,
  280
Giaquinto, Marcus, 336n58
Gibbon, Edward, 163
Giordano, Luca, 242–43, Figure 8.2
Glass Bead Game, The (Hermann Hesse), 72,
  350n39
Goethe, Johann Wolfgang von, xii, 82, 108,
  172, 255, 321
Golden Goose, 55–56, 70, 285
goods, internal and external, 87; external, xii,
  103,105–6, 282, 284; internal, 71, 74, 103,
  105, 282, 298
Good Will Hunting, 141, 255
Goodyear, Gary, 54, 58, 63
Google, xx, 128, 189, 236
Gopnik, Alison, 254, 301
Göttingen, 128, 332n31
Graham, Loren, 169

Grand Emprunt National, 296
Grant, Nicholas, 7, 14
graph, 14, 89, 119–20, 135, 191–93, 255, 312
Grattan-Guinness, Ivor, 237
Gravity's Rainbow. See Pynchon
Gray, Jeremy, 4, 57–58, 67, 241, 316, 319,
  327n6
Grebenshchikov, Boris, 251–52
Greed. *See* pleonexia
Greek mathematics, 34, 43, 46, 68, 84, 232,
  310, 348n20, 372n35, 374n45
Greiffenhagen, Christian, 30, 336n60
Grosholz, Emily, 198, 336n58
Grothendieck universe, 212
group theory, 137–38, 154, 238, 266
Grundzüge der Mengenlehre, 323
Guiding Problem. *See* problem
Gustin, Bernard, 36

Hacking, Ian, 343n40, 363n18
Hagstrom, W. O., 21
Hamlet, 123–24, 354n14
Hand (Invisible), 100
Harris, Michael: Pynchon specialist, 137;
  actor in *Suture*, 166
Harris's tensor product trick. *See* trick
Harvard, 10, 23, 27–28, 158, 265, 277,
  360n51, 374n54, 384n4, 391n62
Hasse's theorem. *See* theorem
Hausdorff Research Institute for Mathematics,
  324
Hausdorff spaces, 323
Heard, John, 288
hedge fund, 10, 72, 85, 98–99, 104, 157
Hedonical Calculus, 152
Heidegger, Martin, 59, 195
Heintz, Bettina, 335n56
Henric, Jacques, 143–44, 162
Hermes, 223, 237
Hero of Our Time. See Lermontov
High John de Conquer, 225
h–index, 21
HNWI (High Net-Worth Individual), 99,
  102–3
Hodge Conjecture, 333n49
Hoffman, Dustin, 141
Homo Ludens. See Huizinga
homotopy, 195, 197, 210–12, 220–21
Hood, Robert, 251